GLASS–TO–METAL SEALS

I. W. Donald

Atomic Weapons Establishment, Aldermaston, UK

Published by the Society of Glass Technology 2009

The objects of the Society are to encourage and advance the study of the history, art, science, design, manufacture, after treatment, distribution and use of glass of any and every kind. These aims are furthered by meetings, publications, the maintenance of a library and the promotion of association with other interested persons and organisations.

Cover image courtesy of Martec.

Society of Glass Technology
Unit 9, Twelve O'clock Court
21 Attercliffe Road
Sheffield S4 7WW
UK

Registered Charity No. 237438

Web site: http://www.sgt.org

ISBN 978-0-900682-62-9

GLASS-TO-METAL SEALS

I. W. Donald
Atomic Weapons Establishment, Aldermaston, UK

CONTENTS

1.   INTRODUCTION

Although many new innovations have been made in the commercial exploitation of glass-to-metal seals and related components, and the science underpinning these systems is now more thoroughly understood, many of the topics covered in Partridge's original monograph on glass-to-metal seals (Partridge, 1949) are still highly relevant today. This is particularly true in the areas of underlying technology, including metal and glass preparation prior to sealing, and certain aspects of stress analysis. The original monograph also continues to provide an excellent introduction to the general area of glass-to-metal systems, as well as providing an historical overview of the early work and technology in this area, particularly in relation to archaic electrical components.

The primary purpose of this new monograph is to provide a thorough review of glass-to-metal seals, with particular reference to the more recent developments in the scientific, technical and commercial fields. Current applications for glass-to-metal seals are extraordinarily diverse, ranging from the humble, taken-for-granted light bulb to complex aerospace and military components developed within the last few years. New applications also continue to emerge where the unique properties of these systems can be exploited, thus making a monograph of this nature timely. It is therefore also the purpose of this monograph to highlight new and emerging fields which are benefiting from the application of glass-to-metal seal and related technologies. In this respect, the scope of the monograph has been broadened to include the related topic of glass-to-metal coatings, coatings in general having found many applications in areas ranging from domestic paraphernalia to the aerospace industry. In addition, the more recent and highly versatile glass-ceramic-to-metal systems are reviewed, glass-ceramics not, of course, having been invented in Partridge's day. Some of the newer ceramic-to-metal, glass-to-glass, glass-to-ceramic and ceramic-to-ceramic systems are also covered briefly, areas very much in their infancy in 1949.

It has of course long been recognised that glasses will, under suitable conditions, bond robustly to many metals and alloys, including copper, silver, gold, and iron-based alloys. Initially, this property of forming a strong, well-adherent bond to metals led to the extensive use of glasses for the coating of metals and alloys (also known as vitreous or porcelain enamelling). In ancient Egyptian times, for example, glasses were applied to metal surfaces in the production of decorative enamelled jewellery and related regalia, whilst the enamelling of iron cooking utensils and baths to give a hard, protective and aesthetically acceptable coating was started in the nineteenth century. More recently, glassy coatings have been applied to metal substrates for the production of a wide range of components, from printed circuit boards to acid-, abrasion- or oxidation-resistant protective coatings for metals subjected to hostile environments; for example, low-pressure turbine blades in jet engines. In 1949, glasses had already found wide scale use in the manufacture of vacuum devices requiring hermetic electrical feed-through connections. Since then, many new applications involving the production of electrically insulating hermetic seals have been developed, including most recently electronic packaging. Since Partridge's era, glass-to-metal seals and coatings exhibiting superior properties can be manufactured using more recently developed metallic alloys and glasses of widely differing compositions. In addition, superior properties, relative to glass-to-metal systems, can now be achieved using the newer glass-ceramic materials, glass-ceramics offering more refractory behaviour and a greater ability to tailor the properties, in particular the thermal expansion behaviour, to meet the requirements of specific applications.

In addition to the newer applications for glass-to-metal seals and related components, a much deeper understanding of the underlying science behind glass-to-metal adhesion and the factors promoting adhesion in specific systems has been acquired. It is now fully appreciated, for example, that as well as the requirement for compatible thermal expansion characteristics, the formation of a chemical bond at the interface between the glass and the metal is essential in order to form a reliable, mechanically strong and adherent seal or coating to a metal.

In this monograph, the basic types of glass-to-metal system are first summarized and the properties of metals and alloys typically employed in the manufacture of components and systems are reviewed. The old and newer theories of glass-to-metal bonding are then outlined. This is followed by a comprehensive review of the more recent glass-ceramic-to-metal systems and recent developments in sealing technology, with particular reference to the influence of materials and processing parameters on the quality of the resultant products. Next a brief review of techniques available for the analysis of materials systems, including thermal analysis, XRD and optical and electron microscopy is provided. Issues relating to the stress analysis of components are covered, with particular emphasis on ageing mechanisms and factors influencing the lifetime behaviour of components. This is followed by a review of glass and glass-ceramic bonding to specific metals and alloys, with particular reference given to the interfacial chemistry involved for each system and how this may affect the bonding behaviour. A brief update is subsequently provided on ceramic-to-metal, glass-to-glass, glass-to-ceramic and ceramic-to-metal bonding, areas which were in their infancy in 1949 and only covered very briefly by Partridge. The monograph continues with a summary of the applications for glass-to-metal seal devices and related systems, and concludes with a discussion of recent advances in glass-to-metal systems and an outlook for the future. Throughout this undertaking, Partridge's work is reviewed for its relevance to today's glass-to-metal seal and related systems and technologies.

2.   BASIC TYPES OF GLASS-TO-METAL SYSTEMS

2.1   Major materials properties of interest

2.1.1   Glasses

Usually when a molten material is cooled a point is reached (the freezing point) where spontaneous nucleation and growth of crystals occurs and the material rapidly solidifies in a crystalline state. This spontaneous nucleation and growth of crystalline phases can, however, be delayed under certain conditions and the material may remain in a liquid state at a temperature below its normal freezing point. As the temperature is further reduced the viscosity of this supercooled liquid increases rapidly and a point may be reached when the viscosity becomes so high that for all practical considerations the material is a solid. It is then known as a glass. The change from a supercooled liquid to a non-crystalline glass occurs over a narrow temperature range referred to as the glass transition and can be detected by a change in slope in a plot of certain properties such as specific volume against temperature, as illustrated in Figure 1. The glass transition temperature, T_g, is usually defined as the temperature at which the viscosity reaches an arbitrary value of around 10^{12} Pa s (10^{13} poise), delineating a change in behaviour from Newtonian liquid to Hookean solid. It should be noted that T_g for a given material is not a constant but is a function of the cooling rate from the liquid to the glassy state and also of the past thermal history of the glass.

Various definitions of glass have been proposed. Probably one of the most widely quoted is the American Society for Testing of Materials, ASTM, offering made in 1945 "Glass is an inorganic product of fusion which has cooled to a rigid condition without crystallizing". This definition is too restrictive today as it only considers glasses prepared from the melt and neither does it take into account organic glasses. One thing all glasses do have in common, irrespective of how they were formed, is that they are x-ray amorphous, that is they do not exhibit long range order as detected by XRD, although even this is being brought into question today for certain categories of glasses. Glasses do, however, exhibit a significant degree of short range order. There have been many theories of glass structure over the years, but the relatively simple three-dimensional random network model of Zachariasen(1932) still holds much favour as a starting point for describing glass structure today. In this model the structure is built up by the union of polyhedra in much the same way as in the corresponding crystal lattice. In the case of a glass, however, there is sufficient distortion of bond angles to permit the structural units to be arranged in a non-periodic way, this giving rise to a random network. In this model oxides can be divided into three main

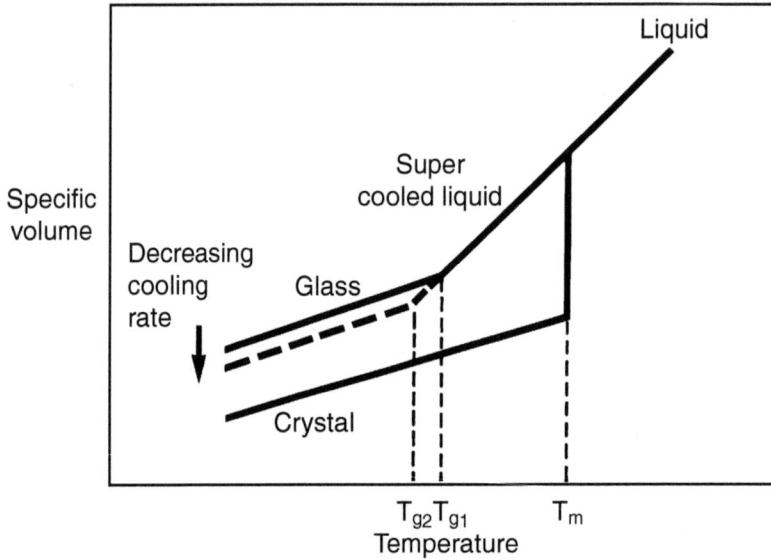

Figure 1: Specific volume as a function of temperature for a glass-forming material

categories. There are those that can form continuous three-dimensional random networks by themselves (network formers), those for which the cation can enter the network substitutionally but which cannot normally by themselves form a network (intermediates), and there are oxides for which the cation can enter the network interstitially (network modifiers). The difference between a regular crystal lattice and a random network structure is illustrated schematically for SiO_2 in Figure 2. Also shown for comparison is the structure of a Na_2O–SiO_2 glass.

It is now generally recognised that any material will, in fact, form a glass, providing that the cooling rate is sufficiently high enough to avoid crystallization and that the final ambient temperature is below T_g. Glass-forming ability is therefore a kinetic phenomenon. In conventional glass technology it is usual, however, still to refer to materials that are 'glass-forming', by which is meant materials that can be formed as glasses at low cooling rates, i.e. generally < 0.1 $K\,s^{-1}$, the most common and practical of which are based on the network forming oxides, SiO_2, B_2O_3 and P_2O_5. The oxide GeO_2 also falls into the category of an easy glass-former, although GeO_2-based glasses are generally of less practical significance. The practical implications of a low cooling rate for glass formation are that these materials can be formed as bulk glasses by conventional casting and related techniques. For the purposes of this monograph only glasses based on the most common of these oxides, SiO_2, B_2O_3 and P_2O_5, are considered in any detail.

Glass has a very long history and throughout the ages its unique combination of properties has made it one of the most useful classes of material known.

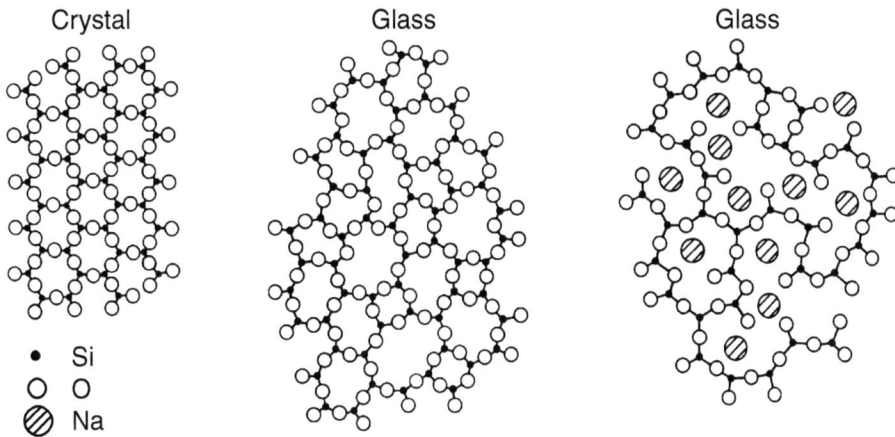

Figure 2: Schematic planar representation of the atomic structure of crystalline and glassy silica and glassy $Na_2O–SiO_2$ (note that the fourth oxygen ion is not shown as this would be perpendicular to the plane of the paper) (after Rawson, 1980)

It is believed that synthetic inorganic glasses, initially in the form of decorative glazes, appeared around 6000 years ago in Mesopotamia and Egypt. Around the beginning of the first century AD the developing art of glass making spread throughout the countries of the Roman Empire, culminating in Venetian glass. In the early seventeenth century glass was brought to the forefront of technology with our perception of the universe forever transformed by Galileo's application of the optical telescope to astronomy. At the present time glasses have many diverse applications ranging from their use in the architectural, food and catering, transportation, engineering, electronics, telecommunications and the aerospace industries, in addition to their artistic and decorative uses. Glasses are undoubtedly highly versatile materials, with their many advantages including ease of formation and fabrication into complex shapes, excellent electrical insulating properties, optical transparency and good chemical durability. For many applications, however, the brittle nature of glass and its susceptibility to catastrophic failure is a serious disadvantage.

The mechanical properties of inorganic non-metallic glasses in which ionic/covalent bonding predominates are dominated by this brittle behaviour. For all practical purposes these materials exhibit fully elastic behaviour up to their breaking points, with no indication of significant macroscopic ductility, at least at temperatures less than the glass transition temperature, T_g, and at ambient hydrostatic pressures. The theoretical strength of glass has been estimated to be of the order of $E/10$, where E is Young's modulus (Kelly, 1973). This suggests that oxide glasses should exhibit strengths of the order of around 7 GPa. In practice, however, useful strengths rarely exceed 100 MPa. This large discrepancy between the theoretical and practical strengths has been explained on the basis of defects in glass, particularly surface defects, which act as stress concentration sites enabling the local stresses generated at the crack tip to exceed the theoretical strength limit for very small applied stresses.

This weakening influence of defects on materials was first highlighted by Griffith(1920; 1924) based on earlier work by Inglis(1913) and Kolosoff(1914). It had been shown by Inglis that cracks, noted in some ships, originated from sharp corners as may be found at doorways and hatchways. Inglis assumed, therefore, that there must be a concentration of stress at sharp irregularities above the normal value. He went on to show that the maximum stress at a crack tip, σ_m, could be given by the expression:-

$$\sigma_m = 2\sigma(c/\rho)^{1/2} \qquad\qquad [2.1]$$

where σ is the applied stress, c is the crack length, and ρ is the radius of curvature of the crack tip, this being a measure of the "sharpness" of a crack.

Griffith extended this idea by developing a theory based on energy considerations for the propagation of pre-existing cracks. He showed that during fracture the net change in energy of the system, ΔH, may be given by the expression:-

$$\Delta H = -(\sigma^2\pi c^2/E) + 4c\gamma \qquad\qquad [2.2]$$

where γ is the fracture surface energy.

For small values of c, ΔH is positive, but for large values the first term in this equation predominates and ΔH becomes negative and decreases as c increases (Rawson, 1980). Differentiation of Equation [2.2] with respect to c and setting the result equal to zero gives a value for c at which the crack becomes unstable at the applied stress σ_f. A flaw will consequently extend under an applied stress and lead to fracture only if the defect is greater than this critical flaw size given by:-

$$c = 2\gamma E/\pi\sigma^2 \qquad\qquad [2.3]$$

Rearranging this equation, the fracture stress of the material (for conditions of plane stress) may therefore be given by:-

$$\sigma_f = (2E\gamma/\pi c)^{1/2} \qquad\qquad [2.4]$$

For conditions of plane strain the corresponding relationship is:-

$$\sigma_f = \{2E\gamma/\pi(1-\upsilon^2)c\}^{1/2} \qquad\qquad [2.5]$$

where υ is Poisson's ratio.

It follows from equations [2.1] to [2.5] that sharp cracks can indeed act as mechanical levers by concentrating the stress at the crack tip. The smaller the value of ρ or the higher the value of c, the greater this stress concentration effect will be, and consequently the lower the strength of the material. By this mechanism, cracks as small as 1–10 μm in size can reduce the effective strength of a brittle material to values normally experienced in practice. This is coupled with a lack of significant stress relieving mechanisms for brittle materials. Cracks, once

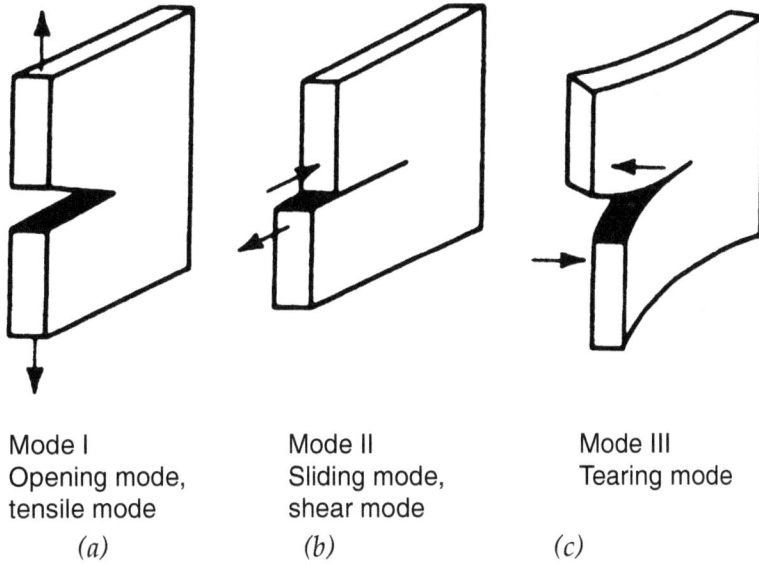

Mode I
Opening mode,
tensile mode
(a)

Mode II
Sliding mode,
shear mode
(b)

Mode III
Tearing mode

(c)

Figure 3: Basic crack extension modes (after Bar-On, 1991)

initiated, will therefore propagate unimpeded under the influence of a critical applied load, leading to catastrophic failure. This is in contrast to ductile metals where dislocation motion can relieve stresses at crack tips, effectively blunting them and therefore making metals less susceptible to catastrophic failure.

Further extension of Griffith's original work has led to the field of fracture mechanics which provides a means of predicting the effective strength and/or lifetime of a particular material in a given structure and environment. Irwin(1958) defined the product $\sigma c^{1/2}$ as the stress intensity factor, K, which is a measure of the intensification of the applied stress due to the presence of a crack. For a crack in an infinitely wide plate it can be shown that:-

$$K = \sigma(\pi c)^{1/2} \qquad\qquad [2.6]$$

whilst for other geometries:-

$$K = \sigma Y c^{1/2} \qquad\qquad [2.7]$$

where Y is a dimensionless constant which takes into account the geometry and loading characteristics of the system.

This approach has led to the concept of a critical stress intensity factor, K_c, which is related to the stress at which a crack can propagate continuously, that is to say catastrophically, in an actual material. Three basic crack extension mechanisms are possible, designated modes I, II and III, with the K_c values corresponding to each mechanism being K_{Ic}, K_{IIc} and K_{IIIc}, respectively, as depicted in Figure 3. Mode Type I applies to a crack opening mode, Type II to a crack sliding mode, and Type III to a crack tearing shear mode. The first mode is the one most universally applied. Values for K_{Ic} can be found experimentally using standard fracture mechanics techniques, as described more fully in Chapter 7.

In addition to exhibiting a very strong dependence of strength on the presence of defects, glasses also exhibit a marked time dependence of strength. If a glass fails at a given stress in a short term test, it will usually fail at a lower stress in a longer duration test. This is due to the phenomenon of subcritical crack growth, or "static fatigue". Static fatigue is caused by the growth of small, pre-existing cracks or other defects present in the material under the influence of an applied stress which may be considerably lower than the normal fracture stress of the material. This occurs at subcritical velocities leading to time-dependent failure of glass under load. During the influence of static fatigue a critical flaw size may be reached after which crack propagation becomes unstable and catastrophic. The mechanical strength of glass therefore depends strongly on both the environment and the length of time under load.

It is clear that brittle behaviour dominates the mechanical properties of oxide glasses and related materials, and that the influence of stress-concentrating defects, coupled with a lack of stress-relieving mechanisms, gives rise both to low mechanical strength and low fracture toughness, together with a strong dependence of the properties on time. The influence of defects and of static fatigue on the properties of brittle materials has extremely important implications when using glass, glass-ceramics, or ceramics for any application that involves an applied or residual stress, or use in an environment in which the effects of static fatigue may be enhanced. This obviously has strong bearing on the lifetime behaviour of these materials when used in seal or coating applications. Further information in relation to the influence that these effects may have on the lifetime behaviour of materials systems is given in Chapter 7.

2.1.2 Metals

In contrast to glasses, most metals and metallic alloys generally exhibit significant ductility and are not normally prone to catastrophic brittle failure. Plastic deformation in crystalline metals occurs by dislocation motion, and it is the high initial dislocation mobility that gives rise to the low yield strengths and extensive plastic deformation associated with ductile metals and alloys. Because the metallic bonding in these materials is non-directional, dislocations can normally move very readily. Unlike brittle materials, plastic deformation in the vicinity of stress concentration sites can lead to crack tip blunting, thereby raising the energy required to propagate a given crack further. In the case of oxide glasses, in which dislocations as conventionally envisaged do not exist, there can be no significant stress relief by plastic deformation, although it may be noted that highly localized deformation of glass, somewhat analogous to plastic flow is observed; for example, during microindentation to give a permanent impression and during abrasion and scratching to yield furrows with raised edges. This implies that limited stress relieving mechanisms are operative, at least at the microscopic level, even in such brittle materials as oxide glasses. The precise mechanism responsible for this behaviour is open to debate, but may be associated with localized densification of the network structure, rather than the movement of "dislocations" of variable Burgers vector.

The ductility of metals can be put to great advantage in glass-to-metal seal design by ensuring that any predominant tensile stresses due, for example, to mismatch in thermal expansions, are only present in the metal components, as described more fully later.

2.2   Glass-to-metal seals

Glass-to-metal sealing is traditionally a fusion technique with the glass melted in contact with the metal parts to be joined or sealed to. In Partridge's day, glass-to-metal seals were widely employed in the electrical industry in the preparation of electrically insulated feed-through connectors, e.g. for lamps and radio components including vacuum tubes and glass-metal terminals for condensers, transformers, etc. These applications evolved from the earlier use of glass-to-metal seals by Edison in the late nineteenth century in the incandescent lamp industry; an industry that is still thriving today. Many thousands of millions of seals in the form of inexpensive light bulbs for domestic and commercial use are produced worldwide annually, confirming the versatility of glass for this application. These components are easy to manufacture in large numbers by either batch or continuous processes and they provide reliable systems for everyday use. Glass-to-metal seals were also used extensively in the construction of laboratory apparatus including metal-glass joints and connectors. Some more recent examples of the use of glass-to-metal seals include high power vacuum tubes, radar magnetrons and microwave connectors, TV tubes, reed and relay switches, telecommunications applications, sensors, automotive electrical components, actuators, aerospace components, electronic systems, and electronics and microelectronics packaging, as discussed more fully in Chapter 10. Glass is a near ideal medium for these types of application. Not only can mechanically strong, hermetic bonding be achieved to metals, but glass is also relatively impervious to gases, it is a good electrical insulator, it is reasonably refractory, it is inexpensive, and the technology is well developed and understood.

Partridge(1949) noted that glass-to-metal seals could generally be assigned to one of four types, and this is certainly still true today. These are: (a) matched thermal expansion seals, (b) unmatched expansion seals, (c) soldered seals, and (d) mechanical joints. Soldered seals and mechanical joints have today largely been superseded by the direct matched and unmatched (compression) seals. Seals can also be classified by a number of alternative descriptions. For example, according to their geometry, e.g. bead seals, strip seals; or by the type of metal or glass employed in their construction, e.g. hard or soft glasses, ductile or brittle metals; or by the method employed in their manufacture, e.g. fusion, solder, etc.

2.2.1   Matched seals

In the case of a matched seal, the thermal expansion characteristics of the glass and the metal are matched as closely as possible over the temperature range of interest. Matched seals are often subdivided according to the thermal expansion of the individual components. Matched seals may be almost free of residual stresses at ambient temperature and are designed so that any stresses that are present do not exceed the strength of the glass at any stage of the manufacturing process, or during use. It is important that any residual stresses are also low enough so that the influence of static fatigue is minimized. Typical examples of matched seals are shown schematically in Figure 4(a).

2.2.2   Unmatched seals

In the case of an unmatched seal, the thermal expansion characteristics of metal and glass may differ markedly. In this instance, it is very important as far as the glass parts of the seal are concerned that the major stresses resulting from thermal expansion mismatch are predominantly compressive in nature. Balancing tensile stresses are kept within the metal parts, which can support them or may deform plastically without failure. Unmatched seals can therefore be subdivided into compression seals and ductile metal or Housekeeper seals.

2.2.2.1 Compression seals

Compression seals are the most popular type of seal, exhibiting high mechanical strength and being ideally suited for applications in which high stresses or pressures may be involved during operation. Typical examples include pin or coaxial connectors. In this design of seal, the so called matched or compression seal, glass may be used with its expansion matched to that of the pin material but having the expansion of the metal housing greater than that of the glass. Alternatively, in addition to the metal housing having a higher thermal expansion than that of the glass, the expansion of the glass itself may be higher than that of the pin material, the so called reinforced compression seal. After cooling from the fabrication temperature the metal housing shrinks onto the glass which itself shrinks onto the pin, and the compressive forces generated at the glass-metal interfaces may be sufficient to provide a hermetic seal without the need for a chemical bond between the metal and glass. This type of seal is often employed in electronic packaging applications. A typical example of a compression seal is shown in Figure 4(b).

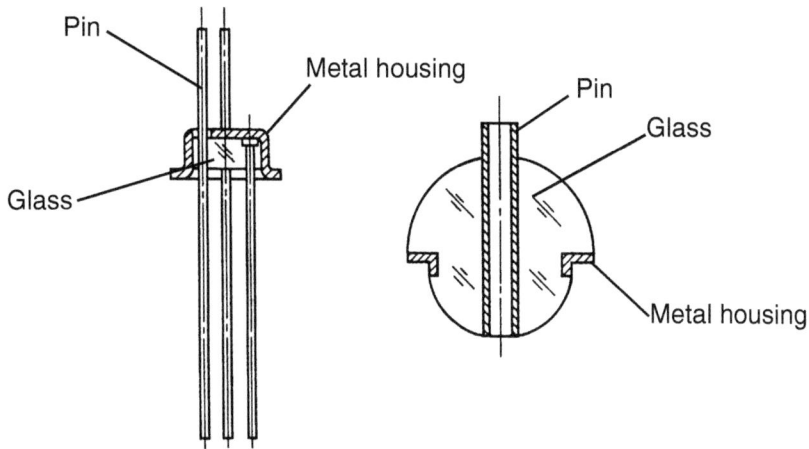

(a) Matched thermal expansion seal

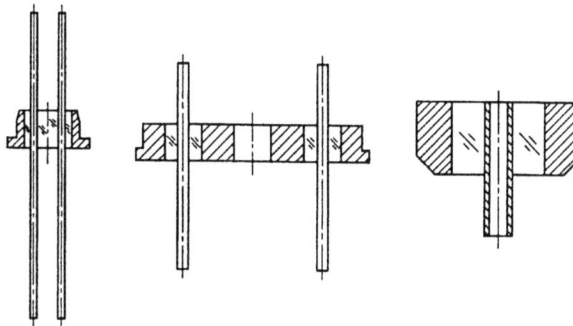

(b) Unmatched (compression) seals

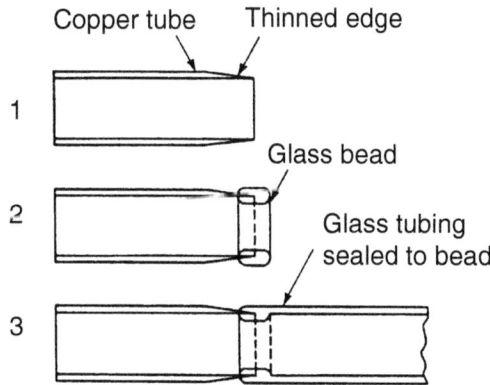

(c) Housekeeper seal

Figure 4: Schematic illustration of some of the basic types of glass-to-metal seal (after Tomsia, Pask and Loehman, 1991)

2.2.2.2 Housekeeper seals

A special case of unmatched seal is the Housekeeper seal, named after G. W. Housekeeper (1923), in which glasses are joined to ductile metals, in particular copper. An example of a typical Housekeeper seal is shown in Figure 4(c). The thermal expansion of the glass does not need to be matched to that of the copper, rather the seal is such that stresses in the glass are minimised through deformation of the copper. In the example shown, a glass tube is sealed to a copper tube by first thinning the edge of the copper tube, applying a thin glass bead, and subsequently bonding the glass tube to this bead. By this means it is possible to bond either low or high expansion glasses to the copper, with all the stress being taken up and relieved by plastic deformation of the metal.

2.3   Glass-to-metal coatings

Much of the early work in the general area of glass-to-metal bonding was concerned with coatings on metals, and specifically with the preparation of enamelled metals. The enamelling of the metals gold, silver, bronze and copper for the production of jewellery was thought to date back to ancient Egyptian times. In fact, most of the artefacts found in Egyptian tombs relied on cold cementing coloured glass or other materials onto metals, rather than by fusion enamelling (The Institute of Porcelain Enamellers, 2004). For example, the gold mask of Tutankhamun was not enamelled but had the coloured inlays cold cemented onto the metal. The earliest objects known to be enamelled using fusion techniques were made in Cyprus around the thirteenth century BC, gold rings decorated with coloured vitreous layers fused to the metal being found in a Mycenæan tomb at Kouklia. Later, in the Byzantine period, the art of enamelling spread to Europe and elsewhere and was broadened to encompass the decorative enamelling of larger items, although it was still generally limited to the preparation of objects of art, rather than for the specific protection of metal surfaces. Some of the most famous vitreous enamelled objects have been produced by Faberge. Many early examples of enamelled works of art are to be seen in churches and museums; for example, The Royal Gold Cup of the Kings of France and England can be found in the British Museum. In the middle of the nineteenth century the large scale preparation of enamelled metals was initiated, originally in Germany, concerned with the industrial production of enamelled cast and wrought iron ware, including cooking utensils, baths and underground piping, the enamel being applied in order to provide a corrosion and abrasion resistant protective surface on the metal. More recently, vitreous enamelled metals have been employed in a multitude of large volume/low technology applications, both in the home and industry. Examples of the use of enamelled metals in the home include their application in cookers, cooking utensils, sink units, dish and clothes washers, water and storage heaters, and gas fires and stoves. Industrial applications include their use in architecture, e.g. building panels and tunnel walls. Other industrial applications include enamelled vessels, pipes, valves, stirrers, etc, used in the chemical industry; for the protection of metals used in agricultural applications including storage silos; and for use in heat-exchangers and related devices (Andrews, 1961; Maskall and White, 1986; Garland, 1986; Anon, 1991).

More recently, glass-ceramic coatings exhibiting superior mechanical and other relevant properties have been applied for the protection of, for example, metal pipes and vessels in the chemical industry.

What could be described as high technology applications for glass-to-metal coatings have emerged in a number of industries over the last 30 years. For example, the use of enamelled steel or copper substrates in the electronics industry for microelectronics applications, including printed circuit boards and multi-layer electronic packaging. In these applications, circuitry is applied to the enamel coating using thin or thick film technology. An enamelled substrate offers a number of advantages over more conventional alumina ceramic or organic polymer substrate materials, including lower dielectric constants (which allow higher operating frequencies), and higher thermal dissipation factors which are important in close-packed electronic systems (Garland, 1986). Enamel coatings have also been employed for the protection of metals against oxidation in the aerospace industry; for example, coating of after-burners and low-pressure turbine blades in jet engines (Garland, 1986). In addition, they have also been assessed as protective coatings for medical and dental prostheses (Krajewski *et al*, 1985). Specific applications are covered more thoroughly in Chapter 10.

2.4 Glass-ceramic-to-metal seals

Glass-ceramic-to-metal seals are a more modern invention used for more arduous applications. McMillan and co-workers were the first to show that glass-ceramics could be employed to produce seals that exhibited superior properties relative to conventional glass-to-metal seals (McMillan and Hodgson, 1963; McMillan *et al*, 1966a, 1966b). Some of the applications for which glass-ceramic-to-metal seals have been utilized include refractory vacuum and laser tube envelopes, vacuum interrupters, high temperature insulation, and high pressure pyrotechnic actuators (Kramer and Massey, 1984; Nash *et al*, 1983). More recently, glass-ceramics have been employed to coat metal substrates used in microelectronics applications. Use of glass-ceramics provides a mechanically stronger system, provides for closer matching of thermal expansion characteristics, and allows higher temperatures to be employed in the application of thick film circuitry; and this higher temperature capability allows the use of conventional screen printing inks to be employed. The topic of glass-ceramic-to-metal seals is covered thoroughly in Chapter 4.

2.5 Types of metal

Glasses and glass-ceramics have been used for sealing to many metals and alloys covering a whole variety of applications ranging from protective coatings to high performance electrical and electronics components. As noted by Partridge (1949), in order to seal successfully to a particular metal or alloy substrate a number of criteria must be met. In particular, the melting point of the metal or alloy must in general be higher than that of the glass it is to be bonded to, although there are some notable exceptions, as described later. The thermal expansion characteristics of glass and metal must also be matched as closely as possible over the temperature range of interest (or be compatible as in the case of compression seals) and the metal must not undergo phase transformations with accompanying changes in thermal expansion behaviour or sharp changes in volume over this temperature range. The metal or alloy must also generally be capable of forming a suitable oxide layer that is strongly bonded to the substrate. Details of some of the more important metals and alloys that have been employed are reviewed and their characteristics are summarized in Table 1, which also gives details of their basic properties and compositions, and in Table 2 which summarizes thermal expansion data. Thermal expansion data are also presented graphically in Figures 5 and 6. Metals that have been successfully sealed to include Cu, Ag and Au, Pt, W and Mo, Fe and Fe-based alloys, Ni and Ni-based alloys, Ti, Ta and Al and their alloys, as summarized below.

2.5.1   Platinum

The thermal expansion of platinum matches that of many of the soft glasses, at approximately $9 \cdot 5 \times 10^{-6} \, K^{-1}$ up to 1100°C, it is very ductile, and its oxide is stable and strongly bonded. Despite these advantages Pt is not, however, now widely used in seal making due to its very high cost.

2.5.2   Copper, silver and gold

These metals have been used extensively since very early times in the production of enamelled jewellery, and more recently in the manufacture of electrical and electronics components. They possess high thermal expansions, with values for the thermal expansion coefficient, α, at ambient temperature of 16·6, 18·6 and $14 \cdot 0 \times 10^{-6} \, K^{-1}$ for Cu, Ag and Au, respectively, rising to 22·5, 26·5 and $18 \cdot 7 \times 10^{-6}$ K^{-1}, respectively at 900°C. Despite the relatively high thermal expansion of these metals they can, in suitable geometries, be sealed to a variety of low, intermediate or high expansion glasses due to their high ductility which allows residual stresses in a finished component to be minimised through plastic deformation of the metal parts.

A special case of a copper-containing material is Dumet wire. This consists of an Fe–Ni alloy core, now usually Nilo-42 alloy although different cores can be employed in order to achieve different thermal expansion characteristics, contained in a outer sheath of copper, giving a duplex wire structure. This system was originally developed by Fink (1915) as an inexpensive substitute for Pt wire and to overcome problems associated with sealing directly to Fe–Ni alloys which often gave porous seals. It was known that copper bonds well to many

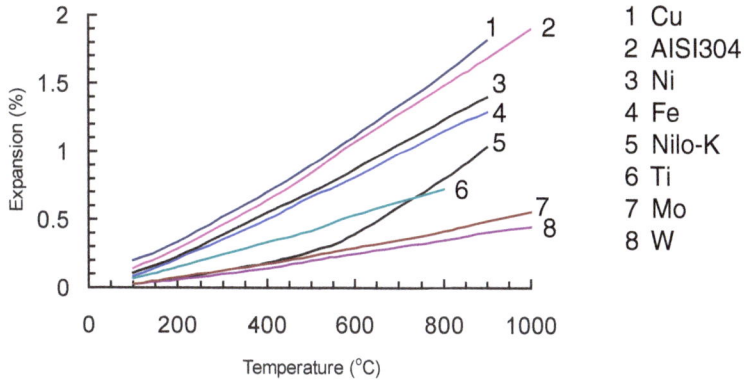

(a) Linear thermal expansion for selection of metals and alloys (data from technical data sheets)

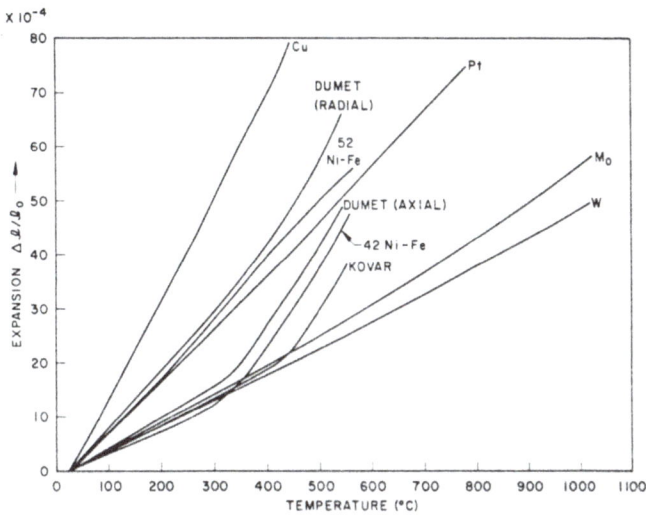

(b) Linear thermal expansion of a selection of metals and alloys
(after Varshneya et al, 1982)

Figure 5: Thermal expansion curves for a selection of metals and alloys

common glasses but its high expansion precluded its use in many applications. Dumet wire allows matching of thermal expansion to common sealing glasses in the radial direction through choice of the core material and thickness of the copper sheath, whilst the mismatch in the axial direction is relieved through deformation of the copper. Dumet wire can be purchased with various cores and in various thicknesses, and also in various forms including pre-oxidized wire and borax-coated wire to aid sealing. Dumet seals are now employed in many electrical applications including cathode ray tubes, vacuum tubes and lamp filaments. Also available commercially is wire with a copper core and a sheath of another metal, e.g. Nilo or stainless steel (Polymetallurgical Corporation). The advantages over Dumet are not clear.

Figure 5: (Continued)

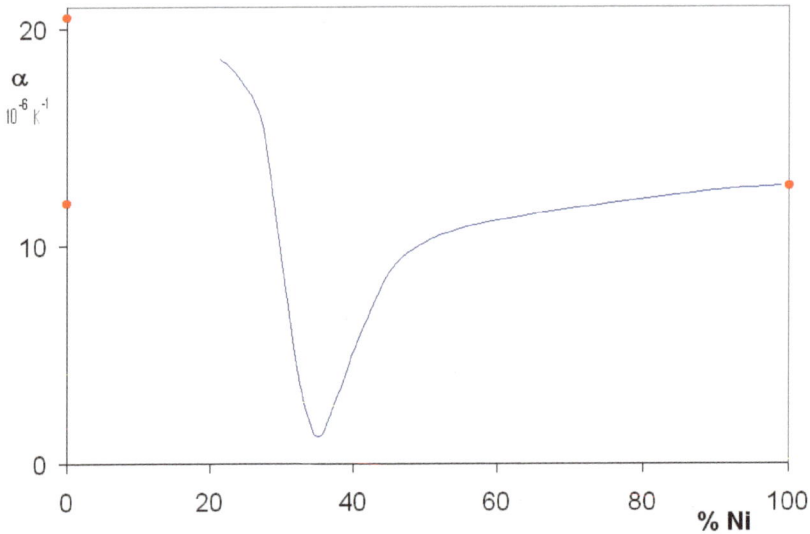

*(c) Thermal expansion as a function of Ni content for binary Fe–Ni alloys
(after Wikipedia)*

2.5.3   Tungsten and molybdenum

Tungsten and molybdenum possess relatively low thermal expansions, around 5.5×10^{-6} K^{-1} for W up to 2100°C and 6.6×10^{-6} K^{-1} for Mo up to 1700°C, and have been extensively used for sealing to hard, low expansion glasses in the manufacture of electrical components requiring refractory properties superior to those that are possible with other metals and alloys; for example, as lamp filaments and lamp filament mandrels and supports. They are, however, relatively brittle metals and are therefore less tolerant of thermal expansion mismatch than is the case for Cu, Ag, Au and Pt. Their oxide films are also more problematic. This is particularly true for Mo for which a thick, non-adherent and volatile oxide layer may be formed. It is therefore essential when forming a seal to these metals that the surface is correctly oxidized before it comes into contact with the glass.

2.5.4   Iron and iron-based alloys

Iron has been used extensively since the middle of the nineteenth century in the production of enamelled washing and cooking utensils. Due to the difficulties associated with working with W and Mo a whole range of Fe-based alloys containing Ni were also developed in the former part of the twentieth century, and these were used as substitutes for Mo and W for sealing to hard glasses. These binary Fe–Ni alloys have now been used for many years specifically in the glass-to-metal seal industry. Specific alloys include the Nilo-series; for example, Nilo-42, a binary alloy containing 42% Ni, Nilo-48 containing 48% Ni, and Nilo-K (also known by various trade names and standard names including Kovar, Nicoseal, Nicosel, Rodar, Telcoseal, Sealvar, Dilver, Pernifer 29-18, Alloy 29-17, ASTM F 15, DIN 17745, AFNOR NF A54-301, and SEW 385), a ternary alloy containing 29% Ni and 17% Co. These simple alloys offer a range of thermal expansion coefficients and are ideal for sealing to many glasses. They are used in many applications. Nilo-42, for example, is employed in sealed beam automobile head lamps and integrated circuit components and is also a core material in copper-clad Dumet wire, whilst Nilo-K with an expansion similar to borosilicate glass is extensively used in the construction of matched seals in metal/glass envelope components, e.g. vacuum and x-ray tubes. These binary alloys sometimes may include a small percentage of Ti which aids in the prevention of gas bubbles during sealing. Pure Fe has an expansion coefficient of $16.7 \times 10^{-6} K^{-1}$ over the temperature range 30–850°C. Alloying with Ni reduces the expansion coefficient considerably, as illustrate in Figure 5(c), with an alloy containing 36% Ni (commonly known as Invar and used in such applications as pendulums and clock springs as well as glass-to-metal seals) exhibiting $\alpha = 2.6 \times 10^{-6} K^{-1}$ over the range 0–200°C. These Fe–Ni alloys do, however, exhibit inflections in their temperature behaviour due to phase transformations, with the result that the expansion coefficient of the same 36% Ni Invar alloy increases to $6.9 \times 10^{-6} K^{-1}$ over the range 200–250°C. The inflection temperature increases with increasing Ni content from 150°C for 36% Ni to 600°C for 55% Ni. The expansion behaviour of a series of Fe–Ni alloys is shown in Figure 6(a). When employed in glass-to-metal seal applications, it is often usual to Ni-plate finished components in order to improve the corrosion resistance.

A whole range of stainless steel compositions based on the Fe–Ni–Cr system have also been developed over many years. These possess a wide range of thermal expansions depending on the precise composition. For example, for the alloy AISI304 over the range 20–400°C, $\alpha \approx 18.0 \times 10^{-6} K^{-1}$, compared to $11.5 \times 10^{-6} K^{-1}$ for AISI 430. Expansion curves are shown in Figure 6(b). The more common austenitic stainless steels have been replaced in some applications by duplex stainless steels which consist of a mixture of austenite and ferrite and which offer superior chloride corrosion resistance (e.g. Davison and Redmond, 1991).

2.5.5   Nickel and Ni-based alloys

Pure nickel, with an expansion coefficient of $13.2 \times 10^{-6} K^{-1}$ up to 420°C, is rarely used by itself in seal applications. There is, however, a series of binary Fe–Ni alloys, as noted above. The more recent Ni-based superalloys were developed in the mid part of the twentieth century to meet the needs of the chemical, engineering and especially the aerospace industries (e.g. Bradley, 1988). The term "superalloy" was first used to describe a series of alloys for use at the high temperatures and in the hostile environments associated with turbosuperchargers and aircraft

TABLE 1(a)
PROPERTIES OF SOME METALS AND ALLOYS

Metal or Alloy	Melting Temp. (°C)	Density (g/cc)	Yield Strength (MPa)	UTS (MPa)	Elongation (%)	Modulus (GPa)	Hardness	Electrical Resistivity (μΩm)	Specific Heat (J/kg K)	Comments
(a) Metals										
Cu	1083	8·93	69	235	49	129	HV 343	0·017	385	Used extensively in the manufacture of electrical and electronic components
Ag	962	10·50	29	177	60	80	HV 251	0·016	238	
Au	1064	19·30	39	137	50	78	HV 216	0·023	130	Has been employed in lead-through seals, but now rarely used by itself
Pt	1772	21·47	–	147	40	170	HV 549	0·098	134	Used in aerospace and medical applications
Ti	1660	4·50	147	316	18	96	HV 970	0·440	523	Used in enamelled ware or alloyed with Ni for feed-through applications
Fe	1535	7·87	157	412	19	211	HV 608	0·100	452	Alloyed with Fe or Ni for feed-through applications
Co	1495	8·79	–	293	3	209	HV 1043	0·065	418	Alloyed with Fe for feed-through applications
Ni	1453	8·90	78	392	30	196	HV 638	0·075	444	Alloyed with Fe or Ni for feed-through applications
Cr	1857	7·19	–	172	–	245	HV 1060	0·130	460	Used extensively in automobile and electronic packaging applications
Al	660	2·70	35	127	50	71	HV 167	0·026	899	Employed in capacitors
Ta	2996	16·62	271	343	12	177	HV 873	0·140	142	Used as lamp filaments and filament mandrels and supports
Mo	2617	10·22	438	638	46	329	HV 1530	0·050	251	
W	3410	19·26	800	1480	4	407	HV 3430	0·050	134	
(b) Alloys										
Nilo-K	1450	8·16	340	520	42	130	HV 160	0·43	–	Fe–Co alloy for sealing to borosilicate glass and alumina
Nilo-42	1435	8·13	250	490	43	150	HV 140	0·61	–	Ni–Fe alloy for glass-to-metal seals and electrical components

Material									Description
Nilo-48	1450	8:20	260	520	43	160	HV 140	0:47	—	Ni-Fe alloy for glass-to-metal seals and electrical components
Nilo-475	1450	8:18	210	520	49	—	HV 140	0:85	—	Ni-Fe-Cr alloy for glass-to-metal seals and electrical components
AISI 316	1399	7.9	241	586	55	200	RB 80	0:74	—	Stainless steel
AISI 430	1510	7.7	310	517	30	200	RB 82	0:60	—	Stainless steel
Ferralium 255	—	7:81	540	825	18	193	BH 230–270	0:8	—	Duplex stainless steel with good mechanical properties to 260°C and good corrosion resistance
Nimonic	1310	8:18	752	1175	47	222	HV 243	1:18	446	Ni-Cr-Co alloy ppt. hardenable for use at temps. up to 920°C
Nimonic	1300	8:36	585	1004	45	224	HV 195	1:15	461	Ni-Cr-Co alloy developed as a sheet material in welded assemblies
Nimonic	1280	8:16	900	1200	—	201	—	1:12	431	Ni-Fe-Cr alloy for use at temps. up to 600°C
Hastelloy	—	9:22	396	914	55	217	RB-98	1:37	373	Ni-Mo alloy with high corrosion resistance to HCl and H_2SO_4
Hastelloy	—	8:64	336	767	58	211	—	1:25	406	Ni-Cr-Mo alloy with high temperature stability and good all round corrosion resistance
Hastelloy	1357	8:69	373	785	62	206	RB-95	1:14	414	Ni-Cr-Mo alloy with good all round corrosion resistance
Hastelloy	1323	8:89	263	682	61	205	RB-87	1:30	427	Ni-Mo-Cr alloy with good all round corrosion resistance
Hastelloy	1260	8:22	380	760	44	196	RB-89	1:18	486	Ni-Cr-Fe alloy for use at temps. up to 1200°C
Inconel 600	1370–1425	8:42	172–345	552–689	35–55	214	HV 115–170	1:03	461	Ni-Cr-Fe alloy for use at temps. up to 1150°C
Inconel 625	1290–1350	8:44	414–758	827–1103	30–60	208	HV 178–300	1:29	410	Ni-Cr alloy for use at temps. up to 1100°C with good all round corrosion resistance
Inconel	—	8:23	1034	1240	12	204	RB-100	1:21	—	Ni-Cr-Fe alloy of very high strength

TABLE 1(b)
CMPOSITIONAL DATA FOR SOME METALLIC ALLOYS

Alloy	Composition (wt%)											
	Ni	Co	Cr	Fe	Ti	Al	Nb and/or Ta	Mo	Cu	Si	Mn	C
Nilo-K	29·0	17·0	-	Bal.	-	-	-	-	-	≤0·2	≤0·3	≤0·02
Nilo-42	42	-	-	Bal.	-	-	-	-	-	0·25	0·50	0·10
Nilo-46	46	-	-	Bal.	0·4	-	-	-	-	0·25	0·50	0·10
AISI 304L	8·0-11·0	-	17·0-20·0	Bal.	-	-	-	-	-	≤0·4	≤2·0	≤0·03
AISI 316	10·0-14·0	-	16·0-18·0	Bal.	-	-	-	2·0-3·0	-	≤1·0	≤2·0	≤0·08
AISI 321	9·0-12·0	-	17·0-19·0	Bal.	≤0·4	-	-	-	-	≤1·0	≤2·0	≤0·08
AISI 430	≤0·5	-	14·0-18·0	Bal.	-	-	-	-	-	≤1·0	≤1·0	≤0·12
Ferralium 255	4·5-6·5	-	24·0-27·0	Bal.	-	-	-	2·0-4·0	1·3-4·0	-	-	≤0·08
Nimonic 90	Bal.	15·0-21·0	18·0-21·0	≤1·5	2·0-3·0	1·1-2·0	-	-	≤0·2	≤1·0	≤1·0	≤0·13
Nimonic 115	Bal.	13·5-16·5	14·0-16·0	≤1·0	3·5-4·5	4·5-5·5	-	3·0-5·0	≤0·2	≤1·0	≤1·0	≤0·20
Nimonic 263	Bal.	19·0-21·0	19·0-21·0	≤0·7	1·9-2·4	0·3-0·6	-	5·6-6·1	≤0·2	≤0·4	≤0·6	≤0·9
Nimonic 901	42·5	≤1·0	12·5	Bal.	2·9	≤0·35	-	5·75	≤0·5	≤0·4	≤0·5	≤0·1
Hastelloy B2	Bal.	≤1·0	≤1·0	≤2·0	-	-	-	26·0-30·0	-	≤0·20	≤1·75	≤0·02
Hastelloy C4	Bal.	≤2·0	14·0-18·0	≤3·0	≤0·70	-	-	14·0-17·0	-	≤0·20	≤1·0	≤0·01
Hastelloy C22	Bal.	≤2·5	20·0-22·5	2·0-6·0	-	-	-	12·5-14·5	-	≤0·08	≤0·50	≤0·015
Hastelloy C276	Bal.	≤2·5	14·5-16·5	4·0-7·0	-	-	-	14·0-17·0	-	≤0·08	≤1·0	≤0·01
Hastelloy X	Bal.	0·5-2·5	20·5-23·0	17·0-20·0	-	-	-	8·0-10·0	-	≤1·0	≤1·0	0·05-0·15
Inconel 600	72·0	-	14·0-17·0	6·0-10·0	-	-	-	-	0·5	0·5	1·0	0·15
Inconel 625	Bal.	≤1·0	20·0-23·0	≤5·0	≤0·4	≤0·4	3·15-4·15	8·0-10·0	≤1·0	≤0·5	≤0·5	≤0·1
Inconel 718	50·0-55·0	≤1·0	17·0-21·0	Bal.	0·65-1·15	0·2-0·8	4·75-5·5	2·8-3·3	≤0·3	≤0·35	≤0·35	≤0·08
Inconel X-750	Bal.	≤1·0	15·5	7·0	2·5	0·7	1·0	-	≤0·5	≤0·05	≤0·05	≤0·08

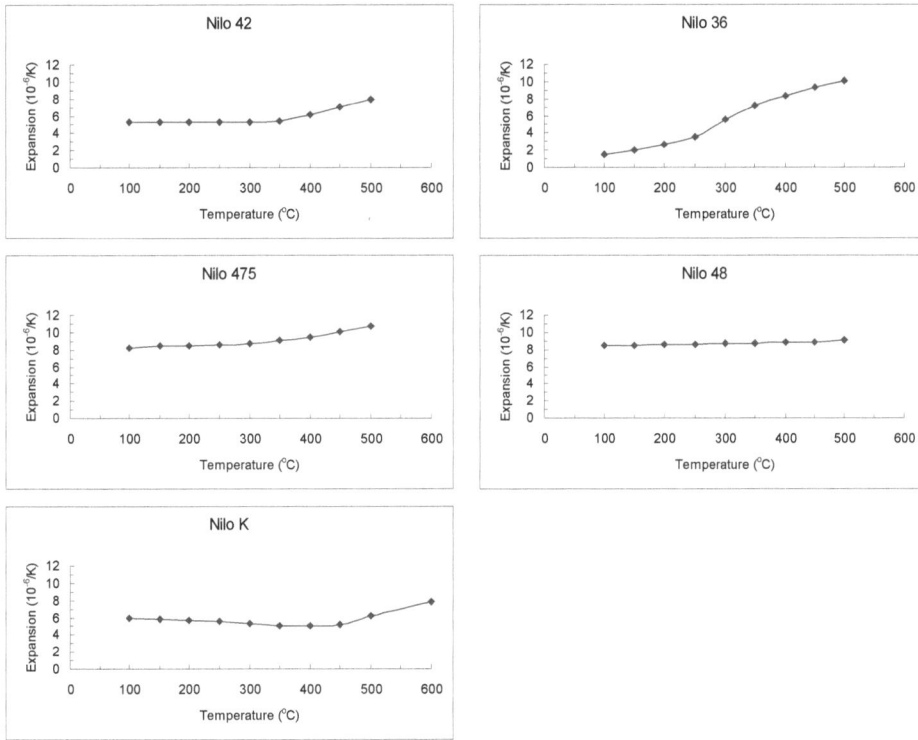

(a) Fe–Ni based alloys

Figure 6: Thermal expansion curves for a selection of metals and alloys. The expansion data points represent the thermal expansion coefficient over the range 20°C to the temperature value given

turbines. Superalloys are generally based on Ni with additions of Fe, Cr and Co, together with smaller proportions of Mo, W, Nb, Ta, Al, Cu and C. Some super-alloys are also based on Co or Fe. They are generally solution and precipitation hardenable, relying mainly on the formation of intermetallic and carbide phases for strength. The thermal expansion characteristics of some Ni-based superal-loys are summarized in Table 2 and are illustrated graphically in Figure 6(c)–(e) which shows the instantaneous thermal expansion coefficient as a function of temperature. It is clear that the expansion behaviour is also far from linear for these alloys, with many of them exhibiting sharp inflections with temperature. Most of these alloys possess relatively similar thermal expansions of around 14×10^{-6} K^{-1} over the range 20–400°C for the Ni-based superalloys, an exception being Hastelloy B2, a Mo-rich alloy, which has an expansion of $11 \cdot 4 \times 10^{-6}$ K^{-1} over the same range. Nickel-based superalloys have been employed recently in the manufacture of ultra-high performance explosive actuators and related devices which utilize high strength glass-ceramic-to-metal seals.

Figure 6: (continued)
(b) Fe-based alloys (stainless steels)

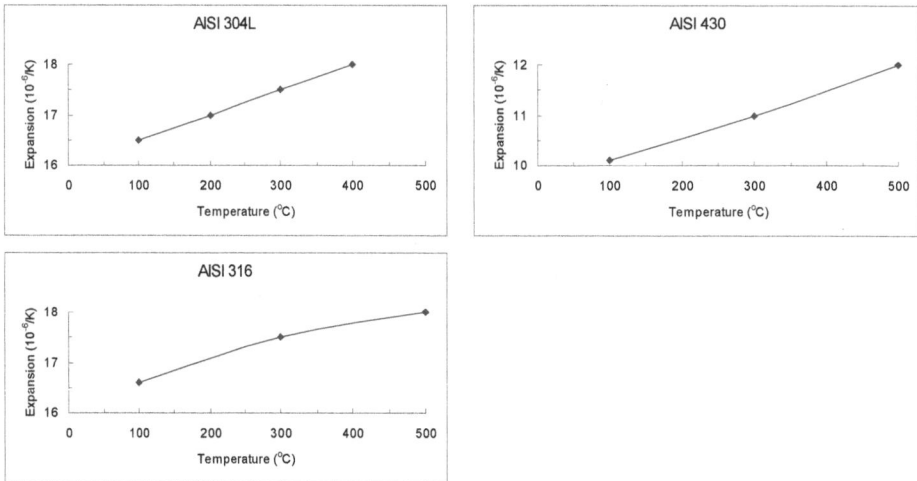

(c) Ni-based alloys (Inconel superalloys)

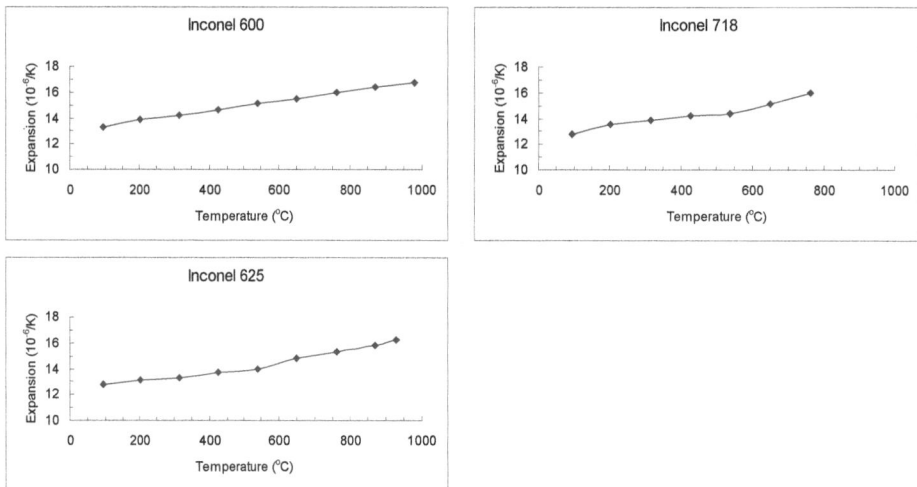

2.5.6   Titanium and its alloys

Pure Ti exhibits a moderate thermal expansion coefficient of $9{\cdot}2\times10^{-6}$ K^{-1} over the range 0–860°C. It undergoes an α to β phase transformation from a low temperature hcp crystal structure to a high temperature bcc phase at 882°C which is accompanied by a volume change. The development of Ti-based alloys in which the α or β phases are stabilized was driven in large part by the aerospace industry in the search for high strength, light-weight alloys for supersonic aircraft engines and structures, including compressor blades and discs (e.g. Polmear, 1988). More recently, Ti and its alloys have found applications in the chemical and biomedical fields where its good corrosion resistance can be put to advantage. The α-stabilised

Figure 6: (continued)
(d) Ni-based alloys (Hastelloy superalloys)

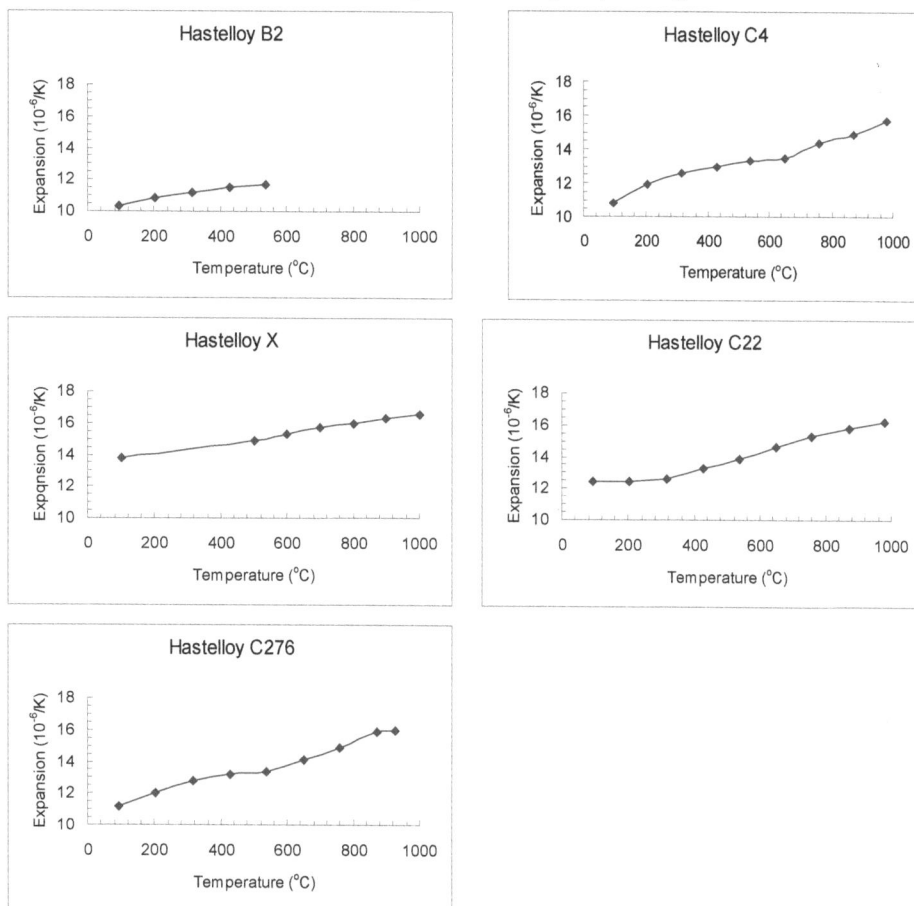

alloys generally contain Cu, Al or Sn alloying additions, whilst the β-stabilised alloys may contain Al, Zr, Mo and Si. Alloys may also contain smaller additions of V, Nb, Pd, Cr, Fe or Bi. Mixed α/β alloys include the widely used Ti–6Al–4V alloy. Sealing of glasses and glass-ceramics to Ti and Ti alloys has been carried out for a number of applications; for example, protective coatings and coatings for biomedical implants.

2.5.7 Tantalum

Pure Ta exhibits a thermal expansion coefficient of $7 \cdot 5 \times 10^{-6}$ K^{-1} up to around 1700°C. One of its major uses is in the electronics industry for the manufacture of capacitors for computers, portable telephones and pagers, and automobile electronics. Its excellent corrosion resistance, particularly to non-fluorinated acids, is also taken advantage of in the chemical industry. It is a particularly difficult metal to seal to or coat, as described more fully in Chapter 8.

Figure 6: (continued)
(e) Ni-based alloys (Nimonic superalloys)

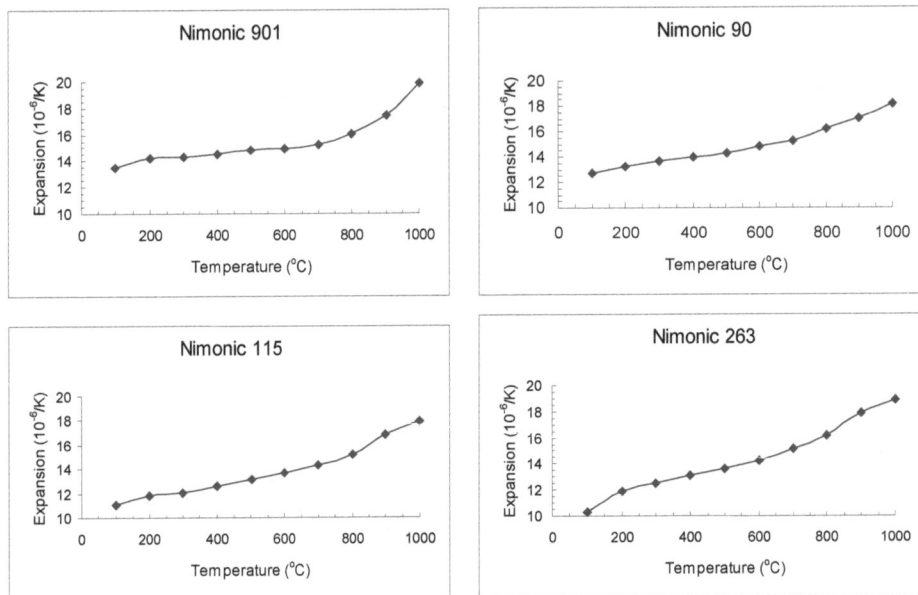

2.5.8   Aluminium and its alloys

Pure Al exhibits a high thermal expansion coefficient of $24 \cdot 9 \times 10^{-6}$ K^{-1} over the range 30–300°C. Aluminium and its alloys are used in a very wide range of applications including use in the automobile, aerospace, building and architectural, packaging, and electrical industries (e.g. Woodward, 1989). The most common alloys are based on Al–Mg, Al–Si, Al–Mg–Cu, Al–Zn–Mg, Al–Mg–Si, Al–Zn–Mg–Cu, and Al–Mg–Li. The enamelling of Al and its alloys has been practised for many years in order to provide a protective finish. More recently, light weight seal components have used aluminium in a number of applications, including electronic packaging.

2.6   Types of glasses

As noted by Partridge(1949), glasses used for sealing and coating applications are often classified according to their thermal expansion and temperature characteristics into "hard" and "soft" categories, and this is still true today. Hard glasses generally have low thermal expansion, i.e. $\alpha < 5 \times 10^{-6}$ K^{-1}, whilst soft glasses possess higher thermal expansions, i.e. $\alpha > 8 \times 10^{-6}$ K^{-1}. In general, glasses of high expansion possess relatively low softening and working temperatures, whilst low expansion glasses have higher softening points. Glasses with low softening temperatures are often referred to as "solder" glasses. This is because they are used for sealing or joining glasses to each other or to other inorganic materials including crystalline ceramics and metals in much the same way as a metallic solder is used to join metallic materials (Takamori, 1979; Singh and

TABLE 2
THERMAL EXPANSION DATA FOR SOME METALS AND ALLOYS

Alloy	Thermal Expansion $(10^{-6}\,K^{-1})$	Temperature Range (°C)
(a) Metals		
Cu[d,e]	16·6	25
	17·2	100
	18·5	300
	19·2	500
	20·7	700
	22·5	900
Al[d,e]	24·9	30–300
	31·3	300–600
Ti[d,e]	8·5	25
	9·2	−120–860
	10·5	860–960
Ni[d,e]	13·0	25
	13·2	−130–420
	17·1	420–990
Ta[d,e]	6·5	25
	7·5	−183–1670
W[d,e]	4·5	25
	5·5	−150–2130
Mo[d,e]	5·0	25
	6·6	−190–1670
(b) Fe-based alloys		
AISI 316 stainless steel	16·6	25–100
	17·5	25–300
	18·0	25–500
AISI 321 stainless steel[c]	16·7	0–100
	17·1	0–315
	18·5	0–540
	19·3	0–650
	20·2	0–815
AISI 430 stainless steel	10·1	0–100
	11·0	0–300
	12·0	0–500

Schukla, 1984; Rabinovich, 1985). The thermal expansion against temperature curves of glasses are usually quite similar in shape, unlike that of many metals and metallic alloys where highly nonlinear behaviour may be observed due to solid state phase transitions occurring as a function of temperature. Some typical expansion curves for common glasses are shown in Figure 7. It can be seen that expansion as a function of temperature is normally near linear in behaviour up to temperatures around T_g at which a pronounced increase is normally observed associated with an increase in volume. A wide variety of sealing and solder glasses

(continued)

Alloy	Thermal Expansion $(10^{-6} K^{-1})$	Temperature Range (°C)
(c) Ni-based alloys		
Nimonic 263[a]	11·9	20–200
	13·1	20–400
	14·2	20–600
	16·2	20–800
	18·9	20–1000
Nimonic 90(a)	13·3	20–200
	14·0	20–400
	14·8	20–600
	16·2	20–800
	18·2	20–1000
Inconel 625(a)	13·1	20–205
	13·7	20–425
	14·8	20–650
	15·8	20–870
	16·2	20–930
Inconel 718(a)	13·5	24–204
	14·2	24–427
	15·1	24–649
	16·0	24–760
Hastelloy C276(b)	12·0	24–204
	13·2	24–427
	13·4	24–538
Nilo-K (kovar)(a)	5·7	20–200
	5·3	20–300
	5·0	20–400
	6·2	20–500

Data from the following sources:- (a) Henry Wiggin & Co. alloy data sheets; (b) Haynes International alloy data sheets; (c) Handbook of stainless steel; (d) Weast, 1974; (e) Samsonov, 1968

are available commercially for bonding to many different metals and alloys. A summary of some of these is given in Table 3(a). Data for experimental sealing and solder glasses are given in Table 3(b). The chemical durability of a selection of sealing and solder glasses is given in Table 4. Glasses can also be broadly classified according to their composition, as summarized below.

2.6.1 Silicate- and borosilicate-based glasses

There are many sealing, solder and enamelling glass compositions based on the silicate and borosilicate systems (e.g. Dalton, 1946; Andrews, 1961; Kataoka *et al*, 1962, 1972; McMillan *et al*, 1970; Ellis, 1971; Kreidl, 1983). The most com-

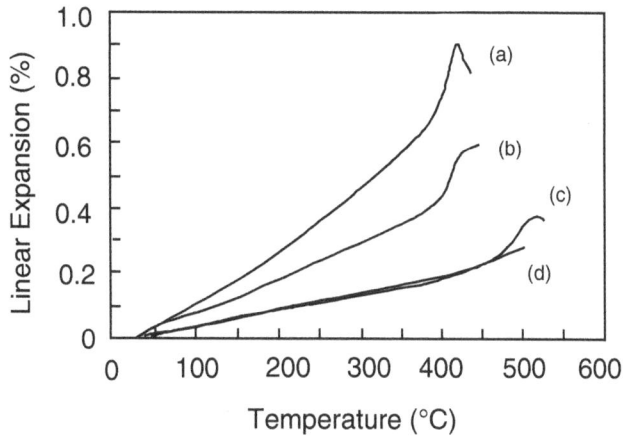

(a) Various experimental glass compositions:-
(a) phosphate glass, (after Chambers et al, 1989)
(b) lithium zinc silicate glass, (after Donald, Metcalfe & Morris, 1992)
(c) lead aluminoborosilicate glass, (after Kataoka & Manabe, 1972)
(d) borosilicate glass, (after Dalton, 1946)

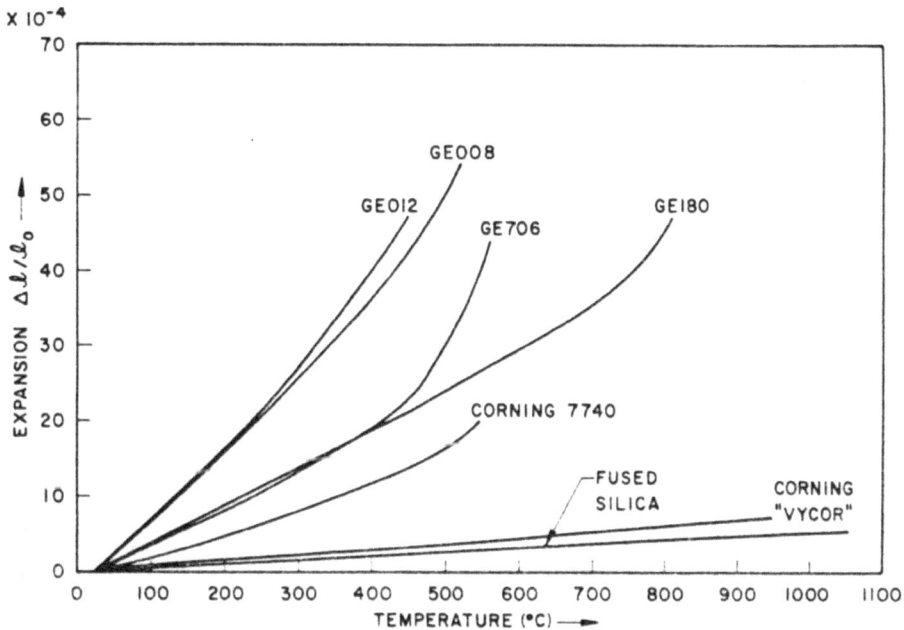

(b) A selection of commercial glasses (after Varshneya, 1982)

Figure 7: Thermal expansion curves for a selection of glasses

mon systems include zinc borosilicates, alkali silicates, alkali borosilicates, lead silicates, lead zinc silicates, and alkali copper silicates. A wide range of thermal expansions is possible in these systems, ranging from around 3 to 14×10^{-6} K^{-1}, as summarized in Table 3.

TABLE 3(a)
COMMERCIALLY AVAILABLE SEALING AND SOLDER GLASSES

Manufacturer and Code	Type of glass	Application	Softening temperature (°C)	Working temperature (°C)	Thermal expansion (10⁻⁶ K⁻¹)	Temperature range (°C)
Schott 8474	alkali phosphate glass	solder sealing	420	512	19·0	20–300
Morgan GBC226	high expansion Pb–free	glass preform	–	–	13·0	0–300
Corning 1990	potash soda lead glass	Fe sealing	500	756	12·4	0–300
Schott 8472	lead borate glass	solder sealing	360	426	12·0	20–300
Schott 8471	lead borate glass	solder sealing	389	456	10·6	20–300
Schott 8470	lead-free glass	solder sealing	570	748	10·0	20–300
Corning 7595	crystallizable solder glass	solder sealing	415 (solder temperature)		9·7	25–set point
Schott 8468	lead borate glass	solder sealing	405	460	9·6	20–300
Corning 7572	crystallizable solder glass	solder sealing	450 (solder temperature)		9·5	25–set point
Morgan GBC400	soda–lime–silica glass	glass preform	720	–	9·4	0–300
Morgan GBC203	–	glass preform	625	850	9·3	0–300
Morgan GBC614	–	glass preform	650	950	9·2	0–300
Schott 8597	crystallizable solder glass	solder sealing	435 (solder temperature)		9·2	20–300
Schott 8467	lead borate glass	solder sealing	418	518	9·1	20–300
Schott 8095	alkali lead silicate glass	Cu sealing	630	982	9·1	20–300
Schott 8512	FeO-containing glass	52Ni/Fe alloy sealing	660	975	9·1	20–300
Schott 8418	alkali alkaline earth silicate glass	51Ni/1Cr/Fe alloy sealing	708	1035	9·0	20–300
Schott 8531	dense lead silicate glass	Cu sealing/ encapsulation of semi-conductors	585	822	9·0	20–300
Corning 7575	crystallizable solder glass	solder sealing	450 (solder temperature)		8·9	25–set point
Corning 9010	potash soda barium glass	–	650	1010	8·9	0–300
GB Glass X88	lead zinc borate glass	solder sealing	550 (solder temperature)		8·8	50–200
Schott 8596	crystallizable solder glass	solder sealing	450 (solder temperature)		8·7	20–300
Corning 7570	high lead glass	solder sealing	440	558	8·4	0–300

GB Glass GS85	alkali alkaline earth borosilicate glass	for graded seals	710	—	8·3	50–400
Corning 7583	crystallizable solder glass	solder sealing	480 (solder temperature)	—	8·3	25–set point
Schott 8465	lead borate glass	solder sealing	461	566	8·2	20–300
Schott 8593	crystallizable solder glass	solder sealing	520 (solder temperature)	—	7·7	20–300
GB Glass X76	lead zinc borate glass	solder sealing	550 (solder temperature)	—	7·6	50–200
GB Glass GS77	alkali alkaline earth borosilicate glass	for graded seals	720	—	7·5	50–400
Corning 7578	crystallizable solder glass	solder sealing	530 (solder temperature)	—	6·8	25–set point
Corning 7556	high lead glass	solder sealing	330	—	6·7	25–set point
Schott 8436	alkali alkaline earth silicate glass	28Ni/23Co/Fe alloy and sapphire sealing	810	1095	6·6	20–300
Schott 8595	crystallizable solder glass	solder sealing	500 (solder temperature)	—	6·5	20–300
Schott 8454	alkali alkaline earth silicate glass	28Ni/23Co/Fe alloy and alumina ceramics sealing	745	1050	6·4	20–300
GB Glass GS65	alkali alkaline earth borosilicate glass	for graded seals	745	—	6·2	50–400
Morgan GBC190	—	glass preform	710	990	5·6	0–300
Morgan GBC593	—	glass preform	720	980	5·3	0–300
Corning 7056	borosilicate glass	kovar sealing	718	1058	5·2	0–300
Schott 8245	high boron glass	28Ni/18Co/Fe alloy and Mo sealing	710	1040	5·2	20–300
Morgan GBC590	borosilicate glass	glass preform - kovar sealing	718	980	5·1	0–300
Schott 8250	high boron glass alloy and Mo sealing	28Ni/18Co/Fe	715	1060	5·0	20–300
GB Glass GS50	alkali alkaline earth borosilicate glass	for graded seals	770	—	4·9	0–400
GB Glass X49BK	zinc vanadium borate glass	solder sealing	610 (solder temperature)	—	4·8	—

Manufacturer and Code	Type of glass	Application	Softening temperature (°C)	Working temperature (°C)	Thermal expansion ($10^{-6} K^{-1}$)	Temperature range (°C)
Schott 2877	alkaline earth borosilicate glass	Mo sealing	790	1170	4·9	20–300
Corning 7040	borosilicate glass	kovar sealing	702	1080	4·8	0–300
Morgan GBC691	borosilicate glass	glass preform - kovar sealing	720	1030	4·6	0–300
Morgan GBC691	borosilicate glass	glass preform - kovar sealing	712	990	4·6	0–300
Corning 7052	borosilicate glass	kovar/Mo or W sealing	712	1128	4·6	0–300
Corning 7574	crystallizable solder glass	solder sealing	750 (solder temperature)	(solder temperature)	4·5	25–set point
GB Glass GS44	alkali alkaline earth borosilicate glass	for graded seals	790	–	4·3	50–400
Corning 1720	aluminosilicate glass	–	915	1190	4·2	0–300
Corning 7593	crystallizable solder glass	solder sealing	650 (solder temperature)	(solder temperature)	4·2	25–set point
Schott 8486	alkaline earth borosilicate glass	W sealing	805	1230	4·1	20–300
Corning 3320	borosilicate glass	W sealing	780	1171	4·0	0–300
Schott 8487	high boron glass	W sealing	770	1135	3·9	20–300
Corning 9700	borosilicate glass	–	805	1200	3·7	0–300
Corning 7720	borosilicate glass	W sealing	755	1146	3·6	0–300
Corning 7740	borosilicate glass	–	820	1145	3·3	0–300
Morgan GBC510	borosilicate glass	glass preform	820	–	3·2	0–300
GB Glass GS30	alkali alkaline earth borosilicate glass	for graded seals	1075	–	3·0	50–400

TABLE 3(b)
EXPERIMENTAL SEALING AND SOLDER GLASS SYSTEMS
(Glass compositions given in Table 9)

Glass	Softening or sealing temperature (°C)	Thermal expansion (α, 10^{-6} K^{-1})	Temperature range (°C)	Reference
Silicate glasses				
S1	480	14·1	20–400	McMillan, Partridge and Ward, 1970
S2	500	11·9	20–400	McMillan, Partridge and Ward, 1970
S3	500	10·9	20–400	McMillan, Partridge and Ward, 1970
S4	–	7·4	not specified	Kataoka and Manabe, 1972
S5	840	2·3	0–300	Ellis, 1971
Borosilicate glasses				
BS1	435	14·4	20–400	McMillan, Partridge and Ward, 1970
BS2	515	8·3	20–400	McMillan, Partridge and Ward, 1970
BS3	–	8·0	not specified	Kataoka and Manabe, 1972
BS4	525	7·7	20–400	McMillan, Partridge and Ward, 1970
BS5	500	7·6	20–400	McMillan, Partridge and Ward, 1970
BS6	465	6·1	20–200	McMillan, Partridge and Ward, 1970
BS7	460	5·4	25–460	Dalton, 1946
Aluminoborate glasses				
AB1	650	3·8	0–300	Malmendier, 1975
AB2	570	3·5	0–300	Malmendier, 1975
AB3	570	3·5	0–300	Malmendier, 1975
Lead borate glasses				
LB1	370	11·0	not specified	Hitachi, 1980
LB2	380	10·4	not specified	Hitachi, 1980
LB3	–	9·3	not specified	Kataoka and Manabe, 1972
Zinc borate glasses				
ZB1	300	12·0	30–250	Malmendier ,1975
ZB2	325	11·5	30–300	Yamanaka and Takagi, 1975
ZB3	336	10·3	30–300	Yamanaka and Takagi, 1975
ZB4	400	7·7	20–300	McMillan, Partridge and Ward, 1970
ZB5	280	6·6	20–200	McMillan, Partridge and Ward, 1970
ZB6	535	5·4	20–400	McMillan, Partridge and Ward, 1970

Glass	Softening or sealing temperature (°C)	Thermal expansion (α, 10^{-6} K^{-1})	Temperature range (°C)	Reference
ZB7	400	5·0	20–400	McMillan, Partridge and Ward, 1970
ZB8	571	5·0	0–425	Pirooz, 1963a, 1963b
ZB9	not specified	4·5	50–350	Suzuki, Nagahara and Ichimura, 1972
ZB10	700	4·2	30–300	Moriguchi, Miwa and Shibuya, 1975
Lead zinc borate glasses				
LZB1	–	12·1	not specified	Sack, Scheidler and Petzoldt, 1968
LZB2	334	11·7	50–250	Ishiyama, Matsuda, Nagahara and Suzuki, 1966
LZB3	410	10·7	20–400	McMillan, Partridge and Ward, 1970
LZB4	–	8·6	not specified	Sack, Scheidler and Petzoldt, 1968
LZB5	435	8·4	20–400	McMillan, Partridge and Ward, 1970
LZB6	450	7·6	20–400	McMillan, Partridge and Ward, 1970
LZB7	490	5·9	20–400	McMillan, Partridge and Ward, 1970
Lead or zinc borate glasses containing a metal halide				
BH1	290	15·2	20–200	McMillan and Partridge, 1973
BH2	295	14·2	20–200	McMillan and Partridge, 1973
BH3	200	13·7	20–100	McMillan and Partridge, 1973
BH4	230	11·5	20–100	McMillan and Partridge, 1973
BH5	295	10·1	20–200	McMillan and Partridge, 1973
BH6	440	7·5	not specified	Hikino and Mikoda, 1971
BH7	510	5·8	not specified	Hikino and Mikoda, 1971
Phosphate glasses				
P1	212	34·7	100–250	Peng and Day, 1991b
P2	221	30·0	100–250	Peng and Day, 1991b
P3	266	26·8	100–250	Peng and Day, 1991b
P4	317	26·0	100–250	Peng and Day, 1991a
P5	303	25·6	100–250	Peng and Day, 1991a
P6	313	24·0	100–250	Peng and Day, 1991a
P7	350	22·0	100–250	Peng and Day, 1991a
P8	374	20·0	100–250	Peng and Day, 1991a
P9	310	17·1	100–200	Asahara and Izumitani, 1975
P10	255	15·3	100–200	Asahara and Izumitani, 1975
P11	345	13·2	100–200	Asahara and Izumitani, 1975
P12	325	12·1	100–200	Asahara and Izumitani, 1975
P13	450	11·6	20–400	McMillan, Partridge and Ward, 1970

Glass	Softening or sealing temperature (°C)	Thermal expansion (α, 10^{-6} K^{-1})	Temperature range (°C)	Reference
P14	330	11·6	100–200	Asahara and Izumitani, 1975
P15	305	11·0	20–200	McMillan, Partridge and Ward, 1970
P16	460	9·6	20–400	McMillan, Partridge and Ward, 1970
P17	540	7·1	20–400	McMillan, Partridge and Ward, 1970
P18	475	7·0	20–400	McMillan, Partridge and Ward, 1970
P19	500	6·4	20–400	McMillan, Partridge and Ward, 1970
Antimonate glasses				
A1	252	21·2	not specified	Eubank and Beck, 1958
A2	250	19·2	not specified	Eubank and Beck, 1958
A3	292	18·2	not specified	Eubank and Beck, 1958
A4	290	15·0	not specified	Eubank and Beck, 1958
A5	312	13·3	not specified	Eubank and Beck, 1958
A6	342	12·2	not specified	Eubank and Beck, 1958
Vanadate glasses				
V1	328	11·8	25–150	Malmendier and Sojka, 1975
V2	308	10·1	25–150	Malmendier and Sojka, 1975
V3	335	7·3	25–150	Malmendier and Sojka, 1975
Devitrifiable solder glasses				
DS1	330	11·3	20–300	Sack, Scheidler and Petzoldt, 1968
DS2	374	10·7	20–300	Sack, Scheidler and Petzoldt, 1968
DS3	356	10·0	20–300	Sack, Scheidler and Petzoldt, 1968
DS4	380	9·6	not specified	Minagawa and Suzuki, 1974
DS5	503	9·3	100–300	Sack, Scheidler and Petzoldt, 1968
DS6	390	8·0	20–300	Sack, Scheidler and Petzoldt, 1968
DS7	600	7·9	100–300	Suzuki and Ichimura, 1968
DS8	424	7·4	20–300	Sack, Scheidler and Petzoldt, 1968
DS9	406	6·6	20–300	Sack, Scheidler and Petzoldt, 1968
DS10	550	6·6	not specified	Martin, 1970
DS11	600	5·4	not specified	Martin, 1970
DS12	440	5·2	20–300	Suzuki and Ichimura, 1968
DS13	620	4·8	not specified	Martin, 1970
DS14	595	4·8	100–300	Sack, Scheidler and Petzoldt, 1968
DS15	620	4·4	20–500	Pirooz, 1963a, 1963b

TABLE 4
CHEMICAL DURABILITIES OF SELECTED SEALING AND SOLDER GLASSES

Glass (mol%)	Durability ($g\ cm^{-2}\ min^{-1}$)	Reference
Soda–lime–silica glass	2×10^{-8}	McLellan & Shand, 1984
$50ZnO–50P_2O_5$	$3 \cdot 4 \times 10^{-4}$	He & Day, 1992
$24 \cdot 5PbO–24ZnO–3 \cdot 5Al_2O_3–6B_2O_3–2CuO–40P_2O_5$	$5 \cdot 1 \times 10^{-7}$	He & Day, 1992
$22 \cdot 5PbO–3Al_2O_3–2CuO–2Fe_2O_3–7Li_2O–40P_2O_5$	$1 \cdot 1 \times 10^{-6}$	He & Day, 1992
$30 \cdot 1K_2O–19 \cdot 9PbO–20 \cdot 3Nb_2O_5–29 \cdot 8P_2O_5$	$4 \cdot 1 \times 10^{-6}$	Aranha & Alves, 1995
$30Na_2O–20BaO–3Al_2O_3–47P_2O_5$ (glass-ceramic)	$5 \cdot 7 \times 10^{-7}$	Wilder et al, 1982
$40NaO–10BaO–Al_2O_3–49P_2O_5$ (glass-ceramic)	$5 \cdot 4 \times 10^{-5}$	Wilder et al, 1982
$34NaO–16CaO–50P_2O_5$	$2 \cdot 3 \times 10^{-4}$	Delahaye et al, 1999
$57 \cdot 7TiO_2–4 \cdot 9Al_2O_3–37 \cdot 4P_2O_5$	$1 \cdot 5 \times 10^{-9}$	Brow et al, 1997
$48 \cdot 8CaO–1 \cdot 4Al_2O_3–49 \cdot 8P_2O_5$	$2 \cdot 3 \times 10^{-8}$	Brow et al, 1997
$45SrO–15Al_2O_3–40B_2O_3$	$2 \cdot 0 \times 10^{-4}$	Brow et al, 1997
$40BaO–20Al_2O_3–40B_2O_3$	$1 \cdot 0 \times 10^{-5}$	Brow et al, 1997
$50CaO–20Al_2O_3–30B_2O_3$	$1 \cdot 0 \times 10^{-6}$	Brow et al, 1997
$39 \cdot 1CaO–29 \cdot 3TiO_2–2 \cdot 8Al_2O_3–28 \cdot 8P_2O_5$	$1 \cdot 8 \times 10^{-9}$	Brow et al, 1997
$50Na_2O–50P_2O_5$	$1 \cdot 1 \times 10^{-3}$	Rajaram & Day, 1986
$50Na_2O–50P_2O_5 + 5$ wt% AlN	$8 \cdot 3 \times 10^{-7}$	Rajaram & Day, 1986
$50Na_2O–50P_2O_5 + 6 \cdot 1$ wt% Al_2O_3	$2 \cdot 9 \times 10^{-6}$	Rajaram & Day, 1986
$38K_2O–6Al_2O_3–56P_2O_5$	$2 \cdot 0 \times 10^{-6}$	Peng & Day, 1991
$30K_2O–10Al_2O_3–60P_2O_5$	$1 \cdot 8 \times 10^{-8}$	Peng & Day, 1991
$30K_2O–10Fe_2O_3–60P_2O_5$	$2 \cdot 0 \times 10^{-7}$	Peng & Day, 1991
$50PbO–10ZnO–40P_2O_5$	$8 \cdot 8 \times 10^{-8}$	Liu et al, 1996
$16Na_2O–23 \cdot 5K_2O–7 \cdot 5Al_2O_3–8B_2O_3–45P_2O_5$	$2 \cdot 5 \times 10^{-7}$	Kilgo et al, 1999
$15Na_2O–18K_2O–9PbO–12Al_2O_3–6B_2O_3–40P_2O_5$	$3 \cdot 2 \times 10^{-9}$	Kilgo et al, 1999

2.6.2   Borate-based glasses

There are also many sealing and solder glass compositions in the aluminoborate, lead borate and zinc borate systems, some with very low softening temperatures, i.e. < 400°C (e.g. Takamori, 1979; McMillan et al, 1970; Dale and Stanworth, 1949; Pirooz, 1963a, 1963b; Suzuki et al, 1972; Malmendier, 1975; Yamanaka and Takagi, 1975; Moriguchi et al, 1975; Hitachi, 1980). The aluminoborate glasses generally have low-to-intermediate thermal expansions, i.e. $\alpha \approx 4$ to 8×10^{-6} K^{-1}, whilst a range of expansions is possible in the other systems, with values from $\alpha \approx 5$ to 15×10^{-6} K^{-1}. In order to lower the viscosity and softening temperatures of borate glasses further, various halide additions have also been incorporated (McMillan et al, 1970; Ishiyama et al, 1966; Hikono and Mikoda, 1971; McMillan and Partridge, 1973). Fluorine is the most useful addition, the other halides, Cl, Br and I, generally lowering the stability of the glasses to an unacceptable degree.

The chemical durability of borate-based glasses, although generally superior to the phosphate systems reported in the next Section, is very much lower than that of most silicate-based systems, and this can limit their useful applications.

2.6.3   Phosphate-based glasses

Zinc, barium and lead phosphate glasses have been reported with properties suitable for sealing, enamelling or solder applications (Fraser and Cianchi, 1953; Ray *et al*, 1973a, 1973b, 1976; Asahara and Izumitani, 1975; Tindyala and Ott, 1978; Kobayashi, 1987; Chambers *et al*, 1989; Abe and Hosono, 1989; Peng and Day, 1991a, 1991b). These glasses exhibit relatively low softening temperatures, often $< 500°C$. Unfortunately, with notable exceptions, phosphate glasses are not very resistant to chemical attack by water. Improved chemical resistance can be achieved, however, by addition of various oxides, e.g. MgO, PbO, Al_2O_3, SiO_2, B_2O_3, etc, although this is often at the expense of an increase in the softening temperature. More recently, addition of significant amounts of Fe_2O_3 has been found to improve durability very substantially, although this is again at the expense of an increase in melting temperature.

Phosphate-based glasses exhibit expansions ranging from around 6×10^{-6} K^{-1} for zinc phosphate based compositions to 15 to 20×10^{-6} K^{-1} for some alkali phosphates containing relatively high concentrations of ZnO.

Alkali aluminophosphate based glasses have also been developed for sealing to low melting point metals such as aluminium. The addition of Al_2O_3 to alkali phosphate glasses leads to improved chemical durability, with the increase in durability being explained on the basis of the formation of $AlPO_4$ groups which strengthen the glass network. It has been noted that the properties can be further improved through the mixed alkali effect (Brow and Tallant, 1997). It has also been reported that the thermal stability and resistance to crystallization of ternary sodium aluminophosphate glasses can be improved markedly by addition of small amounts of B_2O_3, typically < 5 mole% (Donald *et al*, 2004).

2.6.4   Glasses based on less common glass-forming oxides

A number of sealing and solder glass compositions have been reported based on some of the less common glass-forming oxides, in particular, Sb_2O_3 or As_2O_3 (Eubank and Beck, 1958), and V_2O_5 (Baynton *et al*, 1957; Hiruma and Shimizu, 1973; Malmendier and Sojka, 1975). Very high thermal expansions have been noted for a number of Sb_2O_3 based glasses, with values for α in excess of 20×10^{-6} K^{-1}. These glasses also exhibit low softening temperatures, often $< 350°C$, but their chemical durability is open to question and can be very poor, with cation valence states often being very critical. Glasses based on the $PbO–B_2O_3–ZnO–CdO–TeO_2$ system containing high concentrations of TeO_2, up to 40 wt%, and said to be suitable as sealing glasses for magnetic recording heads, have been reported by Mizuno *et al* (1992). The glasses were noted to be chemically compatible with ferrite recording materials and to exhibit good chemical durability.

2.6.5 Devitrifiable solder glasses

The glasses described in the preceding sections remain vitreous during and after sealing. A class of solder glass has been developed, however, which crystallizes either completely or partially during the sealing or soldering operation. These glasses, originally described by Claypoole (1969), are referred to as *devitrifiable* solder glasses. Devitrifiable solder glasses are normally based on either the relatively high thermal expansion lead zinc borate system, with $\alpha > 10 \times 10^{-6}$ K^{-1}, or the lower expansion zinc borosilicate system, with $\alpha < 8 \times 10^{-6}$ K^{-1}, (Pirooz, 1963a, 1963b; Sack *et al*, 1968; Suzuki and Ichimura, 1968; Martin, 1970; Minagawa and Suzuki, 1974; Lee *et al*, 1990). Crystallizable solder glasses are usually employed by application of a mixture of glass particles in a suitable binder, with crystallization occurring during sealing by a surface nucleation mechanism. More comprehensive details on glass-forming systems in general are given elsewhere (e.g. Kreidl, 1983; Volf, 1984).

2.6.6 Composite systems

Composite systems in which a variety of ceramic materials are used as fillers in order to change or improve the properties of a base glass have been developed or proposed (Geodakyan, 2005). The glasses chosen are usually solder glass compositions, typically with a high PbO content. Fillers may include alumina, silica, certain glass-ceramic compositions, lead titanate, zircon, and a variety of alternative ceramics, added as particles in various quantities and in particle sizes ranging from around 8 to 45 μm. Very low expansion solder glass systems have also been developed for sealing materials of near-zero expansion, e.g. fused silica (Morena and Francis, 1998). In this instance, low melting temperature glasses from the SnO–ZnO–P$_2$O$_5$ and PbO–ZnO–B$_2$O$_3$ systems were employed with addition of low expansion fillers including (Co, Mg)$_2$.P$_2$O$_7$. These systems were purported to be suitable for fibre optic sealing applications.

2.6.7 Lead-free sealing and solder glasses

In general, low temperature sealing or solder glasses contain significant amounts of PbO. Currently, in line with environmental directives, there is a drive to reduce the amount of toxic materials such as PbO in these glasses, ideally without detriment to their sealing and chemical properties. Studies aimed specifically at reducing the amount of PbO include the investigation of copper phosphate based glasses for solder applications (Matusita *et al*, 2001) and systems from the SnO–ZnO–P$_2$O$_5$ system (Morena and Francis, 1998). Depending on the application, it is important, in the case of the copper phosphate glasses, to control the Cu$^+$/Cu^{2+} ratio in these glasses, T_g decreasing with increasing Cu$^+$ to below 300°C.

3. GLASS-TO-METAL BONDING THEORIES

The formation of a strong, robust and if necessary hermetic seal between two dissimilar materials depends strongly on both physical and chemical factors. In this respect, the nature of the bond between a glass (enamel) and a metal has been the subject of considerable controversy over the years, but a clearer understanding of the factors involved has gradually emerged. A number of approaches have been applied in an attempt to elucidate the major factors responsible for attaining good bonding characteristics, as outlined below.

Following from the pioneering work of King *et al* (1959), and later work by Pask and co-workers (Pask and Fulrath, 1962; Borom and Pask, 1966; Brennan and Pask, 1973), which was aimed at establishing the factors responsible for the formation of strong bonding between metals and glasses, it is now recognised that strong *chemical* bonding can be obtained between a metal and a glass only if the conditions are such that the glass at the interface can become and can remain saturated with an appropriate substrate metal oxide. This may be accomplished by dissolution into the glass of an oxide layer already present on the metal or, alternatively, by suitable redox (reduction–oxidation) reactions directly between the glass and the substrate. Under ideal conditions a well-defined transitional region between the glass and the metal may be absent, and bonding occurs via a "mono-oxide layer" where there is a very rapid switch in atomic bonding between the metallic substrate and the covalent–ionic glass. Under less favourable conditions, when excess metal oxide is present at the interface, a microstructurally defined oxide-rich transitional layer may be formed. Bonding is then directly via this oxide layer, and the resulting seal or coating will only be strong if the oxide layer adheres strongly to the metal substrate.

Over the years a number of approaches have been applied in an attempt to elucidate the major factors responsible for attaining good bonding characteristics between a metal and a glass, culminating in the chemical approach, as outlined. The steps along the way to the chemical theory are summarized below.

3.1   Thermodynamic approach

In order to form a strong bond between two dissimilar materials the materials must first be brought into intimate contact. In most instances involving sealing, one of the materials will be liquid (i.e. molten) at the sealing temperature. One of the first practical requirements for the formation of a viable glass-to-metal seal or coating is that the molten glass wets the surface of the metal and therefore spreads over it, as depicted in Figure 8. When a liquid is brought into contact with a solid,

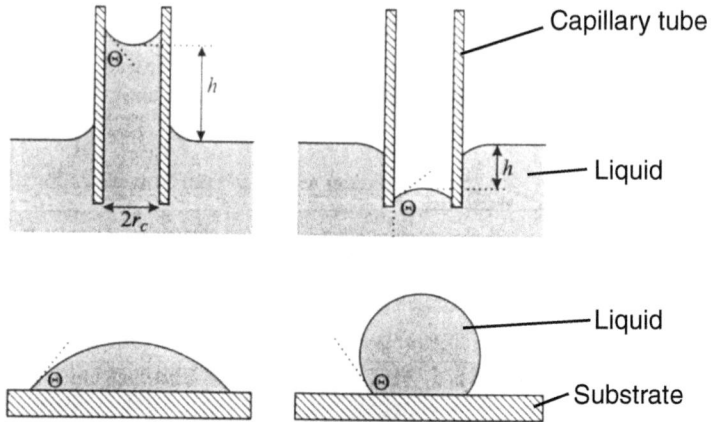

Figure 8: Wetting and non-wetting behaviour (after Butt, Graf and Kappl, 2003)

a new interface is created between the liquid and the solid. This behaviour is often studied experimentally using the sessile drop method, depicted schematically in Figure 9. As shown in this figure, the liquid may or may not spread across the solid surface. According to classical thermodynamic theory the liquid will spread only if the resultant energy of the new solid–liquid interface is less than that of the corresponding solid–vapour interface. The greater this difference, the greater will be the extent of spreading or wetting. The driving force for wetting is therefore related to the difference in energy between the solid–vapour and solid–liquid interfaces; however, because the lowest surface energy configuration for a liquid is a sphere, there is an additional force resisting this spreading. The shape of a sessile drop, as described by its contact angle θ, is therefore a function of three energy terms, as described by Young's classic equation (Young, 1805):-

$$\gamma_{SV} = \gamma_{SL} + \gamma_{LV}\cos\theta \qquad\qquad [3.1]$$

where, γ_{SV}, γ_{SL} and γ_{LV} are the interfacial energies of the solid–vapour, solid–liquid and liquid–vapour interfaces, respectively. A liquid will wet a solid ($\theta < 90°$) when the net energy of the system is lowered by forming a solid–liquid interface. The greater the degree of wetting, the smaller the angle θ. Conversely, the larger this angle, the lower the degree of wetting. It should be noted, of course, that under normal circumstances no liquid will exhibit the maximum contact angle of 180°, even in the case of no wetting, due to the distorting influence of gravity on the liquid droplet. A non-wetting liquid is unlikely to form a strong bond due to its inability to form an intimate interface on a microscopic scale, i.e. it will not penetrate the grain boundaries or other surface irregularities of the material to be bonded to. It may be possible, however, to increase the bonding tendency if reaction occurs between the liquid and the solid material that leads to an increase in the driving force for wetting, as described later.

The concept of the *work of adhesion*, W_A, was subsequently introduced by

(a)

vapour

(b)

vapour

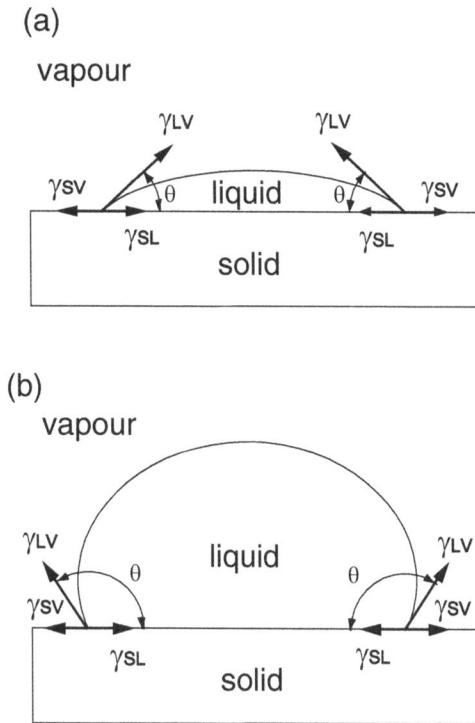

Figure 9: Sessile drop method for determining contact angle: (a) wetting; (b) non-wetting

Dupré (1869). This is related to the surface energies by the equation:-

$$W_A = \gamma_{SV} + \gamma_{LV} - \gamma_{SL} \qquad [3.2]$$

In order to separate the liquid and the solid, work has to be done that is equal to the lowering of the free energy through wetting. An interface is stable when W_A is positive, i.e. when forming the interface results in a decrease in the total free energy of the system. Introducing the contact angle, this equation may be written as:-

$$W_A = \gamma_{LV}(1 + \cos\theta) \qquad [3.3]$$

It should therefore be feasible experimentally to obtain quantitative values for the bonding strength and behaviour of different glass-to-metal systems.

It is recognised, however, that this simple thermodynamic approach to bonding should be treated with caution and should only be employed as a preliminary guide as to whether or not bonding conditions may be favourable for a given system. One of the main conclusions that could be drawn from this approach is that if the liquid has a lower surface energy than that of the solid substrate, the system can lower its energy by spreading the liquid phase over the solid surface, i.e. by wetting the solid. This conclusion can be grossly incorrect because it suggests, for example, that a change in the solid surface that lowers its energy will also lower its wettability. For glass-metal systems this is clearly not the case, as the presence of an oxide layer on the metal (which reduces its surface energy) will almost invariably enhance the wettability. More comprehensive details concerning the wetting and spreading of liquids on solids are given elsewhere (e.g. Adamson, 1982; Blokhuis, 2004; Katoh, 2004).

The simple thermodynamic approach given above assumes no reaction between the liquid and the solid. If a chemical reaction does occur at the interface, a new compound may be formed. Wetting and adhesion may subsequently be enhanced (as in the case of oxidation of a metal surface enhancing adhesion to a glass), and equation [3.3] requires modification to take this into account (Loehman, 1989):-

$$W_A = \gamma_{LV}(1 + \cos\theta) + \gamma_{Ri} - C\Delta G° \qquad [3.4]$$

where γ_{Ri} is the interfacial energy between the solid and the reacted interfacial layer, $\Delta G°$ is the Gibbs free energy of formation of the new interfacial compound, and C is a constant. A spontaneous reaction is characterized by a negative value for $\Delta G°$ which increases W_A and decreases θ. To give an idea of the most likely reactions to occur under bonding conditions, $\Delta G°$ can be calculated for different potential reactions from available thermodynamic data. Reactions with the greatest negative $\Delta G°$ values will be more favourable thermodynamically, although data so derived must again be treated with caution as a given reaction may be kinetically unfavourable. Further details in relation to chemical bonding are given in Section 3.4.

3.2   Mechanical bonding

It was originally thought that adherence between an enamel and a metal substrate occurred via mechanical keying effects. Mechanical theories of adherence originally gained acceptance due to the fact that the interface between a metal and a porcelain enamel often appeared rough, even if the substrate was originally quite smooth. Bonding was thought to occur due to the mechanical keying effect between the roughened substrate and the glass. A number of theories arose based on the mechanism by which this roughening was thought to occur. The dendrite theory, for example, favoured the precipitation within the glass of a metallic phase which provided anchor points, whilst the alternative electrolytic theory proposed that electrolytic attack roughened the surface. In either case the result was an interlocking metal-glass structure, as described below.

3.2.1   The dendrite theory

Expounders of the dendrite theory (King, 1933; Healey and Andrews, 1951; Harrison *et al*, 1952) noted that after sealing a glass to a metal, dendrites of a metallic phase, e.g. Co or a Co–Fe intermetallic phase, were often found within the interfacial region. It was suggested that these dendrites held the coating in place

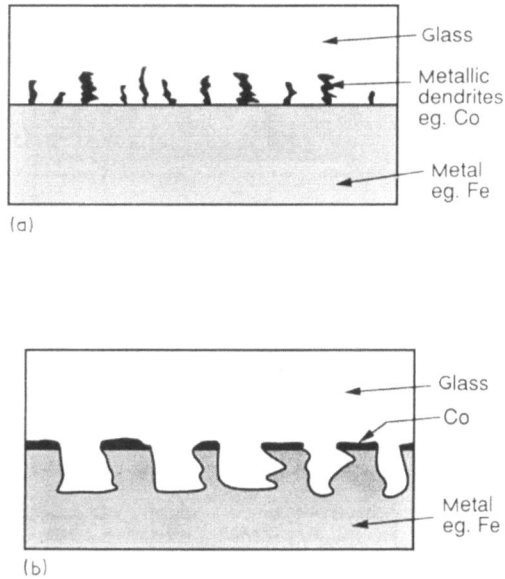

Figure 10: Schematic illustration of bonding between a glass and a metal
(a) bonding according to the dendrite theory; (b) according to the electrolytic theory

due to a mechanical keying effect; i.e. the dendrites acted as anchor points between the metal substrate and the glass, as illustrated in Figure 10(a). The dendrites were believed to be formed due to the reaction of a metal oxide present in the glass with a metallic element in the substrate. For example, for a glass containing CoO bonded to an iron substrate, the following reaction was believed to occur:-

$$CoO_{(glass)} + Fe_{(substrate)} \rightarrow Co_{(dendrite)} + FeO_{(glass)}, \ (\Delta G^\circ = -43 \text{ kJ mol}^{-1}) \quad [3.5]$$

3.2.2 The electrolytic theory

The electrolytic theory (Staley, 1934; Dietzel, 1934, 1935) proposed that roughening of the metal substrate occurred during coating due to an electrolytic or galvanic corrosion mechanism. For example, bonding of a CoO-containing glass in air to an iron substrate results in the precipitation of metallic Co, as proposed in the dendrite theory. In contrast to the dendrite theory, however, it was believed that those precipitates which were formed in contact with the Fe substrate formed the basis of localized galvanic or electrolytic cells. The net effect was the dissolution of Fe into the glass and the formation of a pitted surface into which the glass could flow, thus forming a number of mechanical keys which hold the glass onto the substrate, as illustrated in Figure 10(b). The reaction sequence proposed for this mechanism commences with equation [3.5], followed by:-

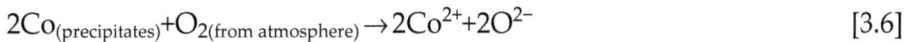

$$2Co_{(precipitates)} + O_{2(from \ atmosphere)} \rightarrow 2Co^{2+} + 2O^{2-} \quad [3.6]$$

$$Co^{2+} + 2e^- \rightarrow Co \quad [3.7]$$

$$Fe - 2e^- \rightarrow Fe^{2+} \quad [3.8]$$

Subsequent work (Harrison *et al*, 1952; Richmond *et al*, 1953; Moore *et al*, 1954) tended at first to support the general concepts of the mechanical keying theories. It was observed by Harrison *et al*, (1952), however, that there appeared to be little correlation between the ease of reduction of metal oxide additions and the strength of the resultant bond, i.e. additions which are easier to reduce than CoO and which should therefore yield a rougher substrate, did not in fact yield stronger bonding. This was one of the factors that eventually led to the mechanical keying theories falling out of favour.

3.3   Intermediate compound formation and mutual solubility

If a new chemical compound is formed by reaction of a liquid with a solid substrate, or mutual solubility occurs, the driving force for wetting may be increased. It has been observed that reactions that involve dissolution of the liquid into the substrate increase the driving force for wetting whilst reactions that involve dissolution of the substrate into the liquid do not.

Another mechanism by which bonding between a metal substrate and a glass could be achieved is via mutual solubility between the glass and an oxide scale on the metal. For example, some of the oxide scale is dissolved by contact with the molten glass, thus forming an intermediate glassy layer which is progressively richer in the metal oxide. Alternatively, bonding could be achieved via the formation of an intermediate compound which links the glass to the metal surface or to an oxide scale. This proposal is closely related to the thermodynamic and chemical approaches, and it is possible to predict compound formation. In this respect much use may be made of available phase diagrams.

3.4   Chemical bonding

It is recognised that from a chemical or molecular approach, bonding must be accomplished by a transitional zone in which the metallic bonding of the metal is gradually substituted for the ionic-covalent bonding of the glass. Strong opposition to the earlier mechanical theories of bonding came from work by King, Tripp and Duckworth(1959) on the bonding of glass to iron. The major conclusion of their studies was that a strong chemical bond was formed at the glass-metal interface only if the conditions were such that the glass could become saturated with an oxide of the substrate metal which, when in solution in the glass, would not be reduced by the metal, e.g. FeO in the case of an iron substrate. By changing the composition of the glass employed in their work, it was possible to alter the solubility of FeO. It was noted that the glass formers, e.g. SiO_2 and B_2O_3, increased the solubility of FeO in the glass, whilst the glass modifiers decreased its solubility, but as long as the glass was saturated with FeO strong bonding occurred independently of the absolute concentration at the interface. A number of alternative metallic systems were investigated, including copper, and it was confirmed that

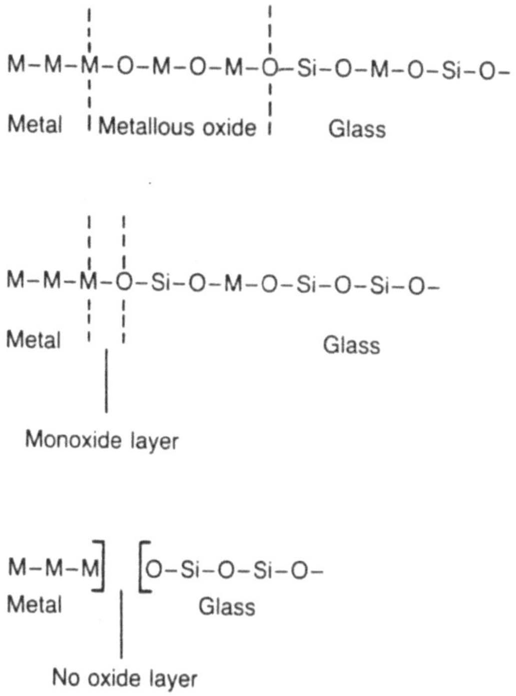

M−M−M−O−M−O−M−O−Si−O−M−O−Si−O−

Metal | Metallous oxide | Glass

M−M−M−O−Si−O−M−O−Si−O−Si−O−

Metal | | Glass

Monoxide layer

M−M−M] [O−Si−O−Si−O−

Metal | Glass

No oxide layer

Figure 11: Schematic representation of bonding according to the chemical theory (after Pask, 1971). Bonding is achieved via a bulk oxide layer, a mono-oxide film or weak van der Waals forces only

good bonding always occurred when the glass was saturated with the appropriate substrate metal oxide. It was suggested that when the appropriate metal oxide is dissolved in the glass up to its saturation point, metal ions will remain at the surface and these will promote metal–metal bonding across the interface, as depicted in Figure 11 (a monoxide layer). At elevated temperatures, when the metal ions in the glass and the atoms of the metal are relatively mobile, there will be a continuous exchange at the glass–metal interface. Metal ions from the glass will diffuse into the metal where they will gain electrons and become zero valent metal atoms, whilst metal atoms will diffuse into the glass and become ionized. A state of dynamic equilibrium will therefore exist at the metal–glass interface which could not be maintained if the glass were not already saturated with the appropriate oxide. At lower temperatures when the atoms and ions will be less mobile, there may exist an exchange of electrons, and ions and atoms at the interface may alternate in properties depending on the concentration of electrons or the degree of ionization. These atoms and ions of intermediate character are therefore expected to provide a transition between the metallic and ionic states, but only so long as the interface remains saturated.

Traditionally, as described by Partridge(1949), the appropriate metal oxide is provided by pre-oxidation of the metal substrate, although at this time it was not appreciated that chemical bonding on a molecular scale was necessary for the formation of a good seal. During sealing this oxide layer dissolves into the glass, and the interface becomes saturated, with favourable bonding conditions prevailing. If, on the other hand, sealing is attempted to a "clean" metal substrate (or, alternatively, sealing conditions are maintained such that all the pre-oxidized layer is dissolved and saturation of the interface is lost) suitable redox reactions must proceed between the metal and the glass in order to achieve or maintain conditions suitable for chemical bonding. The overall reaction between a glass and a *clean* metal substrate will normally involve two steps. Firstly, an oxide reaction product is formed on the metal at the interface by a suitable redox reaction. Secondly, this oxide dissolves into the glass. Saturation at the interface can therefore be achieved, even in the absence of an oxide layer formed first by pre-oxidation. If, however, reaction between a glass and a metal is not favourable, i.e. values for $\Delta G°$ of potential oxide products are positive, the reaction will not proceed spontaneously, at least under standard conditions. It may, however, still be possible for a reaction to proceed under certain ostensibly unfavourable conditions (Pask, 1971, 1987a, 1987b; Pask and Tomsia, 1981). In one particular example, the reaction between a clean Fe substrate and a sodium silicate glass, Tomsia and Pask(1990) noted that the following sequence of reactions was possible:-

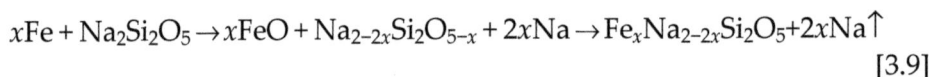

$$x\text{Fe} + \text{Na}_2\text{Si}_2\text{O}_5 \rightarrow x\text{FeO} + \text{Na}_{2-2x}\text{Si}_2\text{O}_{5-x} + 2x\text{Na} \rightarrow \text{Fe}_x\text{Na}_{2-2x}\text{Si}_2\text{O}_5 + 2x\text{Na}\uparrow$$

[3.9]

The free energy relationship for the first step in this reaction will be:-

$$\Delta G = \Delta G° + RT\ln\{[a(\text{FeO})_{(\text{interface})}]^x[a(\text{Na})]^{2x}/[a(\text{Na}_2\text{O})]\}$$

[3.10]

where a is the chemical activity of each component. Under standard conditions, the chemical activities are equal to unity, and $\Delta G = \Delta G°$.

The net reaction is:-

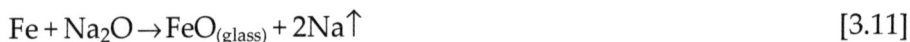

$$\text{Fe} + \text{Na}_2\text{O} \rightarrow \text{FeO}_{(\text{glass})} + 2\text{Na}\uparrow$$

[3.11]

This reaction consists of two steps; first, the formation of the metal oxide at the interface:-

$$\text{Fe} + \text{Na}_2\text{O}_{(\text{glass})} \rightarrow \text{FeO}_{(\text{interface})} + 2\text{Na}\uparrow$$

[3.12]

And secondly:-

$$\text{FeO}_{(\text{interface})} \rightarrow \text{FeO}_{(\text{glass})}$$

[3.13]

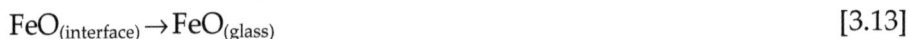

The overall free energy relationship for this type of reaction is:-

$$\Delta G = \Delta G° + RT\ln\{[a(\text{FeO})_{(\text{interface})}][a(\text{Na})]^2/[a(\text{Fe})a(\text{Na}_2\text{O})_{(\text{glass})}]\}$$

[3.14]

The first stage of the reaction (equation [3.12]) does not take place spontaneously at a sealing temperature of 1000°C because no cation in this glass has an oxidation potential less than that of the Fe substrate atom and the value for $\Delta G°$ is positive for this particular reaction (+33·7 kJ mol^{-1}). The reaction may proceed, however, if the value for the activity quotient, $\{[a(FeO)_{(interface)}][a(Na)]^2/[a(Fe)a(NaO)_{(glass)}]\}$, is sufficiently less than one. This is favoured by a low partial pressure of Na, a low activity for FeO, and a high activity for Na_2O. Suitable conditions can be achieved in practice by control over the process parameters, e.g. sealing atmosphere and pressure. The pressure of the Na vapour produced must exceed ambient pressure to form bubbles at the interface and so escape. The reaction is therefore potentially more favourable if sealing is carried out at low ambient pressure. Tomsia and Pask(1990) stressed the importance of being aware of such step reactions and how they may affect the processing parameters. It is important to note, of course, that sealing at reduced pressure may not be desirable, particularly if the starting glass has been produced under normal ambient pressure conditions. This is because gases dissolved in the glass will be more likely to be released at reduced pressure and this would lead to extensive bubble formation in the resultant seal.

The properties of the so-called adherence promoters require special mention in relation to the chemical approach to bonding. It has been appreciated for many years, since the early days of porcelain enamelling, that certain oxides with easily reducible ions, e.g. CoO and NiO, will promote adhesion between an enamel and an Fe substrate. During bonding to a pre-oxidized substrate the oxide scale dissolves into the glass. If, during this operation, all of the oxide is dissolved, the metal surface will come into contact with the glass. When this happens, the adherence promoting oxide is reduced by the metal substrate, thus forming more substrate metal oxide so that interfacial saturation with this oxide is achieved or maintained during the enamelling operation (e.g. equation [3.5] for CoO and an Fe substrate). Nickel oxide is also well known to act in a similar manner to that of CoO. Adherence promoting oxides can therefore play a very important and even essential rôle by creating and maintaining interfacial saturation with the appropriate substrate metal oxide. It is also possible to use other oxides, depending on the glass–metal system employed, e.g. CuO and MnO, which behave in a similar fashion to that of nickel.

It is therefore now clearly recognized that strong chemical bonding can be achieved between a metal and a glass only if the conditions during bonding are such that the glass at the interface becomes *and remains* saturated with the appropriate substrate metal oxide. In practice, and as noted earlier, this is achieved traditionally by pre-oxidizing the metal substrate prior to sealing or coating so that during bonding the oxide layer dissolves into the glass. Ideally, the bonding operation should be terminated at the precise stage when essentially all the oxide (except for a single mono-atomic layer) has dissolved into the glass, the interface remaining saturated with the oxide. This ideal condition is rarely if ever met in practice. Usually, a finite thickness of oxide remains at the interface so that bonding is achieved, not directly between the metal and the glass, but via this discrete oxide layer. In this instance, the overall properties of the seal or coating will be dictated by the properties of the transitional oxide layer and its bonding characteristics both to the metal substrate and to the glass or glass-ceramic. On the other hand, if sealing conditions are maintained after total dissolution of the oxide layer has occurred, saturation at the interface will subsequently be lost due to further diffusion of the oxide into the bulk of the glass away from the interface (under the influence of a chemical gradient). The formation, or otherwise, of strong chemical bonding at the interface will then be controlled by the occurrence of suitable redox reactions between the metal substrate and the glass. For example, reaction between Cr and ZnO to give CrO or Cr_2O_3 at the interface between a Ni-based superalloy and a glass-ceramic, or reaction between Ti and SiO_2 to give an interface composed of $Ti_5Si_3 + TiO_2$, as described later.

It is important to note that although saturation of the glass with the appropriate metal oxide within the immediate interfacial region is necessary for satisfactory chemical bonding to be achieved, too high a concentration must not generally be allowed to build up within the bulk of the glass or glass-ceramic as this will influence the thermal expansion characteristics and other properties of the glass and this may affect the overall behaviour, in particular the long term stability. In this respect, it is highly desirable to have a substrate metal oxide formed that is not too soluble in the glass. On the other hand, in the case of bonding to Fe, Ni or Co metal substrates, it has been noted that the addition of relatively high concentrations of substrate metal oxides can improve bonding to a sodium silicate glass by providing a closer match of the expansion coefficients of glass and metal (Mayer et al, 1974). For example, it was found that addition of 40 wt% Fe_2O_3 to a sodium disilicate glass lowered the expansion from around 16×10^{-6} K^{-1} to 13×10^{-6} K^{-1}, a much closer match to Fe, whilst addition of 20 wt% CoO gave a close match to Co and 8 wt% NiO gave a close match to Ni. This emphasises the dual beneficial influence that such additions can have on the bonding characteristics of metal/glass systems.

Desirable reactions can be promoted by the use of appropriate additives to the glass, as noted earlier. In addition to the classical adherence promoting oxides, CoO and NiO, which react preferentially with the metal to promote the formation of a suitable substrate metal oxide favourable to bonding, various alternative additions may be utilized to promote reactions favourable to chemical bonding. For example, the addition of CuO, MoO_3 or WO_3 to a glass to be bonded to a Cr-containing alloy could favour the formation of Cr_2O_3 at the interface. Not all redox reactions are desirable, of course, and care must be exercised to avoid reactions that create gaseous or unstable reaction products that can disrupt the bonding process. For a given system, it is possible to make rule-of-thumb predictions concerning the most likely reactions using thermodynamic data and phase diagram information, where relevant data are available. Some examples are given in Table 5. Thermodynamically favourable reactions may not necessarily proceed, however, due to kinetic considerations, and some apparently thermodynamically unfavourable reactions may proceed under certain processing conditions (e.g. eqn. [3.11]).

TABLE 5
FREE ENERGY CHANGES, ΔG°, FOR THE REACTION OF A NUMBER OF GLASS AND ALLOY COMPONENTS (calculated at 1300 K)

Reaction	$\Delta G°$ (kJ mol^{-1})
(a) Reaction favourable	
$Ti + 2CoO \rightarrow 2Co + TiO_2$	−421·7
$Cr + 3CuO \rightarrow 1/2Cr_2O_3 + 3/2Cu_2O$	−380·6
$Cr + 3/2CuO \rightarrow 1/2Cr_2O_3 + 3/2Cu$	−330·8
$Cr + K_2O \rightarrow CrO + 2K\uparrow$	−271·4
$Cr + 1/5P_2O_5 \rightarrow CrO + 2/5P\uparrow$	−214·6
$Cr + ZnO \rightarrow CrO + Zn\uparrow$	−190·8
$Cr + 3/2K_2O \rightarrow 1/2Cr_2O_3 + 3K\uparrow$	−182·7
$Cr + 1/2MoO_3 \rightarrow 1/2Cr_2O_3 + 1/2Mo$	−176·7
$Ti + 2/5B_2O_3 \rightarrow 2/5TiB_2 + 3/5TiO_2$	−146·8
$Cr + 1/2WO_3 \rightarrow 1/2Cr_2O_3 + 1/2W$	−137·3
$Cr + 3/10P_2O_5 \rightarrow 1/2Cr_2O_3 + 3/5P\uparrow$	−128·6
$Nb + 5/2Na_2O \rightarrow 1/2Nb_2O_5 + 5Na$	−116·4
$Cr + 3/2ZnO \rightarrow 1/2Cr_2O_3 + 3/2Zn\uparrow$	−92·9
$Fe + NiO \rightarrow FeO + Ni$	−63·1
$Cr + 3/2Na_2O \rightarrow 1/2Cr_2O_3 + 3Na\uparrow$	−61·7
$Cr + 1/2SiO_2 \rightarrow CrO + 1/2Si$	−59·9
$Fe + CoO \rightarrow Co + FeO$	−43·4
$Al + 3/2Na_2O \rightarrow 1/2Al_2O_3 + 3Na\uparrow$	−2·5
(b) Reaction unfavourable	
$Ni + 1/3MoO_3 \rightarrow NiO + 1/3Mo$	+20·3
$Fe + Na_2O \rightarrow FeO + 2Na\uparrow$	+35·0
$Cr + 3/4SiO_2 \rightarrow 1/2Cr_2O_3 + 3/4Si$	+103·0
$Cr + 3/4TiO_2 \rightarrow 1/2 Cr_2O_3 + 3/4Ti$	+136·7
$Ni + 1/3Cr_2O_3 \rightarrow NiO + 2/3Cr$	+138·1
$Cr + 3/4ZrO_2 \rightarrow 1/2 Cr_2O_3 + 3/4Zr$	+244·6
$Cr + 3/2Li_2O \rightarrow 1/2Cr_2O_3 + 3Li\uparrow$	+336·9
$Nb + 5/2Li_2O \rightarrow 1/2Nb_2O_5 + 5Li\uparrow$	+394·9

3.5   Other factors affecting bonding

Many additional factors may influence bonding between a metal and a glass. These include, for example, the presence of surface contamination (oil, grease, water, carbon form graphite jigging), gases dissolved in the metal, water present in the sealing atmosphere or dissolved within the glass, and reactions between the glass and the metal, to be described later.

3.6   Ceramic-to-metal bonding

Ceramics generally exhibit higher melting temperatures than the most common metals to which bonding is desirable. In addition, pure metals do not normally wet ceramics and therefore, unlike glass-to-metal systems, fusion processes cannot normally be employed to bond a ceramic to a metal. The bonding or joining of ceramics to metals has therefore traditionally been accomplished through the application of a metallized coating to the ceramic followed by brazing, or by the use of reactive metal brazes that will wet both the ceramic and metal parts to be joined. This is considered more fully in Section 9.1.

For the limited instances where the metal component does wet the ceramic directly, the bonding processes can be described in terms of the thermodynamics of the system and the products formed at the interface through chemical reaction, in much the same way as for glass-to-metal systems. For example, it has been noted (Naidich, 2003) that high wettability can be achieved using certain transition metals with partly occupied d-electron bands. These transition metal atoms can play the role of bridging the bonding between metallic and ionic/covalent materials. In the case of diffusion bonding between Cu and alumina, $CuAlO_2$ and Cu_2O are formed at the interface, depending on the applied oxygen partial pressure. High resolution TEM, HRTEM, has revealed that only a few monolayers of these reaction products are required in order to increase the interfacial bond strength significantly (Dehm et al, 1997). In the bonding of Ag to MgO it has been shown (Kotomin and Maier, 2003) that point surface defects increase the adhesion energy and cause a redistribution of the electron density across the interface, whilst the wetting of TiC by non-reactive metals has been improved by enhancement of the interfacial reactions through alloying with aluminium (Froumin et al, 2003). In this instance it was noted that a further improvement in the wetting characteristics could be achieved by oxidation of the TiC surface. In both cases, the solubility of TiC in the metal phase is enhanced, thus leading to an improvement in the bonding behaviour. A theoretical study of bonding between ZrC and Fe has been carried out by Arya and Carter(2004). It was shown that a good lattice match can be achieved between the Zr(100)/Fe(110) interfaces, with adhesion achieved through a mix of covalent and metallic bonding interactions.

As has already been noted for glass-to-metal seals, these studies on ceramic-to-metal systems emphasise the importance of achieving strong chemical bonding through careful choice of materials and processing conditions.

3.7   Ceramic-to-ceramic bonding

Recent studies on ceramic-to-ceramic bonding at the atomic level have shown the importance of dislocation structures at the interface in controlling diffusion kinetics. Bonding at the atomic level between ceramics has received attention recently through application of HRTEM. For example, HRTEM has been used to investigate the reactions between MgO films 35–40 nm in thickness vacuum deposited onto sapphire single crystals and subsequently heated to 1100°C (Hesse et al, 1994). Figure 12(a) shows the interface between MgO and Al_2O_3 prior to the heating stage where it is observed that the MgO film is epitaxially orientated. On heating, it was observed that reaction occurred between the MgO and Al_2O_3 to form a layer of $MgAl_2O_4$ spinel at the interface, with the position of the original MgO/ Al_2O_3 interface given by the Pt markers shown in Figure 12(b). In more

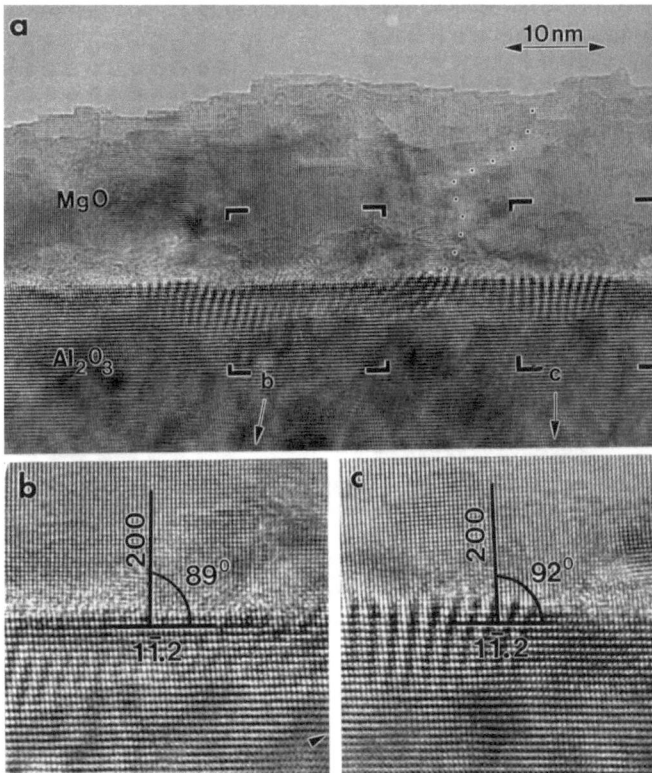

(a) Original interface – MgO is epitaxially orientated ('b' and 'c' are magnified images of the areas marked)

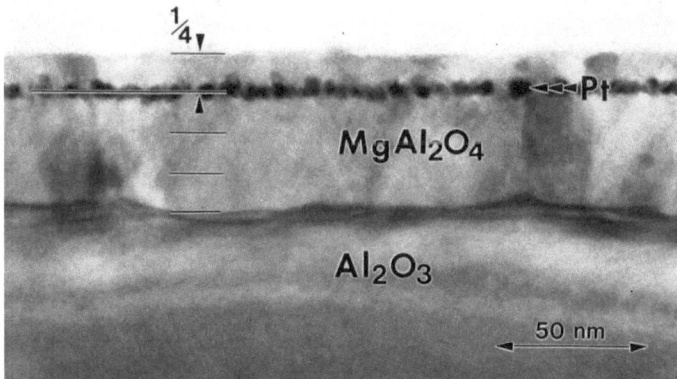

(b) After heating to 1100°C – an $MgAl_2O_4$ spinel phase has formed at the interface (the original MgO/Al_2O_3 interface is shown by the Pt markers)

Figure 12: High resolution transmission electron microscope images of MgO/Al_2O_3 interface (after Hesse et al, 1994)

recent studies the bonding between single crystal MgO and various spinel phases including Mg_2TiO_4, $MgCr_2O_4$ and $MgIn_2O_4$ and between ZrO_2 and $La_2Zr_2O_7$ has been followed and it has been noted that the reaction kinetics depend on the atomic structure at the interface, and in particular on the mobility of interfacial misfit accommodating dislocation networks (e.g. Sieber *et al*, 1997; Hesse and Senz, 2002; Lu *et al*, 2002). During bonding the dislocations must move together with the advancing interface and the mobility of these dislocations will depend on the mode of dislocation movement and on the Burgers geometry of the misfit dislocations. This work may have implications for glass-ceramic-to-metal seals and coatings, but this is an area not yet investigated.

4. GLASS-CERAMIC-TO-METAL SEALS

4.1 Formation of glass-ceramics

Glass-ceramics are a relatively new class of material, originally discovered by Donald Stookey at Corning Glass Works in the 1950s (Stookey, 1956, 1960; Anon, 1957), and subsequently defined as *polycrystalline ceramic materials prepared by the controlled bulk crystallization of suitable glasses* (McMillan, 1979; Strnad, 1986; McMillan and Partridge, 1965; Wada and Kawamura, 1981; Paul, 1982; Grossman, 1982; Hlavac, 1983; Beall, 1983, 1985, 1989; Sarkisov, 1989; Donald, 1993; Smith, 1989; Höland and Beall, 2002). Early work in the UK was pioneered by Peter McMillan at English Electric and later at Warwick University (McMillan, 1979).

Crystallization of conventional glasses normally occurs by the nucleation of crystals at external surfaces. This gives rise to coarse microstructures with large anisotropic crystals which grow inwards from the surfaces. Such materials are usually very weak mechanically because the large crystals can act as powerful stress concentration sites. Control over the crystallization process by the introduction into suitable glass compositions of a large number of small heterogeneities which can act as efficient bulk nucleating centres can, on the other hand, lead to the formation of isotropic microstructures consisting of small, interlocking, randomly orientated ceramic crystals, usually < 10 μm in size and more recently in the nm range, bonded together by some residual glassy phases, usually not more than 2–10%. The resultant glass-ceramic materials can exhibit very good mechanical and other properties.

Bulk glass-ceramics are normally prepared employing a two-stage heat-treatment schedule, as illustrated in Figure 13. During the nucleation stage a large number of small heterogeneities are formed, usually < 100 nm in size, within the glass, at a temperature which is below that at which the major crystal phases can grow at a significant rate (if the crystal growth rate is high at the temperature of nucleation, rapid crystallization will occur from a limited number of nuclei, giving rise to a coarse microstructure). High nucleation rates are achieved in practice by the use of suitable nucleating agents added to the glass batch. These include various elemental metals, certain oxides, and various metallic halides and sulphides. The most common nucleating agents for silicate systems include Group IV transition metal oxides, in particular TiO_2 and ZrO_2 together with MoO_3 and P_2O_5 (McMillan, 1979; Strnad, 1986), as summarized in Table 6.

The precise manner in which a given nucleating agent acts has not been determined unambiguously for all possible glass-ceramic systems, but a number of mechanisms have been identified (e.g. Donald, 1992). These include amorphous (glass-in-glass) phase separation, the precipitation of small crystallites of a compound formed by reaction between the nucleating agent and constituents of the

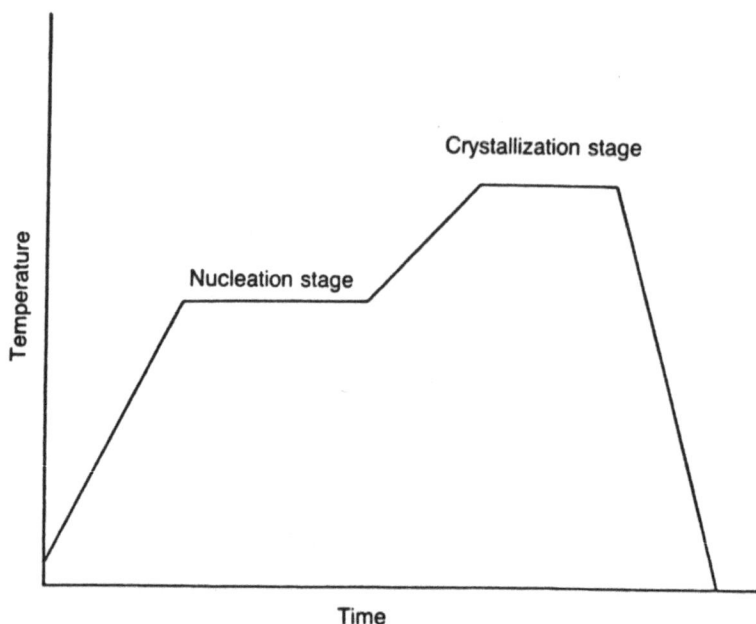

Figure 13: Conventional two-stage heat treatment schedule for producing a bulk glass-ceramic

glass, e.g. reaction of P_2O_5 with Li_2O to give Li_3PO_4 nuclei; reaction of TiO_2 with MgO to give $MgTiO_2$; and *in situ* reaction between mixed TiO_2/ZrO_2 nucleating agent to give $ZrTiO_4$ crystallites. It has also been noted that TiO_2 may first react with other glass constituents to form intermediate phases which act as nucleation sites (e.g. Salman *et al*, 2001); for example, with Li_2O and SiO_2 to form Li_2TiSiO_5, with MnO to form $LiMnTiO_4$, and with FeO to form $FeTiO_3$. Some examples of nucleated glasses are shown in Figure 14. This figure shows a phase separated lithium silicate glass nucleated by P_2O_5. The addition of P_2O_5 stimulates the formation of silica-rich droplets in a lithia-rich matrix. Also shown is a particularly large lithium phosphate nucleus present in a lithium aluminosilicate glass. In this instance bulk crystallization occurs by epitaxial growth of lithium silicate, lithium disilicate and cristobalite onto the nucleus. Crystallization by epitaxial growth was postulated in the 1960s (McMillan, 1964, 1979) with Harper and McMillan (1972) subsequently noting that there is a close similarity between the 111 and 210 d-spacings of lithium phosphate and the 111 d-spacings of lithium disilicate, and suggesting that the lithium phosphate could therefore provide centres for epitaxial growth of lithium disilicate. This type of growth was only observed experimentally relatively recently (Headley and Loehman, 1984) with epitaxial growth of cristobalite, lithium silicate and lithium disilicate all being directly observed for lithium phosphate nuclei.

After successful nucleation, the temperature is raised during the crystallization stage so that crystal growth can occur from these nuclei. It is important that

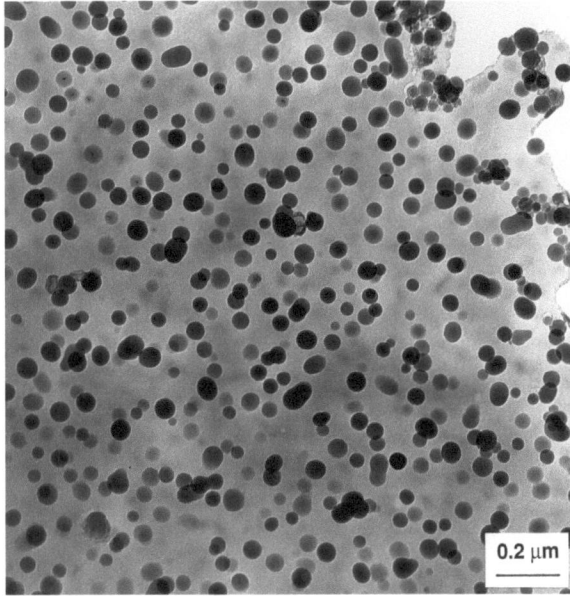

(a) Amorphous (glass-in-glass) phase separation of a lithium silicate glass (after McMillan, 1979)

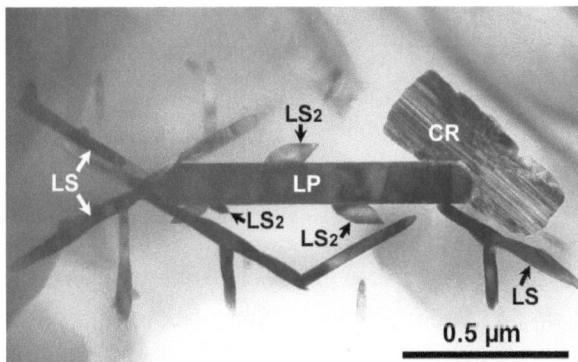

(b) Example of "Swiss army knife" illustrating epitaxial crystallization of lithium silicate, lithium disilicate, and cristobalite on a lithium phosphate nucleus present in a lithium aluminosilicate glass (after Headley and Loehman, 1984)
Figure 14: Examples of nucleated glasses

the nucleation and crystallization temperature regimes do not overlap to any great extent, as depicted in Figure 15, otherwise crystal growth will occur rapidly as nuclei form, leading to a coarse microstructure. Some typical glass-ceramic microstructures are shown in Figure 16. The overall result is a unique family of materials with many useful properties, and with applications ranging from kitchen ware, biomedical implants and telescope mirror blanks, to hosts for the immobilization of toxic and radioactive wastes (McMillan, 1979; Strnad, 1986; Lewis, 1989; Donald, 1993; Donald *et al*, 1997; Beall and Pinckney, 1999; Höland and Beall, 2002).

TABLE 6
TYPES OF NUCLEATING AGENT EMPLOYED IN THE PRODUCTION OF GLASS-CERAMICS

Type of Nucleating Agent	Examples	Mechanism
Metals	Cu, Ag, Au and Pt-group	Limited solid solubility giving rise on cooling the melt to a colloidal dispersion of metallic crystals
Oxides	TiO_2; ZrO_2, HfO_2 V_2O_5; Nb_2O_5; Ta_2O_5 Cr_2O_3; MoO_3; WO_3 FeO; Fe_2O_3, Ag_2O P_2O_5	(a) Limited solid solubility giving rise on cooling the melt to a colloidal dispersion of crystals or causing the glass to phase separate (b) Reaction with a constituent of the glass to form another compound which precipitates out as a colloidal dispersion:- $TiO_2 + ZrO_2 \rightarrow ZrTiO_4$ $TiO_2 + MgO \rightarrow MgTiO_3$ $TiO_2 + FeO \rightarrow FeTiO_3$ $TiO_2 + Li_2O + SiO_2 \rightarrow Li_2TiSiO_5$ $TiO_2 + Li_2O + MnO \rightarrow Li_2MnTiO_4$ $3Ta_2O_5 + Li_2O \rightarrow 2LiTa_3O_8$ $MoO_3 + Li_2O \rightarrow Li_2MoO_4$ $WO_3 + Li_2O \rightarrow Li_2WO_4$ $3Li_2O + P_2O_5 \rightarrow 2Li_3PO_4$
Metallic halides	NaF; CaF_2; MgF_2; Na_3AlF_6; Na_2SiF_6	Causes the glass to phase separate, or, alternatively, on nucleation (re-heating) a colloidal dispersion of the halide is formed
Metallic sulphides	ZnS; CuS; FeS; MnS	On nucleation (re-heating) a colloidal dispersion of the sulphide is formed
Metallic selenides	$CdSe$	Causes the glass to phase separate
Metallic sulphates	Na_2SO_4	Causes the glass to phase separate

It is important to note that different nucleating agents can promote quite different crystallization behaviour in ostensibly the same glass. This is illustrated by Figure 17 for a crystallized lithium zinc silicate glass (Donald *et al*, 1992). Choice of a suitable nucleating agent for a given system is therefore driven strongly by the application in question. Nucleation can also be influenced by the presence of impurities in the precursor glass. It is therefore usual to employ high purity starting materials and to melt in platinum crucibles in order to keep contamination to a minimum, particularly when developing experimental systems. An example of some typical glass starting materials and the resulting precursor glasses are shown in Figure 18. It is important to note that crystallization at different temperatures

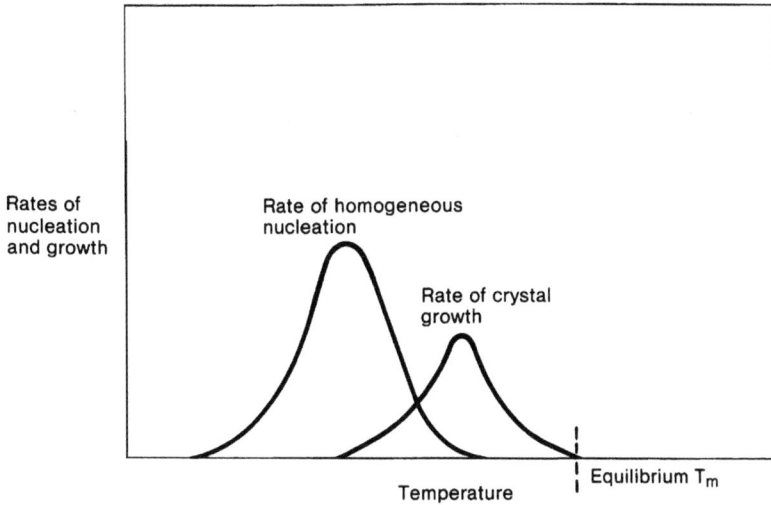

Figure 15: Rates of nucleation and crystallization as a function of temperature. For successful nucleation these regimes should not overlap unduly (after McMillan, 1979)

can also affect the microstructure, with higher temperatures usually resulting in coarser microstructures, as illustrated in Figure 19 (Metcalfe *et al*, 1991).

A major advantage offered by glass-ceramics is that a greater range of practical thermal expansion coefficients can be obtained, relative to their glassy counterparts and, in addition, complex non-linear thermal expansion characteristics can be achieved. This, in principle, enables very close thermal expansion matching to a wide variety of metals and alloys, including those which exhibit nonlinear behaviour due to phase transformations and related phenomena. Data for the thermal expansions of a number of ceramic crystalline phases are summarized in Table 7. The main silica polymorphs, quartz, cristobalite and tridymite, can

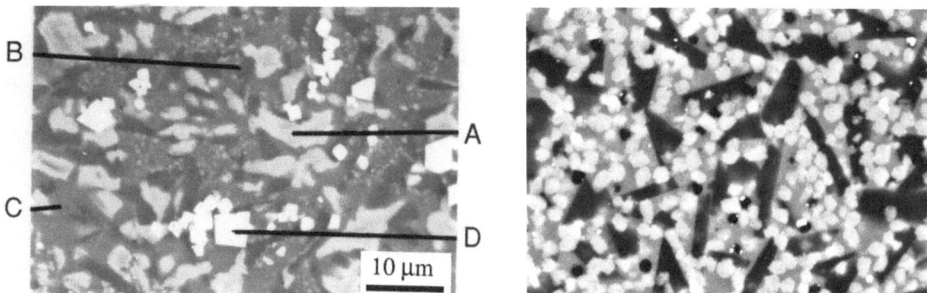

(a) Lithium zinc silicate glass-ceramic containing Cr_2O_3 addition
A = Li_2ZnSiO_4, B = cristobalite,
C = residual glass, D = $ZnCr_2O_4$

(b) Lithium zinc silicate glass-ceramic containing WO_3 addition (same scale as (a))

Figure 16: Typical glass-ceramic microstructures

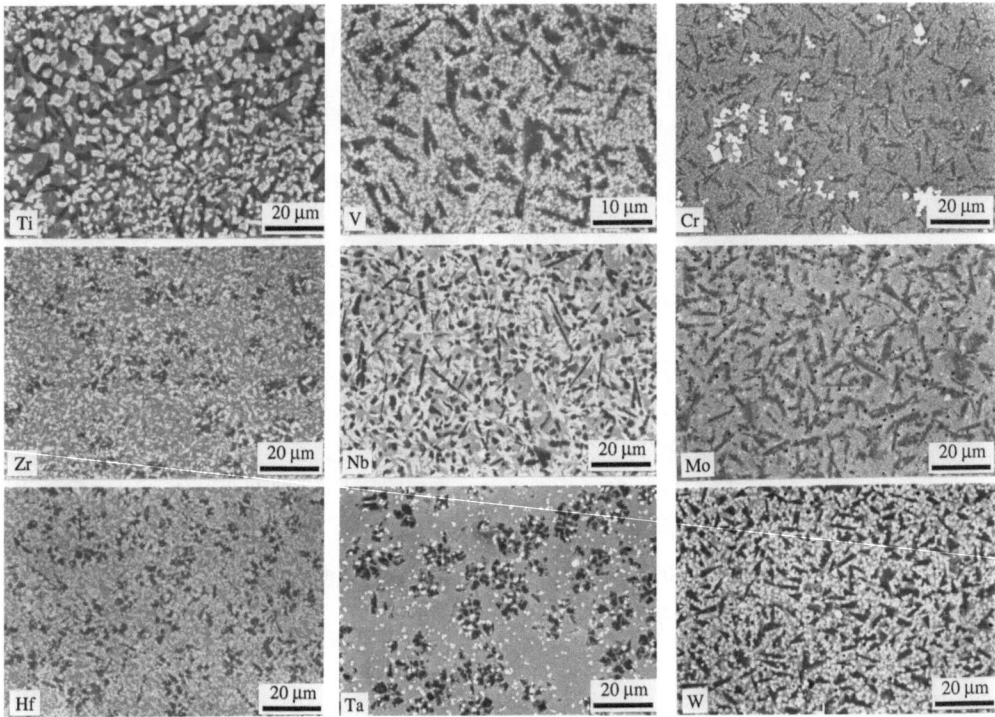

(a) TiO_2; ZrO_2; and HfO_2 (b) V_2O_5, Nb_2O_5 and Ta_2O_5 (c) Cr_2O_3; MoO_3; and WO_3

Figure 17: SEM micrographs showing the effect of different nucleating agents on the resultant microstructures of a lithium zinc silicate glass-ceramic ceramic (after Donald et al, 1992)

also play a major role in determining the thermal expansion characteristics of a given glass-ceramic system. The presence of cristobalite and quartz in particular can lead to high expansion glass-ceramics due to the phase inversions and the accompanying volume increases associated with these phases, as illustrated in Figure 20.

It has been shown to a first approximation (Kingery, 1960) that the thermal expansion coefficient, α, of a glass-ceramic material is an additive function of the thermal expansions of the various phases present, and can be represented by the following relationship:-

$$\alpha_{(glass-ceramic)} = \{\alpha_1 K_1 W_1/\rho_1 + \alpha_2 K_2 W_2/\rho_2 + ...\}/\{K_1 W_1/\rho_1 + K_2 W_2/\rho_2 + ...\}$$

$$[4.1]$$

where α_1, α_2, etc, are the thermal expansion coefficients of the various phases present in the glass-ceramic, K_1, K_2, etc, are the bulk moduli of these phases, ρ_1, ρ_2, etc, are the densities of the phases, and W_1, W_2, etc, are the weight fractions of the phases. The ability to tailor the thermal expansion characteristics of glass-ceramics is therefore a direct consequence of the ability to control (through selection of the starting glass composition and heat-

Figure 18: Glass-ceramics are usually prepared from high purity starting constituents and melted in Pt crucibles in order to minimise contamination

treatment schedule) the type and proportion of crystalline phases present in the final glass-ceramic article.

A more complex thermal cycle is required for the preparation of seals and coatings, normally consisting of three stages, as illustrated in Figure 21. In the sealing process, the glass is initially heated to a temperature high enough to melt the glass and allow it to flow into the metal parts where it wets the surface and reacts to form an interface in much the same way as in the preparation of a conventional glass-to-metal seal. This is usually followed by the standard two-stage nucleation and crystallization cycle associated with the preparation of bulk glass-ceramics. A three-stage sealing and heat-treatment schedule may not always be necessary, however, particularly in the case of coatings. Sometimes, for example, nucleation and crystallization may occur simultaneously during cooling from the high temperature sealing stage, and in this case separate nucleation and crystallization stages are not required. In many instances, the initial sealing cycle does not affect the later crystallization of the glass, i.e. the same phases are produced and in essentially the same proportions regardless of whether the glass-ceramic is part of a seal or is made by the conventional two-stage heat-treatment cycle for producing bulk materials; however, sometimes following a sealing cycle does not produce the same results. For example, in the case a lithium aluminosilicate glass it has been observed that a two-stage treatment yields lithium disilicate as the major crystalline phase, with a thermal expansion coefficient of $11 \cdot 0 \times 10^{-6} \, K^{-1}$, whereas following a sealing cycle produces a mixture of lithium silicate, lithium disilicate and cristobalite with a significantly higher thermal expansion, $14 \cdot 5 \times 10^{-6} \, K^{-1}$ (McCollister and Reed, 1983). The thermal expansion of this glass-ceramic as a

Figure 19: Microstructures of a lithium zinc silicate glass-ceramic crystallized at different temperatures in the range 800–900°C illustrating the larger crystal size obtained at higher temperatures (after Metcalfe et al, 1991)

function of the time at 1000°C is shown in Figure 22. Also shown is the amount of cristobalite formed as a function of time. It is clear that the cristobalite content increases with heat treatment time, thereby giving rise to a higher thermal expansion product. The thermal expansion behaviour of a glass-ceramic may also be strongly dependent on the crystallization temperature as well as the time. It cannot be assumed that a constant or linear variation of thermal expansion coefficient with temperature or time will be obtained. Some examples of the variation in thermal expansion coefficient with crystallization temperature (for a constant time of 1 hour at temperature) are given in Figure 23, where it may be noted that the behaviour is distinctly nonlinear. Other examples where the crystallization behaviour depends on whether or not a sealing cycle has been carried out include a number of lithium zinc silicate glass-ceramics (Donald *et al*, 1992), as noted from the thermal expansion plots given in Figure 24 and the micrographs shown in Figure 25.

TABLE 7
THERMAL EXPANSION DATA FOR SOME CRYSTALLINE PHASES FOUND IN GLASS-CERAMICS

Crystal/Mineral phase	Composition	Thermal expansion coefficient (10^{-6} K^{-1})	Temperature range (°C)
Aluminium titanate	$Al_2O_3.TiO_2$	−1·9	25–1000
Anorthite	$CaO.Al_2O_3.2SiO_2$	4·5	100–200
Beryl	$3BeO.Al_2O_3.6SiO_2$	2·0	–
Calcium orthosilicate	$CaO.SiO_2$	10·8–14·4	–
Calcium zirconate	$CaO.ZrO_2$	10·4	–
Celsian	$BaO.Al_2O_3.2SiO_2$	2·7	20–100
Chromium oxide	Cr_2O_3	7·0–9·6	–
Clinoenstatite	$MgO.SiO_2$	7·8	100–200
		13·5	300–700
Cordierite	$2MgO.2Al_2O_3.5SiO_2$	2·6	25–700
Cristobalite	SiO_2	12·5	20–100
		50·0	20–300
		27·1	20–600
Diopside	$CaO.MgO.2SiO_2$	5·0	–
Enstatite	$MgO.SiO_2$	9·0	20–400
		12·0	300–700
Eucryptite	$Li_2O.Al_2O_3.2SiO_2$	−8·6	20–700
		−6·4	20–1000
Forsterite	$2MgO.SiO_2$	9·4	100–200
Geikielite (magnesium titanate)	$MgO.TiO_2$	7·9	25–100
Kalsilite	$KAlSiO_4$	15·0	–
Lithium metasilicate	$Li_2O.SiO_2$	13·0	20–300
Lithium disilicate	$Li_2O.2SiO_2$	11·0	20–600
Lithium zinc silicate	$Li_2O.ZnO.SiO_2$	9·0	20–200
Magnetite	$FeO.Fe_2O_3$	7·0	–
Mullite	$3Al_2O_3.2SiO_2$	5·3	–
Nephaline (sodium)	$NaAlSiO_4$	9·0	–
Nephaline (natural)	$Na_3K(AlSiO_4)_4$	12·0	–
Osumilite	$MgAl_2Si_4O_{12}$	2·0	–
Pollucite	$CsAlSi_2O_6$	1·5	–
Protoenstatite	$MgO.SiO_2$	9·8	300–700
Quartz	SiO_2	11·2	20–100
		13·2	20–300
		23·7	20–600
Rutile	TiO_2	7·8	–
Spinel	$MgO.Al_2O_3$	8·8	–
Spodumene	$Li_2O.Al_2O_3.4SiO_2$	0·9	20–1000
Stuffed keatite	$ZnO.(ZnO.SiO_2)$	3·0	–
Tridymite	SiO_2	17·5	20–100
		25·0	20–200
		14·4	20–600
Willemite	$2ZnO.SiO_2$	3·0	–
Wollastonite	$CaO.SiO_2$	9·4	100–200
Zircon	$ZrO_2.SiO_2$	4·2	–
Zirconia	ZrO_2	9·8–11·0	–
	$ZrO_2.TiO_2.Y_2O_3$	6·5	–

Data from a number of sources including McMillan, 1979; Hlavac, 1983; Paul, 1982; Höland and Beall, 2002

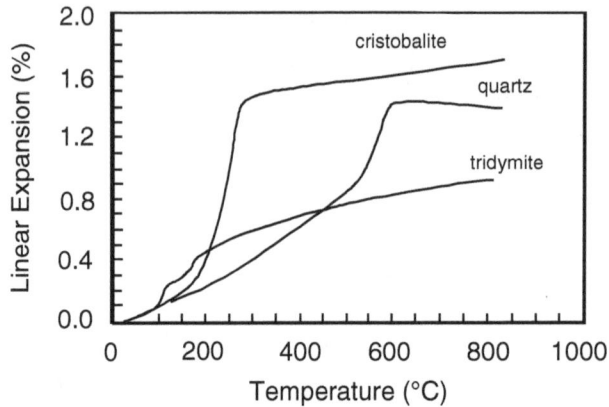

Figure 20: Thermal expansion curves for silica polymorphs

In view of these characteristics it is essential that each glass-ceramic and heat-treatment schedule is tailored to the application in question, i.e. it is not justified to assume that the same product will be obtained regardless of the precise sealing cycle, with some glass-ceramic compositions being considerably less process tolerant than others.

Like glasses, glass-ceramics useful for sealing or coating applications can also be broadly classified according to their composition, as summarized below. Data illustrating the wide range of *practical* thermal expansions that can be achieved are summarized in Table 8, and typical thermal expansion curves for a selection of glass-ceramic materials are shown in Figure 26. Individual glass compositions are given in Table 9. The very wide range of thermal expansions possible for glass-ceramic systems makes them ideal candidates for sealing to most common metals and alloys.

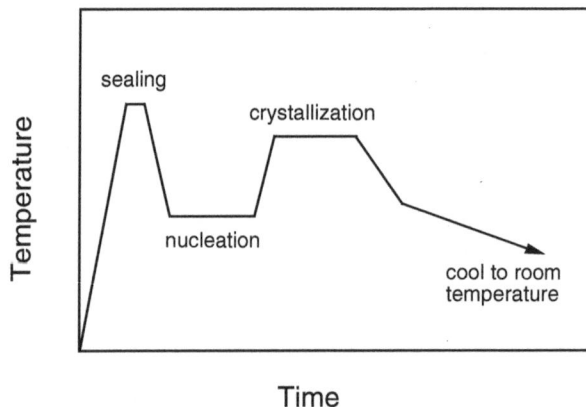

Figure 21: Modified three-stage heat treatment schedule for the manufacture of glass-ceramic-to-metal seals

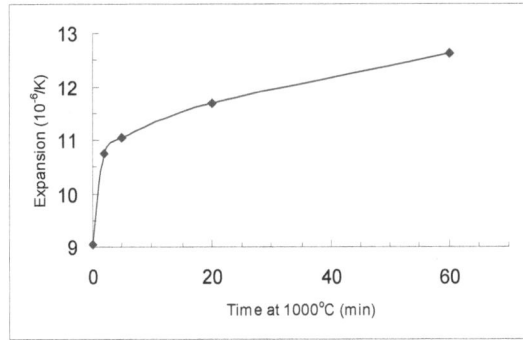

Figure 22(a): Thermal expansion of a lithium aluminosilicate glass-ceramic as a function of nucleation time at 1000°C (SNLA 'S' glass-ceramic, after McCollister and Reed, 1983)

4.2   Silicate-based glass-ceramics

4.2.1   Alkali and alkaline earth silicate glass-ceramics

Glass-ceramic compositions in the alkali metal silicate category are mainly based on lithium silicate, usually in combination with small additions (i.e. <5 wt%) of other oxides, e.g. Na_2O, K_2O, B_2O_3, Al_2O_3, ZnO, etc, together with a suitable nucleating agent, which in the case of lithium silicate glass-ceramics is normally P_2O_5. The major crystalline phases formed in this system are lithium metasilicate and lithium disilicate, together with various silica phases, e.g. cristobalite, quartz or tridymite. Depending on the proportions of phases present, moderate-to-high thermal expansion glass-ceramics can be produced, with α normally in the range ≈ 8–19×10^{-6} K^{-1} (e.g. McMillan and Partridge, 1963; McMillan *et al*, 1966; Kramer *et al*, 1982, 1985; Henderson *et al*, 1984; Hammetter and Loehman, 1987).

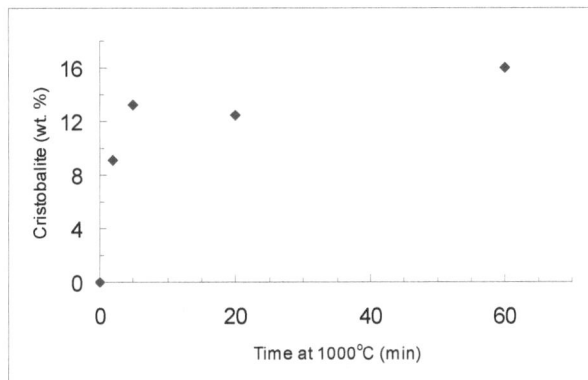

Figure 22(b): Corresponding concentration of cristobalite present in the final glass-ceramic

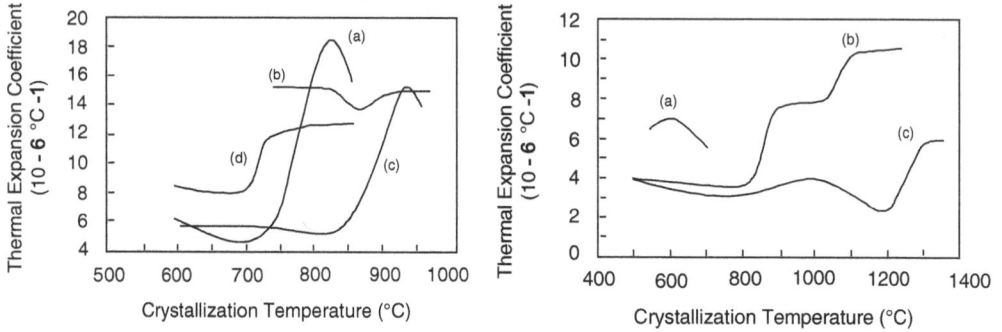

(a) lithium zinc silicate compositions
(a) AZS1; (b) AZS6;
(c) AZS5; (d) AS9

(b) alternative compositions
(a) ACS3; (b) AZS11;
(c) TAS4

Figure 23: Thermal expansion of glass-ceramics as a function of crystallization
temperature (specific compositions are given in Table 9)

4.2.2   Alkali and alkaline earth aluminosilicate glass-ceramics

Glass-ceramics in this category are mainly based on the lithium, magnesium, calcium or barium aluminosilicate systems. The nucleating agent normally employed for these compositions is TiO_2, ZrO_2 or a mixture of the two, although P_2O_5 has also been successfully employed with some compositions. The major crystalline phases produced include β-spodumene, β-eucryptite, β-quartz solid solutions, cordierite, enstatite, wollastonite, sphene and celsian, together with silica polymorphs. Depending on the precise phases, low-to-moderate thermal expansions can be achieved, ranging form zero (or even negative) to ≈9×10^{-6} K^{-1} (e.g. McMillan and Partridge, 1963, 1966; English Electric, 1963; Smith, 1968; Donald et al, 1990; Metcalfe and Donald, 1991).

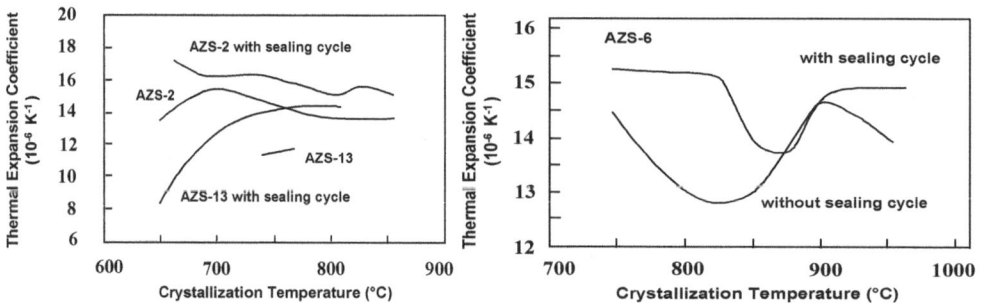

Figure 24: Thermal expansion of glass-ceramics as a function of the crystallization
temperature for samples with and without the inclusion of a high temperature
simulated sealing cycle (after Donald et al, 1992)

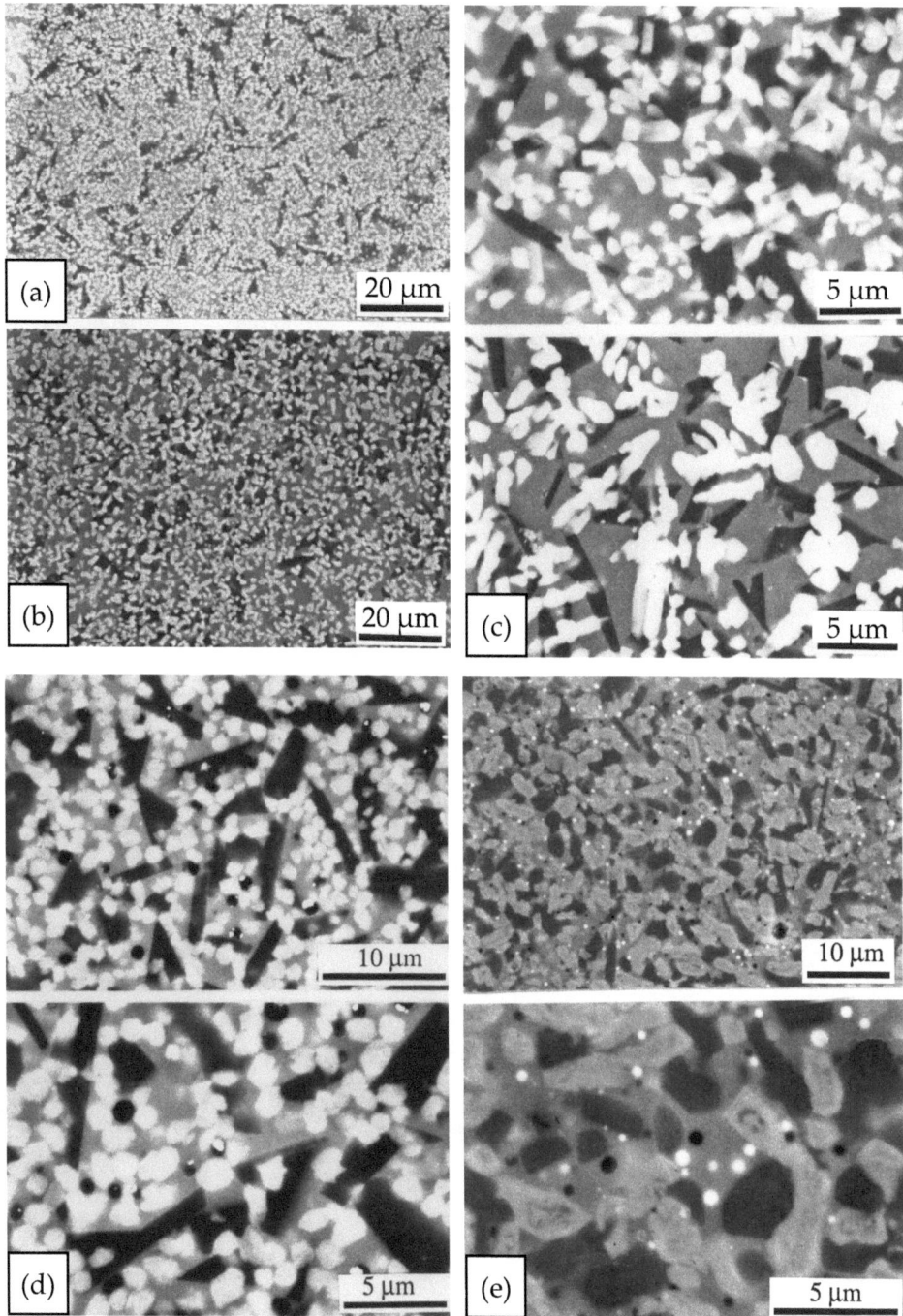

(a) NiO addition, high temperature stage not included;
(b) NiO with a high temperature stage
(c) NiO, higher magnification with high temperature sealing stage
(d) WO₃ addition, high temperature stage not included
(e) WO₃ addition, with a high temperature stage

Figure 25: SEM micrographs of lithium zinc silicate glass-ceramics doped with 2
mol% of NiO or WO₃ with and without the inclusion of a high temperature simulated
sealing stage (after Donald et al, 1992)

4.2.3   Zinc aluminosilicate glass-ceramics

The preparation of alkali-free glass-ceramics in the zinc aluminosilicate system was first reported by Corning Glass Works(1960), using TiO_2 as the nucleating agent. The glass-ceramics so produced were of relatively low thermal expansion and contained gahnite, willemite and rutile as the major crystalline phases. Further work on this system was reported by McMillan and Partridge (McMillan and Partridge, 1972; Partridge et al, 1989a, 1989b) and Vargin et al(1968), who also employed TiO_2 as the nucleating agent, and Strnad et al(1976) and Holleran and Martin(1987), who used ZrO_2. Depending on the precise glass composition and heat-treatment schedule a range of thermal expansions is possible, with values for α from ≈ 3–15×10^{-6} K^{-1}. The higher expansion materials contain a significant proportion of cristobalite.

Alkali-containing zinc aluminosilicate glass-ceramics have also been reported. For example, Omar et al(1991) prepared glasses in the lithium zinc aluminosilicate system containing up to 13 wt% Li_2O and up to 28 wt% ZnO. These glasses contain TiO_2 or ZrO_2 as nucleating agent and can be crystallized to yield glass-ceramic materials with low-to-moderate thermal expansion coefficients in the range 3·6–10·0$\times10^{-6}$ K^{-1}. The predominant crystalline phases are β_{II}-Li_2ZnSiO_4, γ_0-Li_2ZnSiO_4, β-spodumene and β-eucryptite, although the higher thermal expansion materials also contain a significant proportion of residual glass.

4.2.4   Lithium zinc silicate glass-ceramics

Glass-ceramics based on the lithium zinc silicate system were first reported by McMillan and Partridge in 1963. These materials contained up to 59 wt% ZnO, although 30 wt% was described as an upper preferred limit. Unlike the earlier zinc aluminosilicates, the lithium zinc silicate compositions are nucleated by P_2O_5, and exhibit lower crystallization temperatures. Subsequently, further work was reported by McMillan and co-workers (McMillan and Hodgson, 1963, 1967; McMillan and Partridge, 1965; McMillan et al, 1965, 1966a, 1966b) on the preparation of a range of glass-ceramic materials from within this system exhibiting a wide range of thermal expansions from ≈ 4–19×10^{-6} K^{-1}.

Since the early work of McMillan and co-workers additional work has been reported on this versatile glass-ceramic system (West and Glasser, 1970a, 1970b; Chen and McMillan, 1985; Lee and Han, 1987; Donald et al, 1989, 1992; Rao et al, 1990). For example, Donald and co-workers (1989, 1992) have examined the influence of a wide variety of transition metal oxide nucleating additives, used in conjunction with P_2O_5, on the crystallization kinetics, microstructures and thermal expansion characteristics of a lithium zinc silicate glass of analysed composition 17·84Li_2O–5·25Na_2O–17·73ZnO–4·31B_2O_3–1·23P_2O_5–53·64SiO_2 (mol%). Transition metal oxide additives (TMO) investigated included TiO_2, ZrO_2, HfO_2, V_2O_5, Nb_2O_5, Ta_2O_5, Cr_2O_3, MoO_3, WO_3, NiO and CuO. It was noted that the nucleating efficiency, as measured by the activation energy for crystallization,

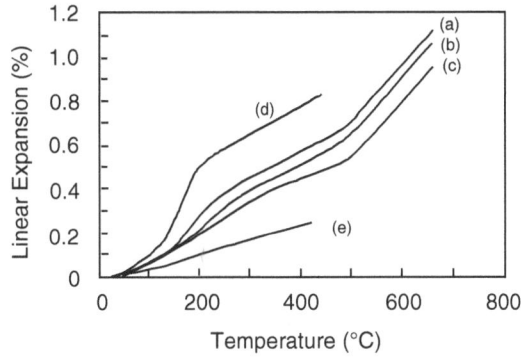

(a) Lithium zinc silicate glass-ceramics
(a) AZS6/750°C; (b) AZS6/850°C; (c) AZS10/850°C; (d) AZS1/810°C;
(e) AZS5/750°C

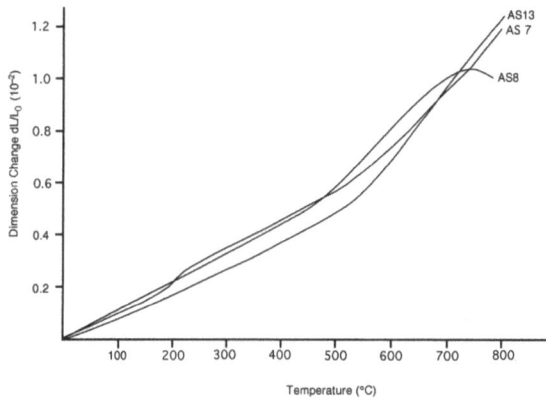

(b) Lithium aluminosilicate glass-ceramics

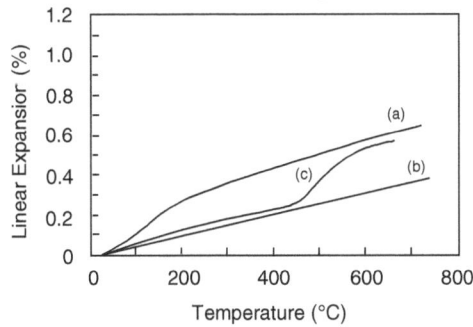

(c) Some alternative silicate glass-ceramics
(a) AAS1/1150°C; (b) AAS1/1100°C; (c) AAS5/900°C
Figure 26: Thermal expansion curves for a selection of glass-ceramics
(specific compositions are given in Table 9)

is directly related to the ionic field strength of the additive employed, with bulk crystallization being favoured by nucleating species of high field strength, in particular Mo, W and V. It was also observed that the TMO additions have a pronounced effect on the thermal expansion characteristics of the resultant glass-ceramics. For a standard heat-treatment schedule it was noted that α could be varied over the range 10·9 to 15·8×10^{-6} K^{-1}, depending on the precise TMO species employed. Variations in the thermal expansion were ascribed mainly to differences in the concentration and ratios of the silica polymorphs cristobalite, quartz and tridymite.

4.2.5   Magnesium zinc silicate glass-ceramics

Limited work has been reported on the preparation of magnesium zinc silicate glass-ceramics. Chen and McMillan(1985) noted that the glass-forming range of this system could be greatly expanded by incorporating small additions of Al_2O_3, together with TiO_2 as a nucleating agent. Crystallized materials contained varying proportions of enstatite solid solution, willemite and cristobalite. Thermal expansions in the range 3·5–10·2×10^{-6} K^{-1} were achievable, depending on the crystallization temperature employed.

4.3   Phosphate-based glass-ceramics

Only limited attention has been devoted to the controlled crystallization of phosphate glasses to produce glass-ceramic materials. Much of this work has been aimed at the production of biomedical materials based on calcium phosphate for applications requiring bone replacement and dental implants (e.g. Vogel and Höland, 1987; Watanabe et al, 1989; Wange et al, 1990; Shi and James, 1992). Of particular interest for sealing work, Wilder et al(1982) examined the crystallization behaviour of a number of phosphate systems, including Na_2O–CaO–P_2O_5, Na_2O–BaO–P_2O_5, Na_2O–Al_2O_3–P_2O_5 and Li_2O–BaO–P_2O_5. The effect of a variety of potential nucleating species was investigated, including TiO_2, ZrO_2, Y_2O_3, La_2O_3, Ta_2O_5, WO_3 and Pt. Of these, only Pt was found to be effective at promoting bulk crystallization. Relatively high thermal expansion coefficients in the range 16·2–22·5×10^{-6} K^{-1} were achieved. A more recent study by Wang and James(1990) of calcium phosphate glasses containing a number of oxide and fluoride additions has shown that TiO_2 in conjunction with Al_2O_3 can be employed to promote bulk crystallization in these systems. Glass-ceramic materials with thermal expansion coefficients in the range 8·6–14·1×10^{-6} K^{-1} were successfully prepared in this study. In addition, Langlet et al(1992) have studied the crystallization behaviour of aluminium phosphate glasses and noted that Li_2O, $NaPO_3$ and AlF_3 can aid in the crystallization process.

4.4   Borate-based glass-ceramics

Very little work has been reported on the preparation of borate-based glass-ceramics. In general, crystallization of borate glasses proceeds from the surface, rather than internally (Simon and Nicula, 1984; Smith et al, 1987; Goktas et al,

1992; Donald *et al*, 1994). In one study, Fahmy and Subramanian(1987a, 1987b) investigated the crystallization of glass of composition $BaO.4B_2O_3$ with and without the addition of TiO_2 as a nucleating agent. They noted that addition of TiO_2 led to the formation of coarse grained glass-ceramic microstructures consisting of $BaTiO_3$, BaB_2O_4 and Ba_2TiO_4 crystalline phases. No mechanical property data were provided, but in view of the coarse crystal structure of the product it is expected that properties would be poor. Crystallization of borate-based glasses to yield crystalline products has also been reported by Pevzner and Klyuev(2000; 2003) for $BaO–Al_2O_3–Ga_2O_3–B_2O_3$ systems, and by Pevzner and Niunin(2000) for $ZnO–PbO–B_2O_3$ compositions, but in these cases crystallization also was observed to proceed from the surface. It is possible that borate glass-ceramics may be of interest for some aspects of sealing work involving, for example, low temperature crystallizable solder glasses, but at the present time it is doubtful that practical high strength systems could be manufactured.

TABLE 8
GLASS-CERAMIC SYSTEMS AND THERMAL EXPANSION DATA
(Glass compositions are given in Table 9)

Glass-ceramic No.	Heat-treatment schedule	Thermal expansion $(10^{-6}\ K^{-1})$	Temperature range (°C)	Major crystalline phases	Reference
(A) SILICATE GLASS-CERAMICS					
Alkali and alkaline earth metal oxide silicates					
AS1	120 min. @ 720°C+120 min. @ 840°C	30·6	0–300	cristobalite	Stookey, 1960a, 1960b
AS2	120 min. @ 720°C+480 min. @ 900°C	24·4	0–300	tridymite	Stookey, 1960a, 1960b
AS1	120 min. @ 720°C+240 min. @ 975°C	23·6	0–300	tridymite	Stookey, 1960a, 1960b
AS3	15 min. @ 1010°C+15 min. @ 670°C +30 min. at 820°C	18·7 / 15·3 / 13·0	50–300 / 50–550 / 50–800	lithium metasilicate+quartz +residual glass	Kramer *et al*, 1985b
AS4	120 min. @ 720°C+480 min. @ 900°C	17·7	0–300	tridymite	Stookey, 1960a, 1960b
AS5	15 min. @ 1010°C+15 min. @ 670°C +30 min. @ 820°C	16·8 / 14·0 / 15·0	50–300 / 50–550 / 50–800	lithium metasilicate+tridymite +lithium disilicate	Kramer *et al*, 1985b
AS6	20 min. @ 1000°C+15 min. @ 650°C +20min. at 820°C	14·5	not specified	lithium metasilicate +lithium disilicate +cristobalite	Hommetter & Loehman, 1987
AS7	15 min. @ 1010°C+15 min. @ 670°C +30 min. at 820°C	14·0 / 12·5 / 14·1	50–300 / 50–550 / 50–800	cristobalite + lithium metasilicate +lithium disilicate +residual glass	Kramer *et al*, 1985b
AS8	15 min. @ 1010°C+15 min. @ 670°C +30 min. at 820°C	13·9 / 13·2 / 16·3	50–300 / 50–550 / 50–800	lithium metasilicate+cristobalite +quartz+residual glass	Kramer *et al*, 1985b
AS9	60 min. @ 500°C+540 min. @ 700°C	12·6	20–500	lithium disilicate +quartz	McMillan *et al*, 1966d

Code	Heat treatment		Temperature range (°C)	Crystalline phases	Reference
AS9	60 min. @ 500°C+60 min. @ 750°C	12·1	20–500	lithium disilicate + quartz	McMillan et al, 1966d
AS10	Nuc. + 60 min. @ 850°C	11·5	20–500	not specified	McMillan & Partridge, 1963
AS11	15 min. @ 1010°C+15 min. @ 670°C +30 min. at 820°C	10·8 9·5 13·2	50–300 50–550 50–800	quartz+ lithium metasilicate +residual glass	Kramer et al, 1985b
AS12	Nuc. +60 min. @ 850°C	10·2	20–500	β-spodumene +lithium disilicate	McMillan & Partridge, 1963
AS13	15 min. @ 1010°C + 15 min. @ 670°C +30 min. at 820°C	10·2 10·8 13·5	50–300 50–550 50–800	lithium metasilicate +residual glass+lithium disilicate	Kramer et al, 1985b
AS14	15 min. @ 1010°C + 15 min. @ 670°C +30 min. at 820°C	10·0 10·4	50–300 50–500	lithium metasilicate +residual glass + quartz	Kramer et al, 1985b
AS9	60 min. @ 500°C+60 min. @ 700°C	8·1	20–500	lithium disilicate +residual glass	McMillan et al, 1966d
Alkali and alkaline earth metal oxide alumino-silicates					
AAS1	30 min. @ 860°C+30 min. @ 1150°C	9·6 8·5	20–460 20–700	$Al_6Si_2O_{13}$+cordierite +magnesium phosphate+zirconium titanate	Donald et al, 1990 Metcalfe & Donald, 1991
AAS2	Nuc. +1230 min. @ 1100°C	9·2	not specified	dibarium trisilicate	Corning, 1960
AAS3	Nuc. +960 min. @ 1200°C	8·5	not specified	anorthite +cristobalite	Corning, 1960
AAS4	Nuc. +960 min. @ 1300°C	6·3	not specified	cordierite +cristobalite+magnesium titanate	Corning, 1960
AAS5	30 min. @ 860°C+30 min. @ 900°C	5·8 8·6	20–460 20–700	$Li_2Al_2Si_3O_{10}$ +β-spodumene +forsterite	Donald et al, 1990;
AAS6	Nuc. +60 min. @ 1250°C	5·6	not specified	cordierite +magnesium titanate +rutile+cristobalite	Metcalfe & Donald, 1991 Corning, 1960
AAS1	30 min. @ 860°C+30 min. @ 1400°C	4·7 5·0	20–460 20–700	$Al_6Si_2O_{13}$ +cordierite+zircon +residual glass	Donald et al, 1990 Metcalfe & Donald, 1991

Code	Heat treatment	CTE	Crystalline phases	Temperature range	Reference
AAS7	Nuc. +960 min. @ 1300°C	4·0	cordierite+rutile +cristobalite	not specified	Corning, 1960
AAS8	Nuc. +180 min. @ 1250°C	3·4	cordierite +magnesium aluminium titanate	not specified	Corning, 1960
Zinc alumino-silicates					
ZAS1	Nuc. +150 min. @ 1200°C	19·3	gahnite +cristobalite+willemite+rutile	not specified	Corning, 1960
ZAS2	Nuc. +120 min. @ 1250°C	18·3	gahnite +cristobalite+willemite+rutile	not specified	Corning, 1960
ZAS3	Nuc. +60 min. @ 1200°C	17·2	gahnite+willemite +cristobalite+rutile	not specified	Corning, 1960
ZAS4	Nuc. +2700 min. @ 1250°C	16·5	gahnite +cristobalite+willemite+rutile	not specified	Corning, 1960
ZAS5	15 min. @ 925°C	5·2	willemite +gahnite+$MgAl_2Si_3O_{10}$	not specified	Holleran & Martin, 1987
ZAS5	120 min. @ 940°C	14·0	as above +cristobalite	not specified	Holleran & Martin, 1987
ZAS6	60 min. @ 555°C+60 min. @ 605°C +60 min. @ 630°C	8·8 / 10·0	β-Li_2ZnSiO_4 +residual glass	20–300 / 20–450	Omar et al, 1991
ZAS7	30 min. @ 1050°C	4·7	not specified	20–600	Nash, et al, 1983; Omar et al, 1991
ZAS8	60 min. @ 540°C+60 min. @ 710°C +60 min. @ 900°C	4·2 / 4·9 / 5·3	β-spodumene +γ_o-Li_2ZnSiO_4	20–300 / 20–450 / 20–700	
Titanium alumino-silicates					
TAS1	Nuc. +120min. @ 1250°C	17·4	not specified	not specified	Corning, 1960
TAS2	Nuc. +960 min. @ 1100°C	17·0	not specified	not specified	Corning, 1960
TAS3	Nuc. +960 min. @ 1100°C	9·4	pyrophanite +tridymite+mullite+quartz	not specified	Corning, 1960
Alkali and alkaline earth metal oxide zinc silicates					
AZS1	60 min. @ 520°C+60 min. @ 810°C	18·8	cristobalite +γ_{II}-lithium zinc silicate +γ_o-lithium zinc silicate	20–400	Chen & McMillan, 1985

Sample	Heat treatment		Temp. range (°C)	Phases	Reference
AZS2	5 min. @ 950°C+60 min. @ 460°C +60 min. at 665°C	17·1	20–460	cristobalite +unidentified lithium zinc silicate +residual glass	Donald et al, 1989
AZS3	60 min. @ 500°C+60 min. @ 850°C	16·5	20–500	not specified	English Electric Co., 1963
AZS4	60 min. @ 500°C+60 min. @ 850°C	16·5	20–500	not specified	English Electric Co., 1963
AZS4	5 min. @ 950°C+60 min. @ 465°C +60 min. at 700°C	16·2	20–460	cristobalite +unidentified lithium zinc silicate +residual glass	Donald et al, 1989
AZS5	60 min. @ 520°C+60 min. @ 920°C	15·5	20–400	quartz +γ_{II}-lithium zinc silicate+cristobalite +γ_{o}-lithium zinc silicate	Chen & McMillan, 1985
AZS2	5 min. @ 950°C+60 min. @ 465°C +60 min. at 850°C	15·2	20–460	cristobalite +quartz+tridymite + unidentified lithium zinc silicate	Donald et al, 1989,
AZS6	5 min. @ 950°C+60 min. @ 585°C +60 min. at 850°C	14·3 / 15·5	20–460 / 20–550	cristobalite+quartz +tridymite +unidentified lithium zinc silicate	Donald et al, 1992
AZS6	5 min. @ 950°C+60 min. @ 465°C +60 min. at 850°C	13·9 / 15·6	20–460 / 20–550	cristobalite+quartz +tridymite+γ_{I}-lithium zinc silicate +γ_{o}-lithium zinc silicate	Donald et al, 1992
AZS4	5 min. @ 950°C+60 min. @ 465°C + 60 min. at 850°C	13·4	20–460	cristobalite +quartz+unidentified lithium zinc silicate	Donald et al, 1989
AZS7	5 min. @ 950°C+60 min. @ 465°C +60 min. at 850°C	13·4 / 15·3	20–460 / 20–550	tridymite +zinc chromate + unidentified lithium zinc silicate	Donald et al, 1992
AZS8	60 min. @ 480°C+60 min. @ 800°C	12·0	20–400	not specified	McMillan et al, 1966b
AZS9	5 min. @ 950°C+60 min. @ 465°C	11·6	20–460	not specified	Donald et al, 1989
AZS10	5 min. @ 950°C + 60 min. @ 465°C +60 min. at 850°C	10·9 / 11·5	20–460 / 20–550	quartz +unidentified Ta-rich phase +unidentified lithium zinc silicate +residual glass	Donald et al, 1991
AZS11	60 min. @ 725°C+60 min. @ 1100°C	10·2	20–400	enstatite solid solution+cristobalite +willemite solid solution	Chen & McMillan, 1985b

AZS12	60 min. @ 680°C+60 min. @ 1100°C	8·7	20–400	enstatite solid solution+cristobalite	Chen & McMillan, 1985b
AZS12	60 min. @ 680°C+60 min. @ 1000°C	7·3	20–400	enstatite solid solution	Chen & McMillan, 1985a
AZS1	60 min. @ 520°C+60 min. @ 750°C	7·0	20–400	γ_{II}-lithium zinc silicate+γ_{O}-lithium zinc silicate+cristobalite	Chen & McMillan, 1985a
AZS5	60 min. @ 520°C+60 min. @ 750°C	5·5	20–400	γ_{II}-lithium zinc	Chen & McMillan, 1985a
AZS11	60 min. @ 725°C+60 min. @ 800°C	4·5	20–400	enstatite solid solution+residual glass	Chen & McMillan, 1985b
AZS12	60 min. @ 680°C+60 min. @ 800°C	3·5	20–400	enstatite solid solution	Chen & McMillan, 1985b

Alkali and alkaline earth metal oxide cadmium silicates

ACS1	240 min. @ 600°C	8·3	?–600	lithium and cadmium silicates+residual glass	Rincon et al, 1987
ACS2	240 min. @ 600°C	8·1	?–600	lithium and cadmium silicates+residual glass	Rincon et al, 1987

Alkali and alkaline earth metal oxide zinc alumino-borosilicates

ZBS1	120 min. @ 500°C+120 min. @ 700°C	7·3	20–500	zinc aluminate+zinc borate+quartz	McMillan & Partridge, 1969
ZBS2	120 min. @ 600°C+120 min. @ 700°C	6·3	20–500	zinc aluminate+zinc borate+quartz	McMillan & Partridge, 1969
ZBS3	120 min. @ 620°C+60 min. @ 900°C	4·5	20–500	zinc aluminate+zinc borate+quartz	McMillan & Partridge, 1969
ZBS4	120 min. @ 500°C+120 min. @ 700°C	3·5	20–500	zinc aluminate+zinc borate+quartz	McMillan & Partridge, 1969

Lead silicates

LS1	120 min. @ 500°C+60 min. @ 725°C	17·0	not specified	not specified	McMillan & Hodgson, 1966

	Heat treatment		Expansion range (°C)	Crystalline phase	Reference
LS2	Nuc. + 120 min. @ 780°C	4·2	not specified	lead titanate	Corning, 1960
(B) PHOSPHATE GLASS-CERAMICS					
PC1	2 h @ 400°C+2 h @ 450°C	22·5	25–250	sodium metaphosphate or sodium trimetaphosphate +sodium barium phosphate or barium phosphate	Wilder et al, 1982
PC2	2 h @ 450°C+2 h @ 500°C	16·2	25–250	-as above-	
PC3	60 min. @ 670°C+60 min. 2 850°C	14·1	20–185	aluminium phosphate +unidentified phase	Wang & James, 1990
		8·6	185–650		
PC3	60 min. @ 700°C+60 min. @ 850°C	12·6	20–185	aluminium phosphate +unidentified phase	Wang & James, 1990
		9·2	185–650		

TABLE 9
GLASS AND GLASS-CERAMIC COMPOSITIONAL DATA

Glass Code	Composition (mole%)															
	Li_2O	Na_2O	K_2O	MgO	CaO	BaO	TiO_2	ZrO_2	ZnO	B_2O_3	Al_2O_3	SiO_2	P_2O_5	Sb_2O_3	PbO	Others
S1	19·50	4·94	3·55	–	–	–	–	–	6·11	53·00	–	–	–	–	–	–
S2	40·05	–	–	–	–	–	–	–	–	–	7·52	52·43	–	–	–	–
S3	35·20	–	–	–	–	–	–	–	–	–	7·48	57·32	–	–	–	–
S4	–	–	–	–	–	–	–	–	–	–	–	59·92	–	–	40·08	–
S5	21·12	–	–	–	–	–	–	–	–	–	8·93	56·02	–	–	–	$11·45Cu_2O+2·48Fe_2O_3$
BS1	–	15·14	5·02	–	–	10·09	–	–	4·98	12·19	–	44·87	–	–	–	$7·21CaF_2+0·25NiO+0·25CoO$
BS2	25·27	–	–	5·76	–	–	–	–	–	63·39	1·71	3·87	–	–	–	–
BS3	–	–	–	–	–	–	–	–	–	26·58	–	30·79	–	–	42·63	–
BS4	4·43	5·34	1·41	3·28	–	–	–	–	–	73·21	1·30	11·03	–	–	–	–
BS5	14·42	–	1·41	4·58	–	–	–	–	–	70·75	–	10·25	–	–	–	–
BS6	4·44	5·35	1·41	1·63	–	–	–	–	–	63·79	1·30	22·08	–	–	–	–
BS7	1·73	2·82	1·58	–	–	1·27	–	–	–	14·14	4·70	72·89	–	–	–	0·87KCl
BS8	7·60	–	–	–	–	–	–	–	–	13·44	10·88	52·70	–	–	15·38	–
AB1	–	–	–	–	–	–	–	–	–	51·48	6·70	–	–	–	–	41·82CuO
AB2	–	–	–	–	–	–	–	–	–	43·43	13·83	–	–	–	–	42·74CuO
AB3	–	–	–	–	–	–	–	–	–	33·26	14·83	–	–	–	–	51·91CuO
LB1	–	–	–	–	–	–	–	–	4·64	21·27	3·31	4·21	–	–	61·02	$4·62CuO+0·93Bi_2O_3$
LB2	–	–	–	–	–	–	–	–	4·99	21·11	3·40	4·33	–	–	60·60	$4·58CuO+0·99Bi_2O_3$
LB3	–	–	–	–	–	–	–	–	–	33·77	3·78	–	–	–	49·74	$12·71Tl_2O$
ZB1	–	–	–	–	–	–	–	–	8·11	28·44	–	2·75	–	–	53·97	$6·73 PbF_2$
ZB2	–	–	–	–	–	–	–	–	–	30·09	1·67	2·83	–	–	65·41	–
ZB3	–	–	–	–	–	–	–	–	4·02	31·01	1·60	2·71	–	–	60·66	–
ZB4	–	–	–	–	–	–	–	–	61·37	38·63	–	–	–	–	–	–
ZB5	–	–	–	–	–	–	–	–	43·78	17·05	–	–	–	–	–	$39·17V_2O_5$
ZB6	4·94	–	–	–	–	–	–	–	46·62	33·50	4·99	9·95	–	–	–	–
ZB7	–	–	–	–	–	–	–	–	66·63	33·37	–	–	–	–	–	–
ZB8	–	–	–	–	–	–	–	–	41·06	24·01	–	18·07	–	–	6·36	10·50CuO
ZB9	–	–	–	–	–	–	–	–	60·13	25·78	–	12·28	–	0·11	0·70	$1·0SnO_2$

	Li_2O	Na_2O	K_2O	MgO	CaO	BaO	TiO_2	ZrO_2	ZnO	B_2O_3	Al_2O_3	SiO_2	P_2O_5	Sb_2O_3	PbO	$Others$
ZB10	–	–	–	–	–	–	–	–	56·34	23·12	–	17·40	1·13	–	–	$0.91Ta_2O_5+0.93CeO_2+0.17Bi_2O_3$
LZB1	–	–	–	–	–	–	–	–	10·85	20·30	–	–	–	–	68·85	–
LZB2	–	–	–	–	–	–	–	–	14·54	21·85	–	–	–	–	63·61	–
LZB3	–	–	–	–	–	–	–	–	9·50	22·21	–	12·89	–	–	55·40	–
LZB4	–	–	–	–	–	–	–	–	16·66	41·67	–	–	–	–	41·67	–
LZB5	–	–	–	–	–	–	–	–	16·72	29·31	–	11·32	–	–	42·65	–
LZB6	–	–	–	–	–	–	–	–	37·67	44·02	–	–	–	–	18·31	–
LZB7	–	–	–	–	–	–	–	–	46·02	33·61	–	7·79	–	–	12·58	–
BH1	–	–	–	–	–	–	–	–	–	19·26	–	8·99	–	–	66·73	$5.02PbBr_2$
BH2	–	–	–	–	–	–	–	–	–	19·34	–	8·78	–	–	66·84	$5.04PbCl_2$
BH3	–	–	–	–	–	–	–	–	–	21·16	–	–	–	–	52·80	$26.04PbF_2$
BH4	–	–	–	–	–	–	–	–	–	19·32	–	8·89	–	–	48·12	$23.67PbF_2$
BH5	–	–	–	–	–	–	–	–	18·16	21·22	–	–	–	–	46·33	$14.29ZnF_2$
BH6	–	–	–	–	–	–	–	–	35·00	35·00	–	–	–	–	–	$30.00PbF_2$
BH7	–	–	–	–	–	–	–	–	37·50	40·00	–	–	–	–	–	$32.50PbF_2$
BH8	–	–	–	30·00	–	–	–	–	40·00	42·50	–	–	–	–	–	$17.50PbF_2$
P1	–	25·00	25·00	–	–	–	–	–	–	–	–	–	50·00	–	–	–
P2	–	35·00	15·00	–	–	–	–	–	–	–	–	–	50·00	–	–	–
P3	–	15·00	–	–	–	–	–	–	–	–	–	–	50·00	–	–	$35.00AgO$
P4	–	–	38·00	–	–	–	–	–	–	–	3·00	–	56·00	–	–	$3.00Fe_2O_3$
P5	–	–	38·00	–	–	–	–	–	–	–	–	–	56·00	–	–	$6.00Fe_2O_3$
P6	–	–	34·00	–	–	–	–	–	–	–	8·00	–	58·00	–	–	–
P7	–	–	34·00	–	–	–	–	–	–	–	4·00	–	58·00	–	–	$4.00Fe_2O_3$
P8	–	–	30·00	–	–	–	–	–	–	–	10·00	–	60·00	–	–	–
P9	–	–	–	–	–	–	–	–	–	–	–	–	50·00	–	50·00	–
P10	–	–	–	–	–	–	–	–	–	–	–	–	58·80	–	41·20	–
P11	–	–	–	–	–	–	–	–	20·00	–	–	–	40·00	–	40·00	–
P12	–	–	–	–	–	–	–	–	20·00	–	–	–	50·00	–	30·00	–
P13	–	–	–	–	–	26·47	–	–	–	9·72	5·63	–	57·18	–	–	–
P14	–	–	–	–	–	8·11	–	–	30·57	–	–	–	61·32	–	–	–
P15	–	–	–	–	–	–	–	–	30·00	–	–	–	50·00	–	20·00	–

	Li_2O	Na_2O	K_2O	MgO	CaO	BaO	TiO_2	ZrO_2	ZnO	B_2O_3	Al_2O_3	SiO_2	P_2O_5	Sb_2O_3	PbO	Others
P16	–	–	–	–	–	15·78	–	–	29·74	–	11·87	–	42·61	–	–	–
P17	–	–	–	–	–	–	–	–	–	–	26·84	–	32·88	–	–	40·28CuO
P18	–	–	–	–	–	–	–	–	–	7·87	5·38	–	46·33	–	–	–
P19	–	–	–	–	–	2·82	–	–	42·48	–	4·77	8·08	41·85	–	–	–
A1	–	–	24·31	–	–	–	–	–	–	–	–	–	–	30·19	–	45·50As₂O₃
A2	–	–	19·99	–	–	–	–	–	–	–	–	–	–	60·03	19·98	–
A3	–	–	22·76	–	–	–	–	–	–	–	–	11·04	–	66·20	–	–
A4	4·88	4·03	4·86	–	–	–	–	–	7·16	14·35	–	–	–	57·54	7·18	–
A5	6·23	3·00	9·88	–	–	–	–	–	–	13·37	–	9·30	–	41·51	16·69	–
A6	–	–	–	–	–	–	–	–	–	42·15	–	–	–	42·95	14·90	–
V1	–	–	–	–	–	–	–	–	–	–	–	–	–	–	47·50	5·00As₂O₃+47·5V₂O₅
V2	–	–	–	–	–	–	–	–	–	–	–	–	–	–	30·00	10·00As₂O₃+60·00V₂O₅
V3	–	–	–	–	–	–	–	–	–	–	–	–	–	25·00	–	50·00V₂O₅+25·00As₂O₃
DS1	–	–	–	–	–	–	–	–	18·75	21·91	–	6·36	–	–	52·98	–
DS2	–	–	–	–	–	–	–	–	18·65	19·61	1·49	6·31	–	–	50·99	2·50CdO
DS3	–	–	–	–	–	–	–	–	18·75	19·73	1·50	6·35	–	–	51·28	2·39Fe₂O₃
DS4	–	–	–	–	–	–	–	–	–	26·99	–	–	–	–	60·24	9·49TeO₂+3·28Tl₂O
DS5	–	–	–	–	–	–	9·83	–	–	33·85	–	–	–	–	56·32	–
DS6	–	–	–	–	–	1·99	–	–	20·67	19·77	0·75	5·09	–	–	51·73	–
DS7	–	–	–	–	–	–	9·12	–	–	41·89	–	–	–	–	48·99	–
DS8	–	–	–	–	–	–	–	–	18·29	21·37	2·92	7·43	–	–	49·99	–
DS9	–	–	–	–	–	–	–	–	18·55	17·34	1·48	2·51	–	0·26	50·37	9·49CuO
DS10	–	–	–	–	–	4·30	8·12	–	23·92	13·98	–	16·20	–	–	37·78	–
DS11	–	–	–	–	–	–	19·79	–	8·10	14·19	–	16·44	–	–	37·18	–
DS12	–	–	–	–	–	–	–	–	19·63	14·91	1·57	6·65	–	–	57·24	–
DS13	–	–	–	–	–	–	19·34	–	15·83	12·02	2·53	13·93	–	–	36·35	–
DS14	–	–	–	–	–	–	17·24	–	–	39·56	–	–	–	–	42·20	–
DS15	–	–	–	–	–	–	–	–	50·86	16·65	–	26·18	–	–	6·31	–
AS1	–	9·68	–	–	–	–	–	–	–	–	1·07	78·93	–	–	–	10·32F
AS2	–	9·19	–	–	–	–	–	–	–	–	–	80·82	–	–	–	9·99F
AS3	24·40	–	–	–	–	–	–	–	–	2·68	2·88	69·02	1·02	–	–	–
AS4	–	14·12	–	–	–	–	–	–	–	–	–	85·88	–	–	–	–

	Li_2O	Na_2O	K_2O	MgO	CaO	BaO	TiO_2	ZrO_2	ZnO	B_2O_3	Al_2O_3	SiO_2	P_2O_5	Sb_2O_3	PbO	Others
AS5	20·29	–	2·50	–	–	–	–	–	–	2·22	2·40	71·75	0·84	–	–	–
AS6	23·70	–	2·80	–	–	–	–	–	–	2·60	2·80	67·10	1·00	–	–	–
AS7	23·46	–	2·90	–	–	–	–	–	–	2·57	2·77	66·35	1·95	–	–	–
AS8	23·09	–	2·85	–	–	–	–	–	–	5·06	2·73	65·31	0·96	–	–	–
AS9	22·3	–	1·50	–	–	–	–	–	3·00	–	–	72·50	0·70	–	–	–
AS10	22·86	–	–	19·31	–	–	–	–	–	–	–	56·77	1·06	–	–	–
AS11	23·04	–	2·85	–	–	–	–	–	–	2·52	5·44	65·19	0·96	–	–	–
AS12	22·53	–	1·48	–	–	–	–	–	–	–	2·13	72·68	1·18	–	–	–
AS13	28·45	–	3·51	–	–	–	–	–	–	3·12	3·36	60·37	1·19	–	–	–
AS14	24·06	–	2·97	–	–	–	–	–	–	2·05	2·84	68·08	–	–	–	–
AAS1	–	–	–	22·50	–	–	4·90	2·10	–	–	19·10	46·50	4·90	–	–	–
AAS2	–	–	–	–	–	16·51	12·70	–	–	–	10·68	60·12	–	–	–	–
AAS3	–	–	–	–	10·97	–	9·60	–	–	–	18·10	61·33	–	–	–	–
AAS4	–	–	–	21·13	–	–	7·08	–	–	–	11·65	60·14	–	–	–	–
AAS5	15·90	–	–	18·90	–	–	3·80	1·20	–	3·40	10·40	41·60	1·00	–	–	–
AAS6	–	–	–	23·07	–	–	6·98	–	–	–	12·16	57·79	–	–	–	–
AAS7	–	–	–	14·34	–	–	9·02	–	–	–	14·14	62·50	–	–	–	–
AAS8	–	–	–	6·07	6·42	–	9·44	–	–	–	17·79	60·28	–	–	–	–
ZAS1	–	–	–	–	–	–	2·58	–	29·75	–	10·20	57·47	–	–	–	–
ZAS2	–	–	–	–	–	–	1·80	–	25·92	–	13·79	58·49	–	–	–	–
ZAS3	–	–	–	7·45	–	–	7·61	–	22·42	–	11·93	50·59	–	–	–	–
ZAS4	–	–	–	–	–	–	6·68	–	24·65	–	13·09	55·58	–	–	–	–
ZAS5	–	–	–	3·50	–	–	5·35	–	32·04	2·02	9·67	52·77	–	–	–	–
ZAS6	22·09	–	–	–	–	–	–	–	12·62	–	9·47	50·47	–	–	–	–
ZAS7	–	4·31	–	–	–	2·24	–	–	28·84	–	6·90	56·60	1·11	–	–	–
ZAS8	23·18	–	–	–	–	–	–	0·74	13·23	–	9·93	52·92	–	–	–	–
TAS1	–	–	–	–	–	–	8·30	–	–	–	18·16	49·61	–	–	–	23·93CoO
TAS2	–	–	–	–	–	–	8·15	–	–	–	15·94	54·22	–	–	–	21·69CoO
TAS3	–	–	–	–	–	–	9·61	–	–	–	18·23	52·33	–	–	–	19·83MnO
AZS1	18·10	–	1·30	–	–	–	–	–	20·10	–	–	59·40	1·10	–	–	–
AZS2	18·11	4·85	–	–	–	–	–	–	18·91	3·54	–	53·33	1·27	–	–	–

	Li₂O	Na₂O	K₂O	MgO	CaO	BaO	TiO₂	ZrO₂	ZnO	B₂O₃	Al₂O₃	SiO₂	P₂O₅	Sb₂O₃	PbO	Others
AZS3	18·16	–	1·28	–	–	–	–	–	21·23	–	–	59·20	–	–	–	1·13MoO₃
AZS4	18·74	4·81	–	–	–	–	–	–	18·67	3·51	–	53·35	0·92	–	–	–
AZS5	18·40	–	–	–	–	–	–	–	20·30	–	–	60·10	1·20	–	–	–
AZS6	17·84	5·25	–	–	–	–	–	–	17·73	4·31	–	53·64	1·23	–	–	2·00Cr₂O₃
AZS7	17·49	5·15	–	–	–	–	–	–	17·37	4·22	–	52·56	1·21	–	–	–
AZS8	22·20	–	1·47	–	–	–	–	–	3·74	–	–	71·85	0·74	–	–	–
AZS9	17·91	1·47	1·61	–	–	–	–	–	17·82	4·26	0·60	55·00	1·33	–	–	–
AZS10	17·49	5·15	–	–	–	–	–	–	17·37	4·22	–	52·56	1·21	–	–	2·00Ta₂O₅
AZS11	–	–	–	21·00	–	–	4·00	–	20·00	–	5·00	50·00	–	–	–	–
AZS12	–	–	–	31·93	–	–	3·24	–	12·66	–	5·04	47·13	–	–	–	–
AZS13	17·90	1·47	1·61	–	–	–	–	–	17·82	4·26	0·60	55·00	1·33	–	–	–
ACS1	20·70	–	–	–	–	–	–	–	–	–	2·60	43·78	–	–	–	32·92CdO
ACS2	21·54	–	–	–	–	–	–	–	–	–	0·90	62·00	–	–	–	15·56CdO
ZBS1	–	–	–	–	–	4·75	–	–	54·30	31·73	4·37	4·85	–	–	–	–
ZBS2	11·71	–	–	–	–	–	–	–	24·50	28·64	13·04	22·11	–	–	–	–
ZBS3	–	–	–	–	–	–	–	–	41·67	21·65	10·35	26·33	–	–	–	–
ZBS4	–	2·37	–	9·12	–	–	–	–	50·46	29·48	4·04	4·53	–	–	–	–
LS1	19·43	–	–	–	–	–	–	–	10·38	–	–	63·55	1·22	–	4·05	–
LS2	–	–	1·37	–	–	–	18·51	–	–	–	5·75	34·60	–	–	41·14	–
PC1	–	40·00	–	–	–	10·00	–	–	–	–	–	–	50·00	–	–	–
PC2	–	20·00	–	–	–	30·00	–	–	–	–	–	–	50·00	–	–	–
PC3	–	–	–	–	40·30	–	5·50	–	–	–	7·40	7·80	39·1	–	–	–
E1Cu	–	2·25	26·07	–	–	–	–	–	–	4·62	–	43·90	–	–	23·16	–
E2Cu	–	–	38·84	–	–	–	–	–	–	–	–	43·81	–	–	17·35	–
E3Cu	–	16·17	4·47	–	5·96	–	–	–	–	12·86	4·46	53·16	–	–	–	–
E4Cu	–	9·23	1·69	–	–	–	–	–	–	2·43	–	55·54	–	–	21·21	2·94As₂O₃+6·96SnO₂
E5Cu	–	4·21	4·98	–	–	2·92	–	–	–	–	–	52·10	–	–	23·38	6·21NaF+6·21AlF₃+2·90As₂O₃
E1A1	–	12·60	–	–	–	–	0·68	0·44	–	4·41	–	55·99	–	–	21·44	4·44V₂O₅
E2A1	9·80	23·46	–	–	–	–	–	–	–	8·13	16·31	–	22·75	–	–	19·55NaF
E3A1	48·63	–	–	–	–	–	–	–	–	–	–	39·38	–	0·60	11·39	–
E4A1	9·90	19·60	–	–	–	–	–	–	–	–	–	54·70	–	0·50	15·30	–

5. THE TECHNOLOGY OF SEALS AND FACTORS AFFECTING SEAL QUALITY

5.1   General bonding requirements

The main requirements for the formation of a high quality seal or coating are the ability to form a strong bond between the components together with the ability either to match as closely as possible the thermal expansion characteristics of the two dissimilar materials, or, alternatively, to ensure that the expansion characteristics are such that predominantly compressive stresses are generated in the brittle glass or glass-ceramic phase. In addition to control over the thermal expansion behaviour, the glass must also meet a number of other requirements. These include an ability to wet and spread over the metal surface(s) at a practical temperature, to bond chemically to the metal but not to promote the formation of undesirable reaction products that may alter the properties at the interface, and to possess acceptable electrical, mechanical, chemical and physical properties. For seals there are also a number of additional technical considerations that may need to be taken into account. For example, in the case of seals for electrical applications, the following may be important: electrical resistivity, dielectric properties, electric flash-over voltage, current carrying capacity, hermeticity, pressure load capacity, mechanical strength, corrosion resistance, temperature capability, etc.

The requirements for the production of strong, dense coatings are similar to those for the production of seals, although it is usual to employ systems in which the coating is in mild compression, rather than being matched in expansion. As coatings are normally initially applied to the substrate in the form of a suspension of glass particles it is, however, more critical that the glass wets and flows sufficiently to cover the substrate, and that the individual glass particles coalesce to form a dense, pore-free coating. The overall requirements may also be strongly dependent upon the specific application. For example, a *refractory* coating impervious to oxygen is required for the protection of a metal from high temperature oxidation. In addition, *specific* requirements for seals and coatings may differ, in particular the manner in which the constituent materials are processed prior to fabrication, as described in the following sections.

5.2   Preparation of glass-to-metal seals

The preparation of a reliable and hermetic glass-to-metal seal can be divided into a number of specific operations. Firstly, selection of the metal and glass component materials, followed by machining or alternative forming into the parts required to make the seal. Next, degreasing and cleaning of the piece-parts prior to their assembly utilizing suitable fixtures or jigs to hold the component parts together in their correct locations. Prior to assembly the metal parts may be subjected to a pre-oxidation stage followed by another cleaning cycle before committing to the furnace, although, as noted later, a pre-oxidation stage is not always necessary. The assembled parts are subsequently subjected to a heat treatment cycle to form the seal, with the furnace operation normally carried out in an inert atmosphere in order to limit the amount of metal oxide formed. After removal from the furnace, components are usually cleaned to remove oxide scale or tarnishing of the exposed metal parts.

5.2.1   Selection and preparation of metal and glass component piece-parts

The choice of specific metal and glass materials is obviously very dependent on the final application required for the finished component and is a topic considered in some detail in Chapter 10 on applications. Particular considerations when selecting the component materials include such factors as material compatibility, thermal expansion and electrical characteristics, temperature and pressure requirements, chemical durability, mechanical reliability and integrity, and lifetime expectancy.

It should also be borne in mind that the quality of the metal parts that are being sealed to can be very important. It has been noted (Geoffrion, 2002) that suppliers of metals may change such conditions as the degree of temper, grain structure, transformation properties, etc. without notification to the customer. Even more critical, the surfaces of metals supplied may be contaminated with debris from rolling mill operations and other fabrication techniques that are difficult to remove. Sintered metal parts may be particularly problematic due to the use of binders in their preparation which have not been adequately removed. Such poor quality products may seriously effect seal quality due to outgassing and other related problems. Geoffrion stresses that high quality vacuum melted metals are best used in the electronics industry, e.g. for semiconductors, computer, aerospace applications, etc, where product reliability is of paramount importance. He also notes that metals which are described as AOD processed (Oxygen Argon Decarburization) are not suitable for seal manufacture as they tend to be inhomogeneous. It should also be noted that metals of ostensibly the same composition can in fact have quite wide variations in composition from batch to batch and supplier to supplier. This is clear from alloy specifications. Taking Nimonic 90 alloy as an example, the specification for this alloy gives the Ni content as between 18 and 21%, Cr between 18 and 21%, Ti between 1 and 2%, Mn as $\leq 1.0\%$, Si $\leq 1.0\%$, Cu $\leq 0.2\%$ and C $\leq 0.13\%$. At AWE we have noted that this can have important consequences on seal quality, with the effects of small concentrations of some elements that may be regarded as impurities being detrimental to the sealing process if their concentration is above a certain limit, which may still be within the specification for the particular alloy in question. Machining of piece-parts is a very important step which must be carried out carefully. In

(a) Courtesy of Schott

(b) Courtesy of Mansol Preforms
Figure 27: Photographs of sintered glass preforms used in glass-to-metal seal manufacture

general, use of sulphur-containing cutting fluids should be avoided as these can lead to problems during sealing if not adequately removed.

The starting glass may be employed as a dense, solid pre-form of suitable shape which is positioned with the metal parts by the use of suitable jigging. Preforms may be made by casting solid blanks and machining to size or, more usually, by powder techniques to prepare sintered preforms. Sintered glass pre-forms are manufactured routinely and are available in many different shapes and sizes from a variety of manufacturers. Some examples are shown in Figure 27. Alternatively, the glass may simply be used as a powder which is packed into the components to be sealed.

TABLE 10
STANDARD PIECE-PART CLEANING PROCEDURE

Metal Components
1. Ultrasonically clean in dichloromethane for 2 minutes
2. Ultrasonically clean in detergent solution* for 2 minutes
3. Rinse in running de-ionised water
4. Ultrasonically clean in de-ionised water for 2 minutes
5. Ultrasonically clean in isopropyl alcohol for 2 minutes
6. Dry under an infra-red lamp

* detergent composition:- 1 part Span 80; 12 parts Triton N101; 1200 parts deionised water; 4800 parts isopropyl alcohol

Glass Preforms
1. Ultrasonically clean in dichloromethane for 2 minutes
2. Rinse in running de-ionised water
3. Lightly etch* for about 10 seconds
4. Rinse in running de-ionised water
5. Ultrasonically clean in de-ionised water for 2 minutes
6. Ultrasonically clean in isopropyl alcohol for 2 minutes
7. Dry under an infra-red lamp

* etchant composition:- 33 ml 47% HF; 40 ml H_2SO_4; 140 ml deionised water

Choice of material for use as moulds or fixtures employed in the manufacture of glass-to-metal seals can be important. In the case of graphite furniture, for example, use of Pb-based glasses should be avoided due to reduction of the glass at the graphite/glass interface to give metallic lead.

5.2 2   Cleaning and assembly of piece-parts

In the preparation of seals it is essential that the components are thoroughly cleaned prior to assembly and fabrication. This is necessary in order to prevent the formation of gaseous reaction products during sealing which may give rise to bubbles within the seal interface. In the case of the metal parts this may involve vacuum annealing and/or solvent cleaning/degreasing procedures and it may also be necessary to decarburize the metal to remove carbide precipitates from the metal surface (see, e.g. Partridge, 1949; Price, 1984). Cleaning and assembly of components is often carried out under clean room conditions or alternatively within a fume cupboard or fume hood in order to eliminate contamination of surfaces by airborne dusts. This is particularly important in the manufacture of electronic components. In any cleaning or assembly operation use of lint-free cloth is also recommended.

Kramer *et al*(1982) have examined the influence of a number of different cleaning procedures for metal and glass component parts prior to sealing on the

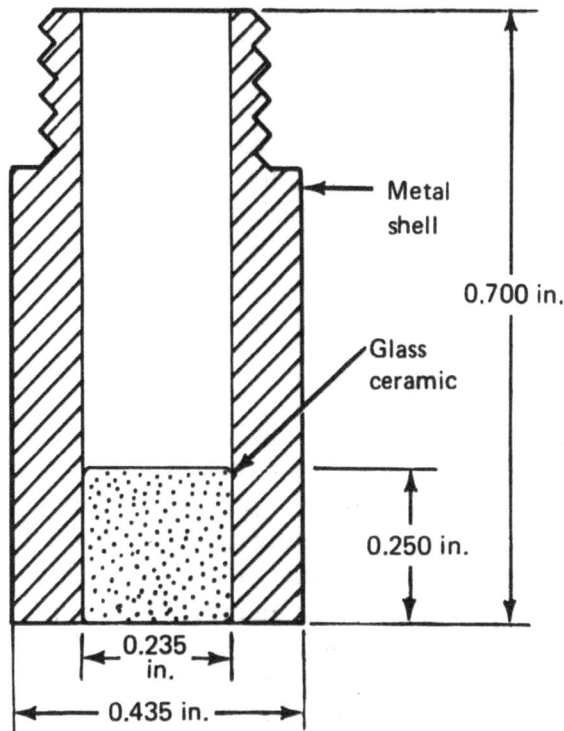

Figure 28: Schematic diagram of shells used by Kramer (1982) for assessing the quality of seals by measuring the hydrostatic burst strength

resultant seal quality. In one such study, metal test shells of Inconel 625, Inconel 718, Hastelloy C276 and stainless steel were first cleaned using a standard four-step solvent cleaning process. This method, which involves the use of chlorinated solvents, is summarized in Table 10. Some of the samples were then sealed to a lithium aluminosilicate glass-ceramic without further treatment. Other samples were subjected to an additional plasma cleaning step using oxygen or argon prior to sealing. In the plasma cleaning process gas, contained in a sample chamber under partial pressure, is ionized at low temperature by application of an external radio frequency field to produce a highly reactive plasma which may react with and volatilize surface contaminants. This leads to a cleaner surface. Variables of this process include choice of gas, power applied and time of application. Seal strength was subsequently assessed by measuring the hydrostatic bursting strength of sealed shells, depicted in Figure 28. It was noted that seals prepared using the metal shells that had been plasma cleaned generally resulted in higher burst strength samples than those which had undergone the solvent cleaning process only, but the results were highly dependent on the metal employed. For example, the highest bursting strength was noted for oxygen plasma cleaned stainless steel (28 MPa compared with 22 MPa for the solvent only sample). In the case of Inconel 718, however, the highest strength was found for the argon

TABLE 11
TRIALS TO ASSESS THE CLEANING EFFICIENCY OF VARIOUS SYSTEMS USED AS SUBSTITUTES FOR DICHLOROMETHANE

Substitute	Comments	OSSE Value
Standard system employing dichloromethane		1223±2
Axarel 4100	Polypropylene glycol in petroleum spirit (Samuel Banner & Co. Ltd)	1110±59
Multisolve MB75	Mixture of aliphatic and chemical solvents (Multisolve Ltd)	1121±32
Acetone	AnalaR (Reagent Grade, BDH)	894±25
Lotoxane (Def-Stan 68-148)	Aliphatic hydrocarbon (Arrow Chemicals)	790±88
MEK	AnalaR (Reagent Grade, BDH)	493±29
MS38	Isoparafinic hydrocarbon (Multisolve Ltd)	501±113

plasma cleaned samples, whilst for Hastelloy C276 reductions in strength were observed for the plasma cleaned samples. In a later study by Salerno *et al*(1990), it was recommended that dichloromethane was replaced by a non-chlorinated solvent in the 4-step cleaning procedure, with N-methylpyrrolidinone being the preferred choice.

For the preparation of seal devices Donald and co-workers have until recently used the solvent cleaning technique developed at SNLA/Mound given in Table 10, employing dichloromethane. More recently, with the increasing drive to more environmentally friendly and less toxic techniques, methods involving the use of non-chlorinated solvents, including aqueous based materials, have been developed at AWE. Surface cleanliness of the metal piece-parts (AISI 304 stainless steel) and glass preforms (lithium aluminosilicate) was monitored using a Patscan Model SQM200 Surface Quality Monitor. Patscan is a non-destructive surface analysis technique utilizing the principle of optically stimulated electron emission, OSEE. It is able to detect particulate, thin film organic and metallic contaminants (Arora, 1985). Any surface layer present, depending on its own photoemission, will either enhance or attenuate the current signal. In general, metals are good photoemitters and give high OSEE values, whereas organic materials are poor emitters and yield low values; therefore, the higher the OSEE value obtained after treatment, the cleaner the substrate surface is likely to be with respect to organic contaminants. Results were compared with data from the original cleaning process and methods giving similar or better cleaning characteristics than the original method were then sealed under standard conditions to assess the resulting seal quality. The highest OSEE value, 1223, was obtained using the current four-stage solvent cleaning process; however, Axarel 4100, which is a mixture of aliphatic and chemical solvents, and MB75, which is propylene glycol ether in petroleum spirit, also gave correspondingly high OSEE values of 1110 and 1121, respectively, which indicated that in terms of surface degreasing these solvents could be the most suitable alternatives to dichloromethane. Seals prepared using these solvents gave results similar to those noted for the original cleaning procedure, thus confirming that these more environmentally friendly cleaning systems are acceptable. The results, which summarise the effectiveness of the various alternatives, are summarized in Table 11. Larger scale cleaning procedures may also employ degreasing and washing stages, the latter being performed using commercial dishwashers.

In addition to the metal parts, the glass components of the seals will normally also be cleaned prior to sealing, and may also be lightly etched in an aqueous solution of hydrofluoric and hydrochloric acids, the HCl being present to dissolve any insoluble fluorides that may be formed by reaction of HF with glass constituents.

5.2.3   Pre-oxidation of metal piece-parts

Traditionally it is also usual, although not always essential, to pre-oxidize metal parts prior to sealing. The pre-oxidation stage can be very crucial, too thick or too thin an oxide layer, or the wrong type of oxide, and the seal may be mechanically weak or not hermetic. In general, an oxide layer must be non-porous, firmly adherent to the metal substrate and must be adequately wetted by the glass. As noted, it is not always necessary, however, nor indeed desirable, to pre-oxidize the metal components prior to sealing. In many instances reliance can be made on redox reactions between the glass and metal providing suitable bonding conditions. It is essential, however, to maintain stable conditions at the interface in order to avoid total dissolution during the sealing process of any oxide that is formed.

There are many anecdotal descriptions of how to pre-oxidize specific metal parts. In the case of Nilo-K, for example, heating in air at a temperature in the range 850-900°C until the parts are a cherry red in colour, i.e. a dull red heat, for around three minutes, followed by cooling at around $10\,K\,min^{-1}$ should give rise to a light grey appearance on cooling to ambient temperature. This is ideal for sealing. If the colour is black, however, the metal has probably been over-oxidized and this may give rise to a poor seal. For the binary Fe–Ni alloys difficulty can be experienced in obtaining an oxide layer that adheres well. These binary alloys are better if alloyed with such elements as Cr, Co, Mn, Si, Al or B (Leedecke *et al*, 2001). Chrome steels are relatively easy to seal to using furnace oxidizing conditions over the temperature range 925 to 1000°C. Other steels can be problematic as scaling may occur, as it also can with copper. In the case of copper, this problem may be overcome by protecting the surface with a thin glassy coating of sodium borate or by nickel plating to avoid scaling. In these instances it is important, of course, that the added layer adheres well to the metal substrate.

5.2.4   High temperature furnace sealing cycle

The furnace atmosphere may also play a very important role in determining the bonding characteristics of the system being sealed, with atmospheres ranging from highly oxidizing to highly reducing depending on whether carried out in air at ambient or reduced pressure, inert gas, reducing gas, etc. Kramer and Osborne(1983) carried out a comprehensive study of the influence of sealing atmosphere on the wetting behaviour of a lithium aluminosilicate glass to the nickel-based superalloys Inconel 718 and Hastelloy C276 using the sessile drop method. They monitored the wetting behaviour as a function of furnace atmosphere which included argon, helium and argon-5% hydrogen at dew points ranging between −40°C to +28°C, together with oxygen concentration in argon at the 10, 300 and 10^4 ppm levels. It was noted that a helium atmosphere consistently produced the lowest contact angles, and therefore the greatest degree of wetting, and that wettability generally increased as the atmospheric dew point was increased. In the case of oxygen in argon it was found that there was little dependence of contact angle for Hasteloy C276 whilst contact angle decreased with decreasing oxygen content for Inconel 718. It is thus clear that furnace atmosphere does indeed play a very important part in

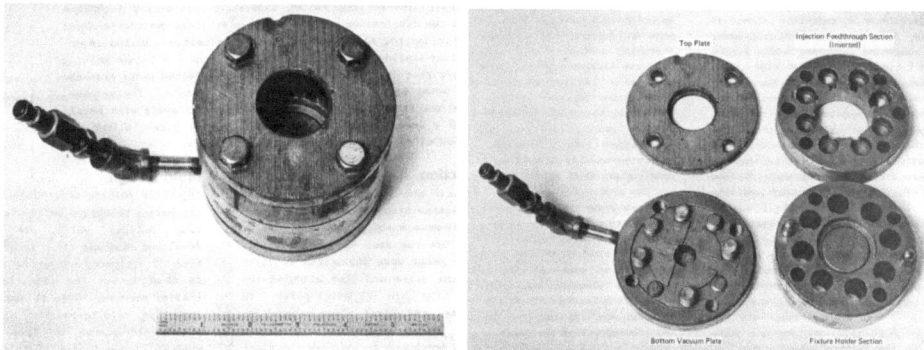

(a) Photograph of the injection moulding tool

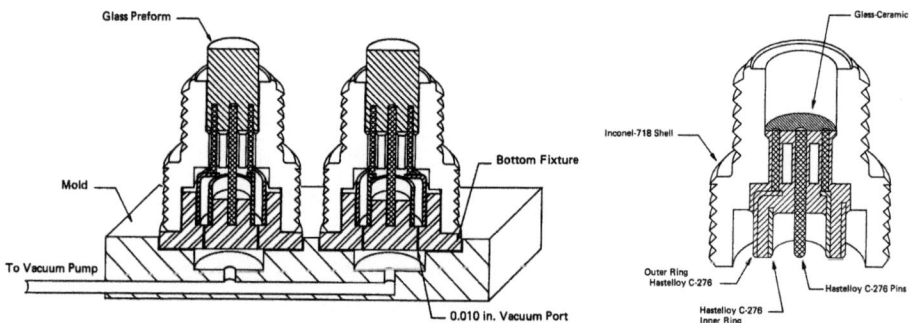

(b) Schematic illustration of typical seal component and fixtures
Figure 29: Vacuum-assisted injection moulding process which can be used with low melting point metals including aluminium for producing glass-ceramic-to-metal seals (after Kramer et al, 1987)

Figure 29: (continued)

(c) An assembled component prior to insertion into the injection moulding tool
(d) Component after sealing showing presence of glass sprue

(e) Sectioned seal illustrating excellent flow of glass into the complex seal geometry

the wetting behaviour of glass–metal systems, but it is equally clear that results noted for a particular system do not necessarily translate to alternative systems, i.e. each system must be judged and assessed independently.

A further method which has been used successfully to prepare glass-to-metal seal components is vacuum-assisted injection moulding (Kramer and Masey, 1984, 1985; Kramer *et al*, 1985). In this process, molten glass is injected under pressure into the metal parts to be sealed held in a suitable mould. A vacuum is applied simultaneously to the opposite side of the mould in order to remove trapped gases and increase the flow of glass. An example of a vacuum-assisted injection moulding rig is shown in Figure 29.

5.2.5   Cleaning and plating of sealed components

After sealing, the exposed metal surfaces of a component are usually tarnished due to the presence of some surface oxide. It is therefore usual to remove this tarnishing in order to improve the aesthetic appearance of the finished component. In the case of chromium-containing metals cleaning and removal of tarnishing may be carried out by immersing the seal in a hot mixture of sodium hydroxide and potassium permanganate. It is also desirable, particularly in the case of electrical components or where good corrosion resistance is required, to gold plate exposed metal surfaces, in particular electrical contacts, pin surfaces and electronic packaging housings and contacts. The gold is normally plated to a thickness of the order of a few micrometres and is of high purity, i.e. ≈99·99%.

5.2.6   Examination and quality control of sealed components

After sealing and cleaning, components may be checked using various techniques ranging from a simple visual check to extensive monitoring of mechanical, electrical and chemical properties. This may include proof testing for pressure components. The most widespread test is a He leak test used to assess the integrity of the seal. These aspects are covered in more detail in Chapter 6 on techniques for the analysis of seals and coatings.

An ASTM standard exists as an aid to the preparation and testing of reference glass-to-metal bead seals, particularly those employed to determine residual stresses in glass-to-metal systems. This standard, together with additional referenced specifications and test methods, is recommended for the development of glass-to-metal systems for use in the electronics industry, in particular (ASTM, 2000). The supplementary tests include measurement of the stress-optic coefficient of the glass and the magnitude of residual stresses, and includes specifications for the glass and metals employed in component manufacture.

5.3   Preparation of glass-ceramic-to-metal seals

In the production of a high strength glass-ceramic component with specific thermal expansion characteristics, e.g. a telescope mirror blank with zero expansion over a given temperature range, it is usual to cast the required shape in glass. This glass blank is then annealed and subjected to the heat-treatment schedule appropriate for the production of a fine grain size material containing the crystal phases necessary to impart the required thermal expansion behaviour. The heat-treatment is generally a two-stage process (Figure 13), consisting of a nucleation stage when a large number of small crystalline or amorphous nuclei form within the glass, followed by a higher temperature crystallization stage when the major crystal phases grow from these nuclei to form a polycrystalline glass-ceramic material. As noted previously, this assumes nucleation and crystallization regimes for the glass in question do not overlap significantly (Figure 15).

As also noted earlier, when selecting a glass-ceramic for bonding to a particular metal or alloy it is essential that a full simulated sealing/heat-treatment schedule, including the high temperature sealing stage, is employed for determining the thermal expansion behaviour of the material (Figure 21). This is because nucleation and growth of crystals can occur during the sealing stage, e.g. during the period required to heat the glass up to the sealing temperature or, alternatively, in the period required to cool the molten glass from the sealing temperature to the nucleation temperature. The proportion and types of major crystalline phases that subsequently form may differ considerably from those obtained when employing a standard two-stage heat-treatment schedule (Donald *et al*, 1989, 1992; Headley and Loehman, 1984; Nash *et al*, 1983); hence, a glass-ceramic which exhibits the required thermal expansion characteristics in bulk form may not necessarily be suitable for sealing or coating applications unless extensive modification of the overall heat-treatment schedule is instigated. Furthermore, use of separate nucleation and crystallization stages may in fact sometimes be unnecessary when using glass-ceramic materials specifically for sealing applications. Headley and Loehman(1984), for example, showed that a separate nucleation stage is superfluous when sealing a lithium silicate glass-ceramic to Ni-based superalloys, whilst Nash *et al*(1983) have noted that a single high temperature heat-treatment stage suffices for sealing a zinc aluminosilicate glass-ceramic to Mo. In addition, for sealing applications, a glass-ceramic is ideally required which is relatively process parameter insensitive over as wide a range of processing conditions as possible. For example, if the required expansion behaviour can only be achieved by crystallizing within the narrow temperature range $T_x \pm 5$ K, this material will be unlikely to make a good practical sealing medium.

Thermal expansion data for a number of glass-ceramic materials are given in Table 8, together with the corresponding heat-treatment schedules. It should be noted that many of the data reported in this Table relate only to a standard two-stage nucleation and crystallization heat-treatment process, this being the only information available for these particular systems. Use of these materials as sealing media would therefore require that the expansion behaviour be re-assessed for samples subjected to an additional high temperature sealing stage. Examples of the influence of process variables on the resultant thermal expansion coefficient of a number of glass-ceramic materials are illustrated in Figure 23. Thermal expansion coefficient is plotted as a function of the crystallization temperature, the other heat-treatment parameters are kept constant for each composition shown in these figures. The influence of a sealing stage is illustrated in Figure 24 which compares α of glass-ceramic samples given standard nucleation and crystallization treatments, with and without the inclusion of a high temperature simulated sealing stage.

In the case of a glass-ceramic seal, it is usual to employ a solid glass pre-form which is positioned relative to the metal parts by the use of suitable graphite or metal jigging; typical assemblies are shown in Figure 30. A heat-treatment cycle

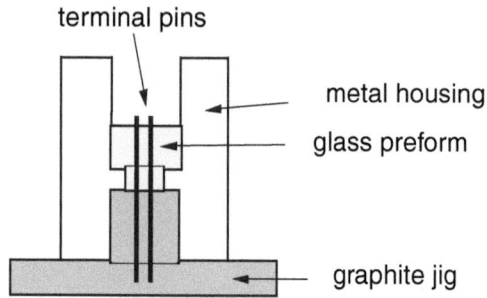

(a) Schematic representation of a simple actuator squib and fixtures

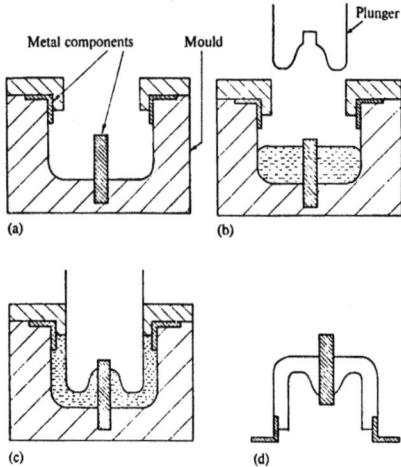

(b) Moulds for pressing glass components using molten glass (after McMillan, Hodgson & Partridge, 1966)

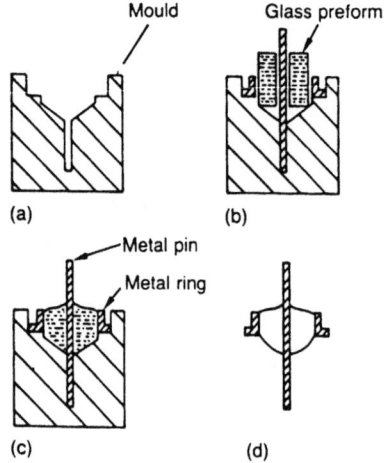

(c) Moulds for producing components using a solid glass preform which is melted in-situ (after McMillan, Hodgson & Partridge, 1966)

(d) Actuator squib components with fixtures

Figure 30: Examples of typical piece-part fixtures used in the manufacture of glass-ceramic-to-metal seal components

will normally include a high temperature sealing stage in order to melt the glass, followed by a lower temperature nucleation stage, and finally a higher temperature crystallization stage, as shown in Figure 21. Hence, a modification of the

Figure 30: (continued)

(e) High vacuum tube envelope consisting of a zinc aluminosilicate glass-ceramic bonded to molybdenum. Figure shows sealed component together with some of the piece-parts and fixtures employed in its construction

(f) 9-pin glass-ceramic-to-stainless steel feed-through seal and metal fixtures

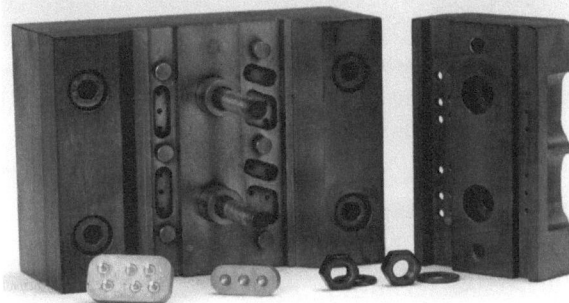

(g) 6-pin and 3-pin glass-ceramic-to-stainless steel feed-through seal and metal fixtures

standard glass-ceramic process is employed, and this must be taken into account when tailoring the thermal expansion characteristics for a specific application, as described earlier. Alternatively, an accurately machined glass-ceramic preform may be sealed to the metal via an intermediate solder glass bond (McMillan *et al*, 1966). The sealing medium can be applied in powder form to the lightly oxidized metal faces to be sealed, and the metal components heated to fuse the glass coating. The metal and glass-ceramic elements are then assembled, using jigs where necessary, and the assemblies are bonded, usually by heating under load in a reducing atmosphere. The glass is partially exuded leaving only a thin film

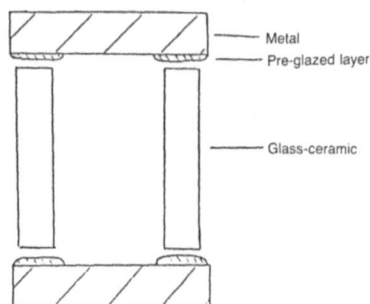

Figure 31: Schematic illustration of a glass-ceramic-to-metal component with butt seals formed using pre-glazed metal parts which are subsequently bonded together using a glass solder (after Partridge, 1990)

between the components. Suitable solder glass compositions can subsequently be heat-treated to induce crystallization and produce a more refractory crystalline intermediate bond, if required. An alternative method is to pre-glaze the metal parts which are then joined to the glass-ceramic using a solder, as depicted for a butt seal joint in Figure 31.

Glass-ceramics have also been sealed to metals employing active metal brazes (Partridge, 1990). In this method, shown schematically in Figure 32 for a typical butt seal configuration, thin foils of titanium and Cu–Ag eutectic alloy are placed next to the glass-ceramic and metal, respectively. Sealing is performed under vacuum at a temperature of around 850°C. The titanium wets and reacts with the glass-ceramic to form a metallized surface whilst the eutectic alloy bonds to the metal part. Alternatively, a single foil of a Cu–Ag–Ti alloy may be used which will bond to both the glass-ceramic and metal parts by itself. Guedes *et al*(2001) have bonded Macor machinable glass-ceramic to titanium using the active metal braze 64Ag–34·5Cu–1·5Ti (wt%). Bonding was carried out under vacuum at temperatures in the range 850–930°C. The interfacial zone was subsequently examined by SEM and EDS. A complex interface up to 150 μm thick and consisting of several different layers was formed. Bond strength was noted to be around 85% of the strength of the bulk glass-ceramic.

Additional details of the technology and types of glass- and glass-ceramic-to-metal seals can be found elsewhere (e.g. Partridge, 1949; Price, 1984; Rulon, 1972; Buckley, 1979; Varshneya, 1982; Partridge and Elyard, 1984; Tomsia and Pask, 1990; Partridge, 1990).

5.4   Novel methods of bonding

In addition to the traditional fusion techniques employed in bonding operations, alternative methods are being developed, including ultrasonic and laser joining. For example, ultrasonic welding has recently been employed for preparing glass-to-metal and glass-to-glass joints. In one variety of the method (Kuckert *et*

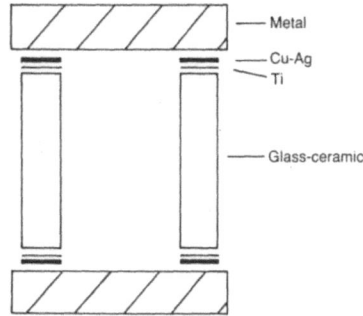

Figure 32: Schematic illustration of a glass-ceramic-to-metal component with butt seals formed by brazing. Separate brazing foils of Cu–Ag alloy and Ti may be used or, alternatively, a single Cu–Ag–Ti reactive alloy braze foil may be employed (after Partridge, 1990)

al, 2004) a borosilicate glass has been sealed to an Fe–Ni–Co alloy using a thin (≈0·1 mm) metal interlayer of Al or Cu between the parts to be joined. An ultrasonic torsion welding system normally used for metal welding was used in these trials. Square 40 mm glass plates 5 mm thick were bonded to metal discs 30 mm diameter by 1·5 mm thick, with He leak tight joints being successfully produced which exhibited tensile strengths of the order of 15 MPa and shear strengths up to 50 MPa. The advantages of using this technique were purported to include very short welding times (< 1 s), coupled with relatively low welding temperatures (< 550°C), high automation capability, and good environmental aspects. The method does, however, suffer from the disadvantage that a static pressure of 20 to 60 MPa is required between the parts to be joined. In addition, it was stressed that the joining surfaces must be parallel to a high degree of accuracy, otherwise imperfect bonding results. These restrictions obviously seriously limit the type and geometry of systems that can be successfully sealed using this technique.

Witte *et al* (2002) have noted that new joining techniques are needed for bonding materials used in the manufacture of microsystems, for example, optoelectronic components including sensors, switches and multicomplexes, and they have employed laser joining to prepare glass-to-silicon seals. Laser bonding was particularly highlighted as a promising alternative to traditional methods for bonding silicon wafer and related materials which include anodic bonding at relatively high temperature which may damage adjacent components. Laser bonding offers the advantage of the ability to apply heat very locally and selectively to very small areas, thereby minimizing damage to materials. Witte *et al* (2002) employed a 1 kW Nd:YAG laser operating at 1·06 µm through a glass fibre to give a spot size of 150 µm and compared the results with a 20 W laser diode. The transmission welding principle was employed whereby one of the parts to be joined is transparent to laser radiation whilst the other is an absorber. It was stressed that special precautions need to be taken to ensure good bonding including low surface roughness (typically < 100 nm), use of high purity materials, and accurate clamping of components. Successful bonding was achieved with both types of laser, but bonding quality was highly dependent on accurate control of the heat input. The advantages of the laser method were considered to outweigh the disadvantages for this type of component and it was considered that the method warrants further development. More recently, Bauer *et al* (2004) have emphasised the importance of new methods for the microjoining of biomedical materials and components including metal-to-glass systems, and have suggested laser joining as a useful way forward.

Electrostatic bonding, also known as anodic or field-assisted bonding, has been employed for joining a range of glasses to metals and glasses to semi-conductors with applications in the areas of pressure sensors, flat screens and fluidic devices. The process relies on application of a voltage to the pieces to be joined held under pressure in the temperature range 300–500°C, a temperature at which the glass becomes an ionic conductor. In the case of bonding pyrex glass to silicon, silicon is made the anode and the glass the cathode. On application of heat and a voltage, Na^+ ions migrate toward the cathode creating a depletion zone in contact with the silicon, this providing the bonding. The precise bonding mechanism is not fully understood, but for glass-metal systems TEM studies by van Helvoort *et al*(2004) on bonding of pyrex glass to aluminium confirms that the process involves the movement of cations and the formation of a thin anodic reaction layer, with Al^{3+} cations reacting with the glass network to form dendritic nanocrystalline structures of γ-Al_2O_3 which grow from the aluminium into the pyrex glass.

Joining of Nilo-K to borosilicate glass suitable for applications involving microfluidic connections has been successfully carried out using a local anodic bonding technique (Blom, 2001). It was noted that reasonably strong bonding was possible in vacuum by applying a voltage of 3 kV for 3 hours at relatively low temperatures in the range 200–300°C. Low temperature metallic solders based on the Sn–Ag–Ti system have also been developed which are reported to wet many glasses, ceramics and metals and are therefore suitable for bonding these materials together in a single step at low temperature (Smith *et al*, 2000). The solders have been used, for example, to bond glass to stainless steel and sapphire to brass at temperatures as low as 250°C.

5.5   Preparation of coatings

5.5.1   Glass coatings

Much of the early work in this area was concerned with the preparation of enam-elled metals. An enamel, often referred to as a vitreous or porcelain enamel when applied to a metal, is a predominantly glassy coating used either to protect the metal substrate, for example when applied to cooking ware, or, alternatively, to provide aesthetically pleasing artefacts, as is the case for jewellery, ornaments, and related regalia. The first major industrial application for glass coatings on metals was in the preparation of porcelain enamelled iron cooking pots and baths, started around the middle of the nineteenth century (Andrews, 1961; Maskall and White, 1986).

The composition of a porcelain enamel coating is usually quite complex, consisting of glass-forming oxides, e.g. SiO_2 or B_2O_3, fluxes, e.g. Na_2O (or B_2O_3 in a dual rôle), and adherence promoters, e.g. CoO and/or NiO, together with colourings and opacifiers, e.g. CuO and TiO_2. Often a glassy ground coat containing NiO or CoO is applied first which possesses good bonding characteristics to the metal substrate. Unlike a seal, the substrate for a coating is often roughened, e.g. by grit blasting, prior to being cleaned and pre-oxidized. Subsequently, further coatings with specific properties may be laid down on top of the ground coat to provide, for example, abrasion or chemical resistance, or an aesthetically pleasing surface. Further details on the rôle of composition, including the use of adherence promoters used in the production of porcelain enamel ware, can be found elsewhere (e.g. Salamah and White, 1981; Klimonda et al, 1981; Wratil, 1984).

Traditionally, enamel coatings have been applied by a number of methods including dipping, painting or spraying. In these methods, the glass is added in the form of a suspension of glass particles in a liquid base, usually water, and often with the addition of a suitable binder to make the suspension adhere to the substrate prior to high temperature firing. The firing operation is carried out either in air or a controlled atmosphere at a temperature high enough to melt the glass particles and cause them to flow over and wet the substrate surface and fuse together. Firing in air leads to the formation of an oxide scale on the metal before the glass has had time to melt and flow. This oxide scale subsequently dissolves in the molten glass, and helps provide the conditions necessary for strong chemical bonding between the glass and the substrate. Alternatively, in the case of metals that are very easily oxidized, the metal substrate may be pre-oxidized prior to firing in an inert atmosphere; more control over the process is possible using the latter method. When metals are pre-oxidized prior to coating, it is essential that the conditions are chosen such that the appropriate metal oxide is formed, e.g. FeO in the case of Fe, and that the oxide scale is highly adherent and non-porous. It is therefore necessary to have access to data for the oxidation behaviour of the metal or alloy in question (see e.g. Stott, 1989).

More recently, other more exotic methods have been employed to coat a metal substrate. These include electrostatic deposition in which a dry glass powder which has been given an electrostatic charge is applied to an oppositely charged substrate, electrodeposition where the component to be coated is immersed in a suitable slurry and made the anode of a galvanic cell, and screen printing in which a relatively high viscosity suspension of glass particles is applied through a wire mesh or mask onto the surface to be coated. This latter method, also known generically as silk screen printing and employing screens or masks made from closely woven stainless steel or nylon, allows complex patterns to be printed onto the substrate.

In addition, sol-gel methods offer potential for coating applications. Sol-gel processing (see, for example, Sakka, 1982; Thomas, 1982; Turner, 1991; Guglielmi, 1997) offers the advantage of being able to produce glasses at reduced temperatures, significantly below that required for preparation from the melt. This means, for example, that many refractory glasses which are very difficult to prepare by fusion techniques can now be made more readily. To date, sol-gel glass and ceramic coatings have mainly been applied to optical and related materials as anti-reflection coatings, and to glass articles for improving the mechanical properties (Klein, 1981; Fabes et al, 1986; Dislich, 1982; Biswas et al, 1989; Floch and Priotton, 1990). The method does, however, offer scope for the coating of other materials, including metals, with a limited number of studies reported in this area. For example, De Sanctis et al(1990) reported sol-gel coatings of silica on a number of different stainless steels, prepared by dip coating. Coatings were applied using an alkoxide route in which the metal samples were dipped into solutions of tetraethoxysilane, TEOS, ethanol and distilled water. After drying, samples were heated at 500–700°C to yield surface coatings up to 1·5 mm thick. Bonding between the coating and stainless steel was believed to be via an intermediate oxide interface. Coated samples were noted to exhibit lower corrosion rates in nitric acid together with improved oxidation resistance. Coatings have also been applied to steel substrates using solution chemistry in which colloidal solutions of silicates are applied by dipping or spraying followed by firing (Vakhula, 2000).

The technique of 'self-propagating high-temperature synthesis' in conjunction with centrifugal casting has recently been applied to coat the inside surface of copper pipes with an abrasion and corrosion resistant ceramic protective layer (Han-Guan, 2004). The method is based on the well known thermite process which utilises a self-sustaining combustible mixture of finely divided aluminium and iron oxide powders, widely used in the past for welding and joining metal parts.

5.5.2   Glass-coated wire

A number of methods have been reported for the coating of thin metal wires with an electrically insulating or protective glass layer. Glass-coated wires may, for example, be drawn through molten glass held on a series of heated metal support loops, as reported by McMillan and co-workers (McMillan et al, 1966). Alternative methods rely on feeding wire through a heated glass tube (Miriam, 1966), or drawing wire through a heated Pt bushing to which molten glass is metered via a Pt tube (Potter and Henning, 1968). Some examples are shown in Figure 33. In addition, a remarkably simple and novel method for the preparation of thin metal wires coated with a glass layer in a single operation *directly* from the melt was first reported by Taylor in 1924 (Taylor, 1924, 1931). In this method, known as the Taylor-wire or microwire process, the metal to be pro-

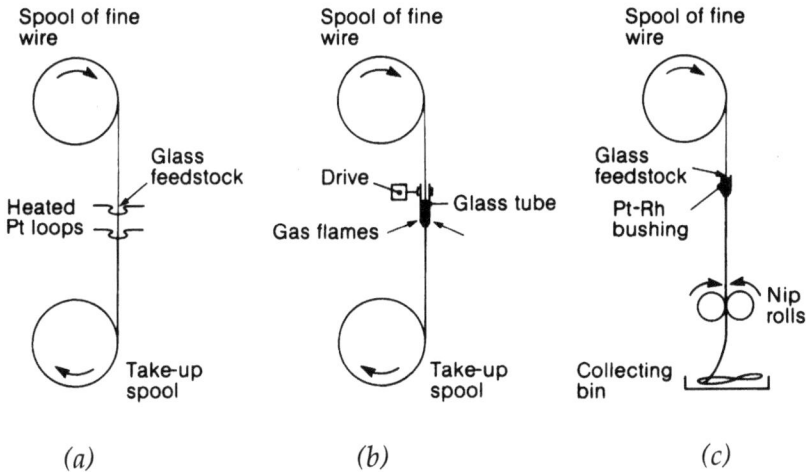

Figure 33: Methods employed in the coating of thin metal wires with glass
(a) method used by McMillan et al, 1969
(b) method used by Miriam, 1966
(c) method used by Potter and Henning, 1968

duced in wire form is held in a glass tube which is closed at one end. The metal is then melted and the glass softened. The end of the tube is subsequently drawn down to produce a glass-encapsulated metal filament. The Taylor-wire process is illustrated schematically in Figure 34(a). An actual production machine for the preparation of microwire (Donald and Metcalfe, 1991) is shown in Figure 34(b). Spools of microwire, each holding around 1 km of product, are shown in Figure 34(c). By suitable choice of the starting glass composition and drawing conditions it is possible to prepare wires of diameter in the range ≈1 to 100 µm coated with a layer of glass of thickness ≈2 to 30 µm. Since the introduction of the Taylor-wire technique, many metals have been prepared as glass-coated filaments, including Cu, Ag, Au, Fe, Co and Pb, together with a number of Fe-, Ni-, Cu- and Pb-based alloys, as reviewed by Donald(1987). Depending on the process parameters employed, and the diameter of microwire obtained, very high cooling rates can be achieved using the Taylor-wire process, with estimates ranging from 4×10^3 K s^{-1} for a core diameter of 100 µm, to 10^7–10^8 K s^{-1} for a 1–2 µm diameter core (Manfre and Vianello, 1987; Pardoe et al, 1978; Nixdorf, 1967; Goto, 1978). These cooling rates are high enough in principle to enable the vitrification of a range of metallic alloys, and the process has indeed been employed successfully to produce glass-coated microwire with an amorphous metal core, as first reported by Weisner and Shneider (1974) for FeP based alloys. Additional alloys have been successfully vitrified since this time including Fe- and Ni-based alloys (Goto, 1977, 1980, 1983; Miroshnichenko et al, 1980; Bashev, 1983; and Donald and Metcalfe, 1991, 1996). The high cooling rates may also promote the formation of nanocrystalline materials and in fact many of the crystalline microwires reported over the years are likely to possess nanometre grain sizes. Such small grain sizes

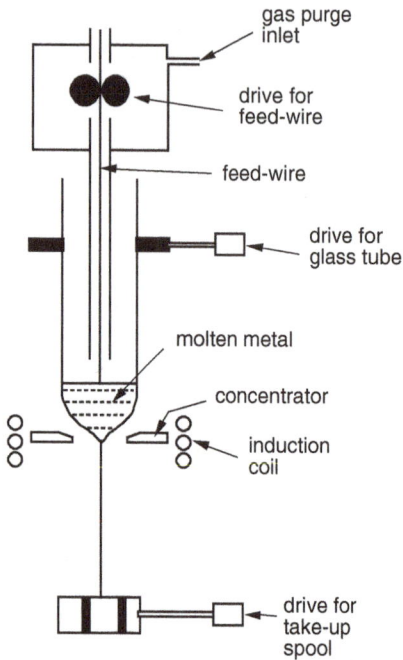

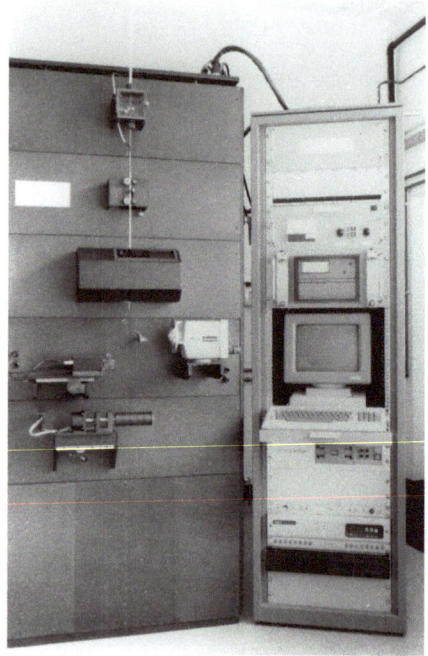

(a, left) Schematic illustration of the Taylor-wire process
(b, right) A Taylor-wire production machine (after Donald, 1993; Donald and Metcalfe, 1996)

(c) Spools of microcrystalline copper and amorphous Ni–Si–B alloy microwires
prepared using the Taylor-wire process
Figure 34: The Taylor-wire (microwire) process for the manufacture of glass-coated
metal filaments

have been observed directly recently for some Cu-based microwires (Del Val *et al*, 2001). Micrographs of glass-coated crystalline and amorphous metal filaments manufactured by this process are shown in Figure 35.

Little interest was initially shown in the Taylor-wire process, although by the 1950s and 1960s significant attention began to be shown in this novel process,

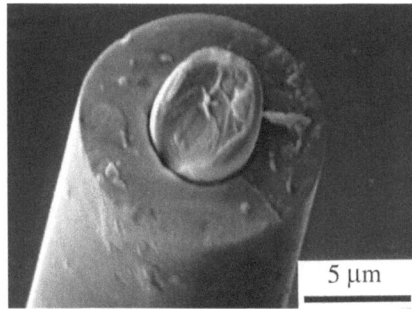

(a) Broken end showing the classic vein pattern associated with ductile failure of an amorphous metallic alloy core

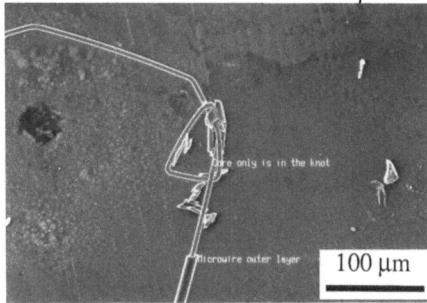

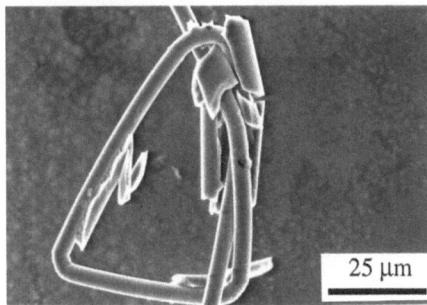

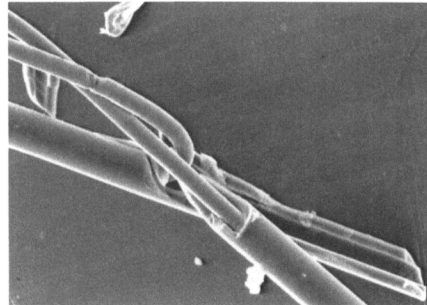

(b) Amorphous metal microwire bent into the form of a knot to illustrate the exceptional ductility of the metal core; note that the silicate glass outer sheath has spalled off
Figure 35: SEM micrographs of glass-coated amorphous Ni–Si–B alloy microwire

particularly by workers in the former USSR (Ulitovsky, 1957), as potential applications for glass-coated microwires emerged; for example, miniature electrical components including wire-wound resistors and electrical motors, in addition to composite materials, screens and catalysts. Although most interest subsequently declined by the 1980s, a further revival of interest has more recently taken place in the Taylor-wire process for the manufacture of Fe-, Ni- and Co-based microcrystalline and amorphous microwires exhibiting useful magnetic properties, with potential applications as sensors and magnetic shielding (Wiggins *et al*, 2000; Zhukov, 2002; Tufescu *et al*, 2003; Deprot *et al*, 2002; Phan *et al*, 2003). This area is covered more fully in Section 10.8.

5.5.3   Glass-ceramic coatings

Glass-ceramic coatings are a more recent addition to the family of coating materials. As in the case of seals, glass-ceramics offer all the advantages of glasses, in addition to the benefits of a more refractory nature and an ability to tailor the thermal expansion characteristics more closely to those of the substrate. Glass-ceramics may be applied, initially as a suspension of glass particles, by any of the standard methods. The heat-treatment schedule is normally more critical, however, and can involve separate fusing/consolidation, nucleation and crystallization stages. It is essential, during the high temperature fusing or firing stage, that the individual glass particles do not rapidly crystallize. If crystallization does occur during this stage, it may not be possible for the individual glass particles to melt and flow adequately over the metal substrate; the production of a continuous and porosity-free coating is then not be feasible. It is also possible, in principle, to manufacture glass-ceramic coated wire by the Taylor-wire method, although this has proved difficult in practice due to the lower thermal stability of glass-ceramic precursor glasses.

5.5.4   Ceramic coatings

There are many methods by which *ceramic* protective coatings may be applied to metals and alloys; for example, chemical or physical vapour deposition, sputter deposition, thermal or plasma spraying, etc. (e.g. Mevrel, 1989; Rickerby and Matthews, 1991). Sol-gel techniques may also be employed (Guglielmi, 1997). Ceramic coatings are covered more fully in Section 9.5.

5.6   Factors affecting seal and coating quality

There are many factors which influence the quality of a seal or coating. As noted earlier, the presence of a suitable substrate metal oxide is essential in order to promote strong chemical bonding. As a general principle, this oxide must be non-porous, of suitable thickness, firmly bonded to the substrate, wetted by the glass so that the glass can flow and coat all metal surfaces, and must not all be consumed by the glass during sealing. In this case, any diffusion into the glass must also be even in order to prevent the formation of areas on the substrate that are denuded of an oxide layer.

   In addition to these requirements, many supplementary factors need to be taken into account if a strong, hermetic and reliable seal is to be produced. For example, the occurrence of bubbles or voids at the metal/glass or metal/glass-ceramic interface is a particularly troublesome occurrence and can not only weaken the interface mechanically, but may also prevent a seal from being hermetic. Bubbles occur for a number of reasons. In coatings, which are normally applied as a particulate suspension, very careful control over the green coating characteristics and subsequent firing and heat-treatment schedules is required in order to prevent the entrapment of gases during the stage when the individual glass particles are melting and undergoing fusion. One of the most common causes of bubbling during sealing or coating is the formation of gaseous products due to the release of volatile contaminants, e.g. traces of cutting oils, fingerprint grease, carbon dust from jigging etc. This can normally be avoided by the use of stringent cleaning/degreasing treatments prior to sealing, and careful assembly of components in "clean" conditions. Another common cause of bubbling during sealing is oxidation of carbon or carbides which are present near the surface of many metals and alloys (oxidation may occur either due to the presence of oxygen in the furnace atmosphere or due to redox reactions between the carbides and the glass). This can be eliminated or minimized by subjecting the metal to a de-carburization treatment prior to sealing, e.g. a high temperature vacuum anneal. Redox reactions between the metal substrate and constituents in the glass can also result in the formation of gaseous products, e.g. liberation of gaseous Na by reaction of Na_2O with Al in Fecralloy, and reaction of SiO_2 or P_2O_5 with Ti to release oxygen.

Another mechanism for the formation of bubbles is diffusion of gases, and in particular hydrogen which may be dissolved in the metal, into the interfacial region during cooling from the sealing temperature (the solubility of hydrogen in metals decreases with decreasing temperature, and hence hydrogen gas may be liberated during cooling of a metal seal from the sealing temperature). This effect can be particularly pronounced in the enamelling of iron, and can lead to an effect called "fishscaling" whereby the enamel surface spalls off under the influence of a build up in hydrogen gas pressure at the interface after sealing. This effect can similarly be reduced by, for example, vacuum annealing the metal prior to sealing. The presence of water can also present a serious problem during sealing. For example, water vapour present in the furnace atmosphere can lead to severe blistering of an enamel (Andrews, 1961), or to the formation of bubbles in glass-to-metal seals (McMillan and Hodgson, 1966). Other potential enamelling defects include blistering, pinholes and dimples caused by organic surface contaminants or high water content in the sealing atmosphere, burn-off and Shiner scales due to localized over saturation of the enamel with FeO, splitting due to localized detachment of the enamel cover coat during the initial stages of firing, together with orange peel, "Shore lining", hairline cracks and "Luder's lines" due to a variety of processing related faults (Wratil, 1984).

Perhaps less well appreciated is the influence that water *dissolved in the glass* can have on the overall sealing characteristics (Craven *et al*, 1986; Haws *et al*, 1985) (It is, of course, well known that dissolved water can have a particularly detrimental effect on the transmission properties of infra-red optical glasses, and precautions are taken to ensure low water contents in these glasses (Donald and McMillan, 1978)). Glasses produced by conventional techniques can contain a significant proportion of dissolved water, either in the form of molecular water or hydroxyl ions. For most glasses the water content is in the range 0·02–0·06%, although higher concentrations of water up to ≈10% have been noted in some glasses prepared at high pressure (Volf, 1984). This dissolved water can react

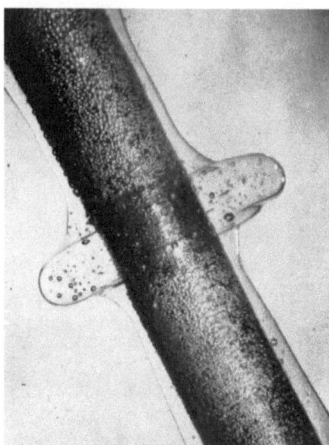

(a) Bubbles in a glass-to-tungsten seal ostensibly caused by release of hydrogen dissolved in the glass (after Partridge, 1949)

(b) Micrograph of a lithium zinc silicate glass-ceramic bonded to Inconel 718 alloy showing presence of voids ≈100 μm in diameter along the interface possibly caused by the evolution of hydrogen gas due to the presence of water in the starting glass
Figure 36: Examples of gas bubbles present in seals

with metallic species diffusing into the glass from the metal substrate to form hydrogen gas, as shown in Figure 36. As noted earlier, and summarized in Table 12, thermodynamic data suggest that hydrogen gas formation may, in fact, be responsible for bubble formation in many metal/glass and metal/glass-ceramic systems, in particular for those alloys rich in Cr, Nb, Ti or Al. Two different approaches have been adopted in order to minimize this effect. In the first method, the glass is produced by melting under "dry" ambient conditions (a maximum of a few ppm of water vapour) in order to produce a glass which is virtually free from dissolved water; sealing is then carried out under similar dry conditions. In the second approach, additives to the glass can be employed which will react preferentially with constituents that would otherwise react with the dissolved water, to form non-gaseous reaction products; e.g. reaction of CuO with Cr to give Cr_2O_3 and Cu_2O. This topic is covered in more detail in Section 7.2.2.

TABLE 12
FREE ENERGY CHANGE, $\Delta G°$, FOR THE REACTION OF WATER DISSOLVED IN GLASSES WITH METALLIC COMPONENTS OF COMMON ALLOYS
(calculated at 1300 K)

Reaction	$\Delta G°$ (kJ mol^{-1})
(a) Reaction favourable	
$Hf + 2H_2O \rightarrow HfO_2 + 2H_2\uparrow$	−524·3
$Y + 3/2H_2O \rightarrow 1/2Y_2O_3 + 3/2H_2\uparrow$	−500·6
$Zr + 2H_2O \rightarrow ZrO_2 + 2H_2\uparrow$	−498·8
$Al + 3/2H_2O \rightarrow 1/2Al_2O_3 + 3/2H_2\uparrow$	−366·7
$Ti + 2H_2O \rightarrow TiO_2 + 2H_2\uparrow$	−354·0
$Ta + 5/2H_2O \rightarrow 1/2Ta_2O_5 + 5/2H_2\uparrow$	−299·7
$Ti + H_2O \rightarrow TiO + H_2\uparrow$	−240·0
$Nb + H_2O \rightarrow NbO + H_2\uparrow$	−229·1
$Nb + 5/2H_2O \rightarrow 1/2Nb_2O_5 + 5/2H_2\uparrow$	−228·0
$Nb + 2H_2O \rightarrow NbO_2 + 2H_2\uparrow$	−211·0
$Cr + 3/2H_2O \rightarrow 1/2Cr_2O_3 + 3/2H_2\uparrow$	−129·6
$Li + 1/2H_2O \rightarrow 1/2Li_2O + 1/2H_2\uparrow$	−124·8
$Mn + H_2O \rightarrow MnO + H_2\uparrow$	−112·9
$V + 5/2H_2O \rightarrow 1/2V_2O_5 + 5/2H_2\uparrow$	−77·2
$Mn + 4/3H_2O \rightarrow 1/3Mn_3O_4 + 4/3H_2\uparrow$	−76·5
$Mn + 3/2H_2O \rightarrow 1/2Mn_2O_3 + 3/2H_2\uparrow$	−46·5
$Zn + H_2O \rightarrow ZnO + H_2\uparrow$	−26·4
$Fe + H_2O \rightarrow FeO + H_2\uparrow$	−9·7
$Mo + 2H_2O \rightarrow MoO_2 + 2H_2\uparrow$	−5·9
(b) Reaction unfavourable	
$Fe + 4/3H_2O \rightarrow 1/3Fe_3O_4 + 4/3H_2\uparrow$	+2·9
$W + 3H_2O \rightarrow WO_3 + 3H_2\uparrow$	+16·0
$Fe + 3/2H_2O \rightarrow 1/2Fe_2O_3 + 3/2H_2\uparrow$	+21·2
$Co + H_2O \rightarrow CoO + H_2\uparrow$	+33·3
$Cu + 1/2H_2O \rightarrow 1/2Cu_2O + 1/2H_2\uparrow$	+50·5
$Ni + H_2O \rightarrow NiO + H_2\uparrow$	+51·7
$Mo + 3H_2O \rightarrow MoO_3 + 3H_2\uparrow$	+94·3
$Cu + H_2O \rightarrow CuO + H_2\uparrow$	+134·0

In addition to the water and other reactions outlined above, many alternative undesirable reactions may occur during glass-to-metal or glass-ceramic-to-metal bonding. For example, reaction of Cr or Fe with P_2O_5 employed as a nucleating agent in some glass-ceramic compositions to form chromium and/or iron phosphide(s). This particular reaction is extremely undesirable because the nucleating agent is removed from the interfacial region resulting, upon crystallization, in the formation of a coarse microstructure which may possess thermal expansion characteristics which are very different from those of the bulk glass-ceramic. These effects can be minimized by careful control over the starting glass composition and heat-treatment schedule, or, in the case of glass-ceramics, by the use of an alternative nucleating agent. It is possible to use thermodynamic data in order to predict possible seal reactions. Thermodynamic data for a number of potential reactions are given in Table 5. These effects are discussed in more detail in Chapter 7.

More extensive details of the technology of glass- and glass-ceramic-to-metal coatings are given elsewhere (e.g. Andrews, 1961; Maskall and White, 1986; Ikeda, 1968; Eppler, 1983; Garland, 1986; Chapman *et al*, 1987).

6. TECHNIQUES FOR THE ANALYSIS OF SEALS AND COATINGS

There are many diverse methods available for studying and characterizing seals and coatings and the materials that make them up. Only a brief summary of some of the more important techniques is given here.

6.1 Thermal analysis

Differential thermal analysis, DTA, and differential scanning calorimetry, DSC, are used extensively for the characterization of glass and glass-ceramic materials; for example, in determining T_g, T_x, T_m, etc. When used in conjunction with dilatometry, which provides information on the thermal expansion behaviour of materials, the data obtained from DTA or DSC can be used in the derivation of appropriate heat-treatment schedules for selected metal/glass or metal/glass-ceramic systems. A technique related to DTA and DSC, namely dynamic mechanical thermal analysis, DMTA, has also been found useful by providing information on phase separation in glasses and in resolving T_g of the residual glassy phase in glass-ceramics (Gilbert, 1989; Hill and Gilbert, 1992; Hill et al, 1992). Great care must be exercised when carrying out thermal measurements using any of the above techniques, however, because the data obtained are very dependent on such parameters as sample condition, type of apparatus, use of appropriate calibration procedures, heating or cooling rates employed, etc. (Mackenzie, 1972; Daniels, 1973; McNaughton and Mortimer, 1975; Pope and Judd, 1973; Sestak, 1984; Brown; 1988; Hohne et al, 2003; Brown and Gallagher, 2003).

Differential thermal analysis has been defined by the Nomenclature Committee of the International Confederation for Thermal Analysis, ICTA, as "a technique for recording the difference in temperature between a substance and a reference material against either time or temperature as the two specimens are subjected to identical temperature regimes in an environment heated or cooled at a controlled rate". Differential scanning calorimetry, on the other hand, has been defined by the same Committee as "a technique for recording the energy necessary to establish a zero temperature difference between a substance and a reference material against either time or temperature as the two specimens are subjected to identical temperature regimes in an environment heated or cooled at a controlled rate".

In practice, in both DTA and DSC, a sample and an inert reference material are heated (or cooled) at a constant rate and thermal differences between them

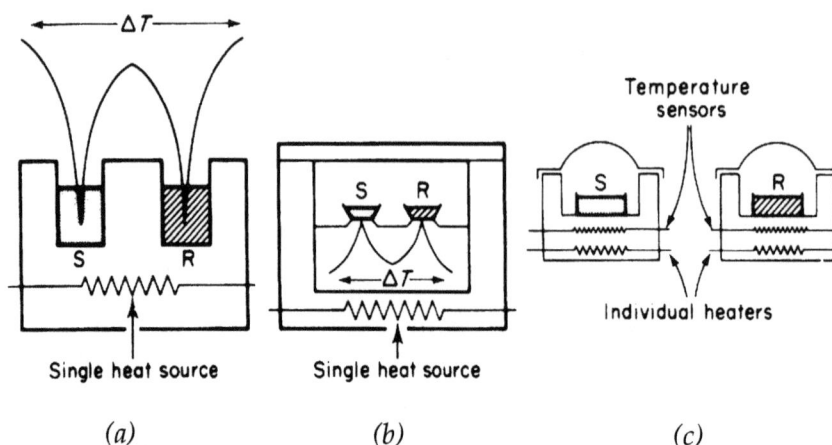

Figure 37: Schematic illustration of the three differential thermal analysis techniques (after Pope and Judd, 1980)
(a) Conventional Differential Thermal Analysis, DTA
(b) Heat-flux Differential Scanning Calorimetry, (HF)DSC
(c) Power-compensated Differential Scanning Calorimetry, (PC)DSC

are monitored, normally as a function of temperature. The different techniques are illustrated schematically in Figure 37.

In classical DTA the temperatures of the sample and reference materials are measured by means of separate thermocouples, with the sample and reference materials being placed in small crucibles which are thermally isolated from one another. Thermal isolation can be achieved, for example, by situating the sample and reference materials in individual wells in a "massive" thermally isolating ceramic block heated externally by a wire-wound resistance furnace. If the sample undergoes a thermal event, e.g. a glass transition, crystallization, crystalline phase change, melting, etc. heat is evolved or absorbed and this is monitored as a transitory change in temperature between the sample and the reference, i.e. the temperature differential between sample and reference is monitored continuously. Conventional DTA systems are available with operating temperatures up to around 2000°C.

In the case of DSC, two fundamentally different operating modes are available, namely power-compensated DSC and heat-flux DSC. In the power compensated mode, which is often regarded as the only true DSC mode, separate platinum heaters and resistance thermometers are provided for both the sample and the reference (there is no external furnace). As heat is evolved or absorbed by the sample during a thermal event the temperature of the sample and reference are kept approximately equal by providing more or less power to the sample heater, i.e. the power required to keep sample and reference materials at the same temperature is monitored. Heat-flux DSC, on the other hand, is more closely related to DTA. An external furnace is provided but, unlike classical DTA, both the sample and reference materials are placed in thermal contact on a conducting metal plate, i.e. they are connected by a controlled thermal resistance. As in classical DTA, the temperature difference between sample and reference is monitored but, unlike DTA, the temperature differential is very much smaller and can be accurately related to the heat flow. Amplification and electronic compensation of the differential temperature signal, coupled with the use of small samples to minimise thermal gradients, then provides a known calorimetric response over a wide temperature range. It is claimed that both the power-compensated and heat-flux DSC modes can provide quantitative calorimetric and kinetic data, provided that adequate calibration and operating procedures have been followed. Due to technical limitations the power compensated DSC mode is limited at present to maximum operating temperatures of around 900°C whilst the heat flux mode can be used to around 1750°C.

6.2 X-ray and neutron diffraction

Extensive use is made of x-ray diffraction analysis, XRD, for determining the types of crystalline phases present in glass-ceramic materials, and in confirming that novel glass compositions are indeed amorphous. High temperature XRD is particularly useful for determining crystal phase formation and phase transformations as a function of temperature, and has been employed by Donald *et al*(1992) in the development of lithium zinc silicate glass-ceramics for sealing to stainless steel and Ni-based superalloys. X-ray and neutron radiography may also be employed in the analysis of seal devices and also in failure analysis. These techniques are particularly useful for the non destructive evaluation of integrated circuit devices to detect encapsulated foreign material, short circuits and misalignment problems. The topic of XRD with respect to its application in glass technology is covered comprehensively elsewhere (e.g. Rao, 2002).

6.3 Spectroscopic techniques

Various spectroscopic techniques may also be employed to probe the detailed atomic structure of glasses and ceramics used in sealing applications. Such methods include infra-red and Raman spectroscopy, Mössbauer spectroscopy, electron spin resonance, ESR, nuclear magnetic resonance, NMR, and neutron diffraction. These topics are also covered fully elsewhere (e.g. Bach and Krause, 1999; Rao, 2002). Other useful characterization techniques include small angle neutron scattering, SANS, and secondary ion mass spectroscopy, SIMS. Further details of these and related techniques, as applied to the materials and systems under review, are given elsewhere (e.g. Chalker, 1991).

Mössbauer spectroscopy, also known as γ–ray spectroscopy, for example, can be employed to investigate the local environment of an atom within certain materials. The method relies on the emission and absorption of γ-rays and is isotope specific. This limits its use to systems containing a limited number of isotopes, the most commonly used being ^{57}Fe. Others include isotopes of K, Cs, Ni, Zn, Ge, Tc, Ru, Ag, Sn, Sb, Te, I, Cs, Ba, La, Hf, Ta, W, Re, Os, Ir, Pt, Au, Hg, the rare earth elements and some actinides, although these isotopes have been far less frequently studied.

6.4   Optical and electron microscopy

Optical microscopy is a widely used surface analysis technique but is limited to a maximum optical resolution of around 0·5 μm. In practice the resolution is limited further by such factors as the specific properties of the surface being examined, and it may often be difficult to resolve clearly artefacts < 5 μm in size. It can be used for the examination of interfaces, after sectioning and polishing of suitable samples, but the resolving power is not high enough for the detection of many of the artefacts of interest; for example, precipitates, reaction products, and thin surface layers. Optical microscopy also suffers from poor depth of focus, particularly at high magnification. Electron microscopes, on the other hand, posses far greater resolving power since the wavelength of electrons is much smaller than that of light.

Use of transmission electron microscopy, TEM, once the reserve of metallurgists, is now used extensively for the study of ceramic microstructures. More recently, high resolution TEM, HRTEM, has been employed to examine interfaces between different materials including ceramic-metal systems (e.g. Pirouz and Ernst, 1989) and ceramic–ceramic systems (Hesse *et al*, 1994; Hesse and Senz, 2002; Lu *et al*, 2003). Resolutions of the order of a few nm are possible. In transmission electron microscopy an electron beam is transmitted through the sample. This requires the preparation of thin samples transparent to an electron beam and with some exceptions, noted above, is not therefore widely used in the area of sealing or coating. Replication techniques, used widely in the past, are now less common but do offer some advantages. In this technique a very thin carbon film is deposited onto the surface to be examined. This is subsequently removed, by etching the sample for example, and examined in the TEM. The film matches the contours of the surface onto which it was deposited very accurately and may also include precipitates and inclusions removed with the film. It therefore provides a reasonably accurate representation and can be employed with sectioned seals. It cannot of course be used to map elemental distributions across interfaces.

Scanning electron microscopy, SEM, on the other hand, is used extensively for the examination of microstructures, including interfaces, and can have an overall resolving power of < 20 nm. Unlike optical microscopy, the SEM also offers great depth of focus making it ideal for the examination of surfaces with some topographical relief. Use of SEM also enables (qualitative) x-ray analysis of individual microstructural features to be performed in-situ in the SEM by means of the energy dispersive spectrometer, EDS, facility which is fitted to many modern instruments. This is particularly useful for the examination of small precipitates and other reaction products within a glass-metal interfacial region. The SEM generates a beam of electrons which is focused through a series of electromagnetic lenses and scanned across the surface of a material. These electrons interact with the material under investigation and generate secondary or backscattered electrons from the surface of the material which are detected and recorded as an image of the surface of the material. Contrast is achieved based on the atomic number of the elements making up the individual phases in the material, with high atomic number elements creating a brighter image. During the process of interaction of the electron beam with the sample surface x-rays are also generated. Each element emits a characteristic x-ray spectrum and the EDS spectrometer detects the x-rays as a function of their intensity and energy. EDS analysis can thus provide information on the elements present in a sample and can give a semi-quantitative estimate of their concentration. SEM with an EDS attachment is also useful in component failure analysis investigations.

As a cautionary note, it should be appreciated that sample preparation can have a pronounced effect on the resulting microstructure as revealed by optical or electron microscopy. Great care should be taken in sample mounting, grinding and polishing operations, as these may introduce stresses into the material which may lead to cracking or de-bonding at interfaces. It may then be impossible to differentiate between defects present in the original seal and those introduced subsequently during sample preparation. If samples are etched prior to examination in order to reveal grain structures, great care should be taken to ensure that all surface debris, including any insoluble products formed by reaction between the sample surface and the etchant, are removed. Similarly, preparation of samples for TEM examination can introduce defects and artefacts not present in the original sample. In addition, the electron beam may cause damage to the sample, and this may give rise to misleading results.

6.5 Mechanical properties

Measurement of the mechanical properties of materials employed in sealing and coating applications and in the properties of the components themselves is of fundamental importance in the design and selection of suitable materials for specific applications. In the case of the glass or ceramic materials themselves the main properties of importance include mechanical strength (tensile, compressive, flexure), elastic moduli (in particular Young's modulus), hardness, ductility (more relevant in the case of metals), fracture toughness, and Poisson's ratio. In the case of seal or coated components the mechanical behaviour and adhesive properties of the seals and coatings can be assessed using a number of techniques. For example, coating adhesion can be evaluated employing pull-off, shear or indentation methods, whilst other tests employ tribological techniques for determining resistance to abrasive, erosive and chemical media (see, e.g. Fairhurst et al, 1985; Bull and Rickerby, 1991).

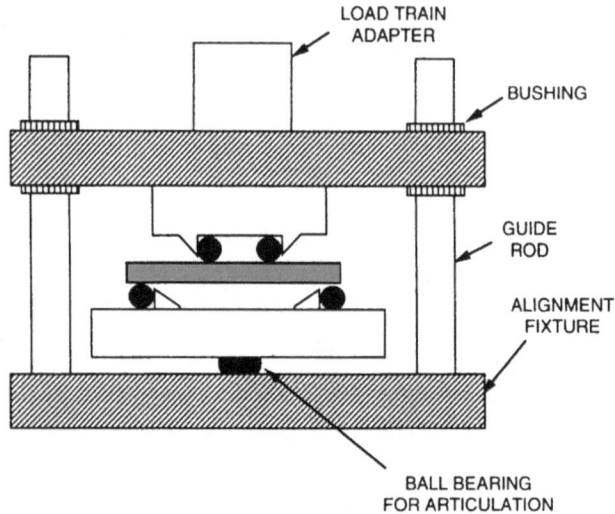

Figure 38: Four-point bending test rig (after Faber et al, *1998)*

Tensile testing of brittle materials such as glasses or glass-ceramics is rarely carried out due to the intrinsic difficulties associated with experimental design. More commonly, standard three- or four-point bend testing is the norm (Quinn and Morrell, 1991; Ferber *et al*, 1998). Unfortunately, these methods suffer from the disadvantage that failure almost invariably occurs from the edge of brittle samples, with flaws in the edges acting as severe stress concentration sites. To overcome this limitation, techniques based on biaxial flexure have been developed, these methods giving a better estimate of the intrinsic strength of the brittle material. Some of the more important testing methods are described below.

6.5.1   Flexural strength of materials

Three- and four-point bending tests are standard methods for determining the flexural strength of brittle materials. A typical arrangement for a four-point bending fixture is shown in Figure 38. Biaxial flexure strength has more recently become a popular testing technique for measuring strength (Vitman and Pukh, 1963; Wachtman *et al*, 1972; ASTM, 1978; Ritter *et al*, 1980; de With and Wagemans, 1989), and may be preferred over the more standard three- or four-point bending methods. Biaxial flexure offers the advantage that edge effects are minimised, thereby giving a more accurate picture of the intrinsic strength of the material. Biaxial testing generally involves the use of disc shaped samples which are supported on three or more points situated equidistant from its centre and near to its periphery, with loading applied through the central portion of the disc. The region of maximum tensile stress occurs near the centre of the lower face of the sample, with the region near the edges subject to minimal stress. A number of different biaxial testing regimes are in use. These include piston-on-ball, ball-on-

Figure 39: Testing rig for measuring the biaxial flexure strength of brittle materials (piston-on-three-ball method shown)

ring, and ring-on-ring. In the piston-on-ball test, shown in Figure 39, a disc or plate specimen is supported by three or more ball bearings equi-spaced on a circle and loaded with a small diameter flat piston near its centre. In the ball-on-ring test the specimen is supported on a ring and loaded centrally with a ball, whilst the ring-on-ring test also employs a ring support but uses a smaller diameter ring to provide the load.

In the case of the piston-on ball method the biaxial flexure strength, σ_{max}, may be given by:-

$$\sigma_{max} = [3P(1+v)/4\pi t^2][\{1+2\ln(a/b)\}+(1-v)/(1+v)][(1-b^2/2a^2)[a^2/R^2] \qquad [6.1]$$

where P is the load at fracture, v is Poisson's ratio for the glass, t is the sample thickness, a is the radius of the support points, b is the radius of the piston, and R is the radius of the sample.

It is now generally accepted that the ball-on-ring test is the most reliable in terms of providing consistent data (De With and Wagemans, 1989), employing equation [6.1] with a value for b being taken as equal to a third of the sample thickness, i.e. $b = t/3$.

The mechanical properties of enamelled float glass have been monitored using a biaxial test method (Krohn *et al*, 2002). It was noted in this study that the ring-on-ring test geometry apparently gave more consistent results than ball-on-ring testing.

6.5.2   Coating-substrate bond strength

The measurement of bond strength between coatings and substrates has been reviewed by Chapman *et al*(1987). It was noted that testing can be subdivided into a number of areas including tensile, shear, strain and fracture mechanics. Tensile test methods include direct pull-off, the topple test, ultracentrifuge testing, and impact deceleration. Shear tests include the lap shear method, the napkin ring test and the rod and ring test. Strain methods involve bending twisting or stretching shear, with stresses generated at the interface being related to the total strain of the body, whilst fracture mechanics tests relate the failure strength to crack size. Test methods are illustrated schematically in Figure 40.

In the direct pull-off tensile test the force applied perpendicularly to the interface required to cause failure is measured. The tensile strength, σ, of the coating/substrate bond is then given by $\sigma = F/A$, where F is the failure force and A is the area over which failure occurs. It is one of the most widely used tests and has been applied to many materials combinations including glass-to-metal systems (ASTM C633-79). Attachment and alignment of test fixtures is very important in this method, and bending moments due to misalignment can be problematic. The tensile test method has been modified to enable direct pull-off without the requirement to attach a pull rod to the surface of the coating, and this has made it suitable also for high temperature testing. The other methods to determine bond strength, which have in general been less extensively used for glass-metal and ceramic-metal systems, are reviewed in detail by Chapman *et al*(1987).

A number of different adhesion tests have been performed by Ashcroft and Derby(1993) on glass-ceramic thick films deposited on metal substrates which were aimed at identifying a reliable method of measuring adhesion. Testing was performed on a lithium zinc silicate and various lithium aluminosilicate glass-ceramic compositions applied to copper and copper–invar–copper laminates as films 20–200 μm thick by screen printing and heat treatment. Tests included the direct pull test, scratch test, interfacial shear test, indentation test, and bend test. The direct pull test relies on increasing the tensile strain applied to the coating–substrate interface through a stud attached (adhesively bonded) to the coating until failure occurs. This method is limited by the strength of the adhesive used to attach the stud to the coating. According to Ashcroft and Derby (1993) this test must, however, be regarded as a qualitative comparative test for similar materials as no quantitative model of interface fracture is currently available that could be used for all types of material. In the scratch test, normally used to measure the adhesive strength of thin films, although it may sometimes be used to estimate the adhesive strength of thicker coatings, the load required by a diamond indenter as it traverses the surface to delaminate the coating is monitored. Coating failure is complex and may be dependent on the coating thickness in addition to actual interface strength. In addition, no universally accepted model is available to analyse the data quantitatively. As in the case of the direct pull test, the scratch test is therefore employed more to rank

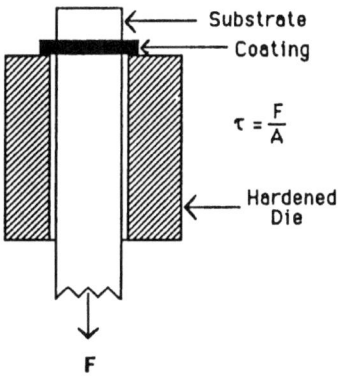

(a) Rod and Ring Test

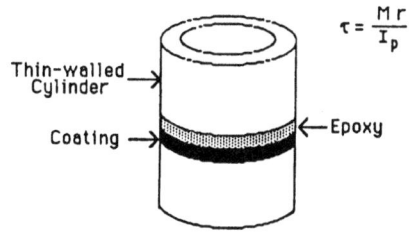

(b) Napkin Ring Test

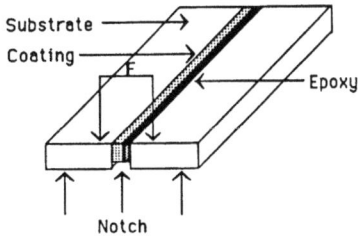

(c) Double Torsion Test

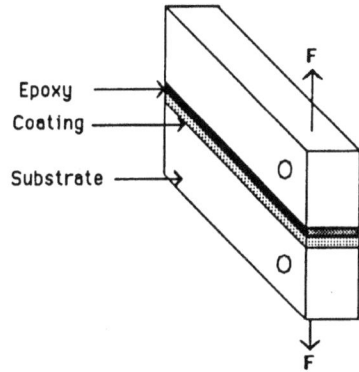

(d) Double Cantilever Beam test

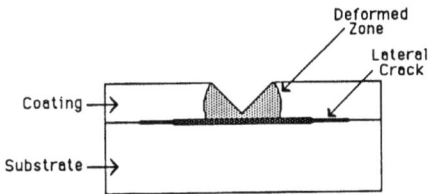

(e) Indentation Crack Zone Test

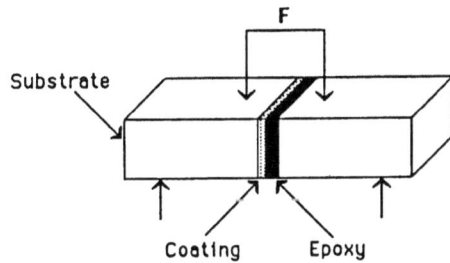

(f) Four-point Bend Test

Figure 40: Schematic illustrations of mechanical testing arrangements for coatings (after Chapman et al, 1987)

the properties of similar materials, rather than to generate quantitative data. In the interfacial shear strength test a coated sample is pulled in tension and the density of cracks formed in the coating at the point where delamination occurs and the coating begins to spall off the substrate is monitored. The maximum crack density has been shown (Agrawal and Raj, 1989) to be dependent on the interfacial shear strength, τ, through the equation:-

$$\tau = \pi t \sigma / \lambda \qquad [6.2]$$

Figure 40: (continued)

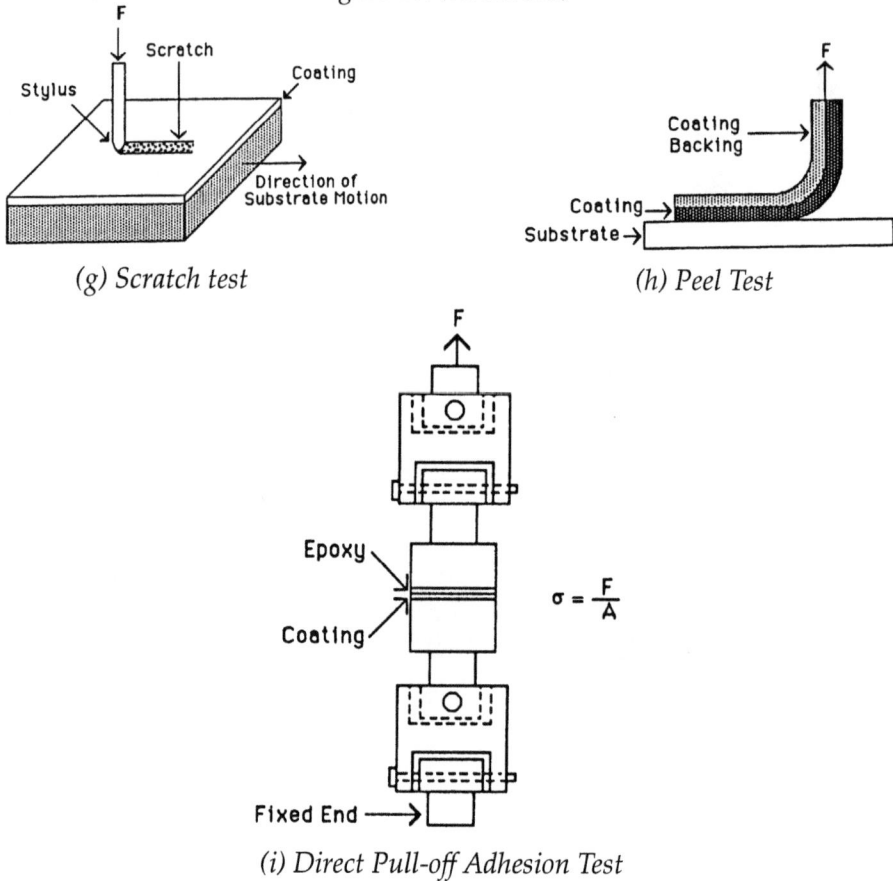

(g) Scratch test

(h) Peel Test

$$\sigma = \frac{F}{A}$$

(i) Direct Pull-off Adhesion Test

where λ is the maximum crack spacing once constant crack density has been achieved, t is the coating thickness, and σ is the tensile strength of the film. It was noted by Ashcroft and Derby that the indentation test developed by Marshall and Evans(1984) can be used to measure the fracture resistance of the interface, as can the four-point bend test. In the case of the indentation test, however, it was found impossible to propagate an indentation crack without deforming the substrate and damaging the coating, thus adding to the difficulty and shortcomings in analysing the resulting data. Similarly, difficulty was experienced in propagating interfacial cracks successfully using the four-point bend test. It was therefore concluded that none of the currently available tests was completely satisfactory, although they did rank the samples studied in the same order of adhesive strength. More work is undoubtedly required before a more universally acceptable test giving consistent data is established.

6.5.3   Coating–substrate fracture toughness

Determination of interfacial fracture energies using four-point bending of strip seals is generally based on crack initiation from a notch tip in a coating. The crack is propagated into an interface using tensile stresses in four-point bending

(Charalambides *et al*, 1989). To promote delamination of the coating, the resistance of the coating to bending is generally increased by incorporating a stiffening layer (the same material as the substrate may be employed for this purpose). The load–deflection curve of this three layer composite is subsequently recorded as the crack propagates, the load becoming independent of the displacement as the crack travels along the interface. The load corresponding to delamination is a measure of the fracture energy or toughness of the interface. When this point is achieved, the magnitude of the fracture energy may be determined as follows:-

The energy, G, under plain strain conditions is given by:-

$$G = \{M^2(1-v^2)\}/2\omega EI \qquad [6.3]$$

where E is the elastic modulus, I is the moment of inertia, and M, the moment, is given by:-

$$M = Pa/2 \qquad [6.4]$$

where P is the load, and a is the distance between the outer and inner loading points. As long as the crack proceeds in a steady state condition, P will be constant and will be noted as a plateau in the load–deflection curve. It can then be shown that the interfacial fracture energy, G_{int}, is a function of P, E, a, ω, v, and the thickness of the substrate, t, according to the following expression:-

$$G_{int} = \{21P^2(1-v^2)a^2\}/16E\omega^2t^3 \qquad [6.5]$$

The interfacial fracture toughness, K_{int}, may be given by (Hutchinson and Suo, 1992):-

$$K_{int} = \{G_{int}E_{int}\}^{1/2} \qquad [6.6]$$

where E_{int} is the interfacial elastic modulus given by:-

$$1/E_{int} = 1/2\{(1/E_1) + (1/E_2)\} \qquad [6.7]$$

where E_1 and E_2 are the elastic moduli of the substrate and the coating, respectively.

The interface fracture energy of metal-to-ceramic joints has been measured by Turwitt *et al*(1986) using a four-point bending method. The system investigated consisted of sandwich seals between alumina and niobium formed by solid state bonding in vacuum. It was found that the fracture energy of the joint increased with increasing Nb grain growth during bonding, but decreased with increasing Nb grain size if aged after bonding had been carried out.

6.5.4 Young's modulus

Young's modulus of materials can be determined by a number of different methods. In the case of a coating on a substrate, the modulus of the coating can be determined using three- or four-point bending or indentation techniques. In the case of three-point bending it is usual to employ a bar sample coated to a similar thickness on both sides. Young's modulus is calculated using the expression (Fawcett, 1998):-

$$2E_cI_c + E_sI_s = P\ell^3/48d \qquad [6.8]$$

where E_c and E_s are Young's modulus of the coating and substrate, respectively, ℓ is the span, P is the load, d is the corresponding displacement, and I is the moment of inertia.

For four-point bending a sample coated on one side only may be employed. Strain gauges are attached to the coating and substrate. It can be shown that (Chui, 1990):-

$$E_c = E_sR\{(KR + 2K{-}R)(2R{-}K + 1)\} \qquad [6.9]$$

where R is the ratio of substrate to coating thickness, and K is an average value for individual strain across the entire range.

Indentation may also be used to determine a value of Young's modulus. The apparent Young's modulus of the sample, E^*, is calculated from the load-displacement curve using standard Hertzian contact theory, when it can be shown that (Oliver and Pharr, 1992):-

$$E^* = 0{\cdot}75Ph^{-3/2}R^{-1/2} \qquad [6.10]$$

where h is the depth of elastic penetration of the indenter, P is the load, and R is the indenter radius.

A value for the true modulus of the sample can be determined using the expression:-

$$E = (1{-}v^2)/\{(1/E^*){-}[(1{-}v_i^2)/E_{int}]\}) \qquad [6.11]$$

A value for the fracture toughness of a coating, K_{Ic}, can also be found using the indentation method. A load is applied which is sufficiently high enough to cause cracking to occur around the indenter. The fracture toughness can then be determined using the expression (Beshish *et al*, 1993; Shetty *et al*, 1985):-

$$K_{Ic} = 0{\cdot}016(E_c/H)^{1/2}(P/c^{3/2}) \qquad [6.12]$$

where H is the hardness of the coating, c is the length of the crack caused by the indentation, and P is the indenter load.

An alternative equation for measuring K_c has been given by Evans and Charles (1976) for bulk glass-ceramics:-

$$K_{Ic} = 0{\cdot}0824(P/c^{3/2}) \qquad [6.13]$$

6.5.5 Novel techniques

A number of novel techniques have been reported for monitoring the behaviour of seals and coatings. For example, the microstructural fracture of yttria stabilized zirconia barrier coatings on nickel-based superalloys for turbine applications has been studied using a novel miniature mechanical testing device mounted in an SEM (Wessel and Steinbrecht, 2002). It was noted that fracture followed pre-existing flaws in the coating. Fracture studies of ceramic coatings on metals have also been performed using four-point bending in-situ and observed by optical microscopy (Takahashi *et al*, 2001). In this study several different coating materials were examined. It was found that the fracture behaviour depends strongly on the top coat microstructure in addition to the heat treatment schedule applied after coating.

The deformation behaviour of model Nb–alumina–Nb sandwich seals has been investigated using a surface strain mapping technique (Liu and Brunner, 2002). Seals were first prepared by diffusion bonding metal and alumina sheets together under pressure at 1400°C. Surface displacements during subsequent compressive loading were monitored using an optical imaging system with data fed to a software driven PC. It was observed that deformation of the joint started in the Nb close to the interface, extending throughout the Nb before interface debonding or fracture of the alumina occurs. It was also noted that the yield strength of the joint increased with decreasing thickness of the Nb layers.

Simple methods for measuring the tensile and shear bond strengths of ceramic–ceramic and ceramic–metal systems, and for determining the elastic modulus and fracture strength of brittle coatings have been developed recently utilising compression tests (Bao and Zhou, 2002). The high temperature fatigue deformation of spray coated steel substrates has also been monitored using an electronic speckle pattern interferometry method (Wang and Kido, 2003).

A miniaturised disc bending test based on biaxial flexure geometry has been devised by Eskner and Sandstrom (2003) to monitor the strength and ductile-to-brittle transition behaviour with temperature of thin nickel aluminide coatings on Ni-based superalloy substrates. In this method, the specimen is positioned between two dies and a punch inserted through the upper die exerts a load on the sample via a ball bearing. The method is therefore related to the ball-on-ring biaxial flexure testing method. The die assembly can be heated in a furnace, thus facilitating measurements at high temperature.

Overall, it must be concluded that there are currently no standardised tests offering universal acceptance for determining the mechanical behaviour of seal components or coated materials. This is therefore an area requiring further attention.

6.6   Acoustic emission

Acoustic emission has been employed to monitor the fracture behaviour of ceramic coatings on metals. Watanabe *et al* (2003) used laser interferometry to detect acoustic emission from a plasma sprayed coating of alumina on stainless steel during thermal cycling between ambient temperature and 1000°C. Scanning acoustic microscopy, SAM, which makes use of the absorption and reflection of ultrasonic waves in a sample can be employed for non-destructive examination of seals or components for delamination and related defects. Scanning acoustic microscopy has also been employed as a non-destructive test to measure Young's modulus of thermally sprayed and thin film coatings (Rats *et al*, 1999; Schneider *et al*, 1993). Acoustic emission is undoubtedly an area worthy of further investigation.

6.7   Modelling studies

Many studies have been aimed at modelling glass-to-metal and related architectures, and in using model systems to derive data.

The phenomenon of fish scaling in enamelled iron and steel is a well known effect. It is caused by a build up of hydrogen gas within the interface during the enamelling process which, if the pressure is high enough, causes the enamel to spall off the metal surface, usually as a function of time. A model has been developed, based on an evaluation of the amount of free hydrogen present in the metal after enamelling, which enables the susceptibility of particular steels to fish scaling to be assessed (Valentini *et al*, 1992a; 1992b).

Extensive use has been made of finite element modelling of metal–glass systems. The simplest models employ a 2-D mesh for which rotational symmetry about a central axis applies. This would be applicable, for example, to a simple cylinder seal with a single central pin. In the case of more complex geometries, for example a cylinder seal with two metal pins for which rotational symmetry does not apply, 3-D meshes are required. A 3-D mesh requires a much larger number of elements and takes up significantly more computing time. In addition to the setting up of an appropriate mesh, finite element analysis also requires the input of reliable materials data, including such properties as yield and/or tensile strength, elastic moduli, Poisson's ratio, and thermal expansion characteristics. The operating conditions of the component being modelled must also be known. In the case of an explosive actuator squib, for example, the maximum pressure and duration of the pressure pulse must also be known in order to predict the magnitude of the stresses expected within the component elements. As discussed later in Chapter 7, the presence of residual stresses due to thermal expansion mismatch and the occurrence of precipitated phases and reaction products in the interfacial zone may play a very large part in determining the overall performance of a component. The properties of precipitated phases in particular may not be known with any degree of accuracy, making it difficult to model with confidence. Not taking into account the presence and properties of interfacial phases may therefore lead to misleading analysis. In view of this, it is often more appropriate to resort to simple analytical techniques to estimate the influence of residual stresses on the bulk system through a general knowledge of the material and processing parameters and use of intuition. Some specific studies involving modelling of specific metal–glass or metal–ceramic systems are summarized below.

The cracking behaviour of functionally graded ceramic coatings on metal has been modelled by Cai *et al*(1998) using finite stress analysis techniques. In this study, the coating was considered to consist of a ceramic material reinforced with ductile metal particles and with the density of particles increasing towards the substrate. It was shown that such coatings can exhibit fracture toughness values very much higher than the unreinforced ceramic alone, a reasonably obvious conclusion that has already been established in the case of ceramic composites (Donald, 1995). Crack propagation in ceramic/metal functionally graded plates has also been modelled using finite element analysis techniques by Noda *et al*(2003). It was noted that crack propagation paths are influenced by such parameters as temperature, the composition and profile of the system, and interaction among multiple cracks. An alumina/niobium system has been used by Kohnle *et al*(2002) as a model ceramic/metal system. The influence that the plasticity of the metal component plays on the interface strength and energy release was examined.

Numerical studies have been employed to assess the influence of existing pre-cracks on the properties of thermal barrier coatings (Zhou and Kokini, 2003). It was found, for example, that the presence of multiple vertical surface pre-cracks was beneficial in reducing the propagation of interface cracks, particularly in the case of short pre-cracks and a large pre-crack density. It is unlikely, however, that this would provide an acceptable method for improving the interfacial performance of practical systems.

A very extensive bibliography of finite element modelling of ceramics and glasses in general has been provided by Mackerle(1999). This deals with such topics as mechanical properties, composites, electroceramics, etc. for a range of glass and ceramic materials. There is undoubtedly a great deal of information available on finite element analysis of such materials, although much less data are available on ceramic-to-metal systems, with only just over 100 references given to this topic as compared to a total of over 1000 in all.

6.8 Optical stress analysis

Residual stresses will invariably be present in seals and coatings at ambient temperature. These stresses arise from a variety of different origins and are fully described in Chapter 7. In the case of glass-to-metal seals where the glass component of the seal will normally be transparent to the visible spectrum, it may be possible to utilize optical stress measurement techniques to monitor the location and magnitude of these residual stresses. These methods are not, of course, suitable for measuring stresses in opaque systems such as glass-ceramic-to-metal seals and most coatings. In addition, many commercial glass-to-metal seals employ deeply coloured glasses for identification purposes, ruling out optical stress analysis as an option. The measurement of stresses using optical stress analysis techniques for use with simple seals and where the glass is transparent is very comprehensively covered by Partridge(1949), and the basic principles and methods used are still relevant today. Partridge noted that the direct measurement of stresses present in seals is generally preferable to analytical calculations which draw on experimental thermal expansion data. Although optical stress analysis can be employed for monitoring stress in simple seal geometries, the present day complexity of many seal designs does, however, make analysis by this method difficult if not impossible. More detailed information on optical stress measurements in glass in general can be found elsewhere (e.g. McKenzie and Hand, 1999; Varshneya, 1982).

As noted by Varshneya, stresses in the glass components of a glass-to-metal seal can be measured by a number of different techniques. For example, methods based on the stress birefringence properties of glasses, or use of strain gauges, with the former method being more popular due to the ease of measurement of the photoelastic properties of glass. Photoelastic methods rely on measuring the optical retardation associated with stressed glasses. The relative retardation, R, between perpendicularly polarized waves upon emerging from stressed glass of thickness d, is given by:-

$$R = (\sigma_1 - \sigma_2)/Cd \qquad\qquad [6.14]$$

where C is the stress optic coefficient of the glass, and σ_1 and σ_2 are the principal stresses.

Measurement of stresses therefore involves measuring optical retardation. Retardation may be measured using compensators, of which the three most common methods are the Babinet–Soleil compensator, the Berek compensator, and Senarmant compensation. As far as stresses in glass-to-metal seals are concerned, it is only the tensile stresses that are of particular importance. An estimate of the stresses present in glass may be made using a strain viewer. In this method, the sample is viewed between crossed polarizers with a white light source. The light which emerges is coloured and the colour can be related to the optical retardation. Stresses corresponding to higher than around 400 nm retardation would normally be high enough to fracture the glass which makes the strain viewer method rather insensitive, as noted in Table 13. It is therefore necessary to employ a suitable tint plate which greatly enhances the sensitivity, as also noted in Table 13. An added advantage of using a tint plate is that both positive and negative retardations can be resolved (and hence both tensile and compressive stresses can be identified). The strain viewer would normally be calibrated using a glass rod which is bent known amounts in order to introduce precise stresses which are then related to the identified retardations and therefore given colours.

6.9   Seal hermeticity

As noted earlier, it is usually highly desirable, if not essential, that a seal system is hermetic. This is particularly true for such components as vacuum systems including vacuum tubes, lamp envelopes, photomultiplier and x-ray tubes, to-

TABLE 13
STRAIN VIEWING DATA RELATING COLOUR TO OPTICAL RETARDATION
(after Varshneya, 1982)

| *Without tint plate* | | *With 565nm tint plate* | |
Retardation (nm)	*Colour*	*Retardation (nm)*	*Colour*
50	iron-gray	300	pale yellow
200	gray-white	290	greenish yellow
300	yellow	180	green
430	orange	93	greenish blue
530	Red	46	blue
565	magenta	23	violet-blue
635	blue	0	magenta
675	blue-green	−23	reddish-violet
740	green	−46	reddish-orange
840	greenish-yellow	−69	orange
880	yellow	−93	yellowish-orange
		−145	gold-yellow
		−180	yellow
		−220	pale-yellow
		−300	white

gether with electronic and microelectronic packaging, certain medical implant and related components, and reed switches and sensors, to name but a few. There are a number of tests that can be carried out to assess the hermeticity of a given seal. These include relatively simple dye penetrant, bubble emission and pressure decay tests, but undoubtedly the test which provides the most useful and quantitative information is the He leak test used in conjunction with a mass spectrometer. In all the tests, including the He leak test, a differential pressure is generated between the internal volume of the seal and the outside environment. The pressure gradient so generated causes the testing medium, e.g. He, to leak through the seal where it can be monitored and measured, e.g. by a mass spectrometer. Leak rates are usually measured for a one atmosphere pressure differential in units of He volume per unit time, e.g. 10^{-x} He cm^3s^{-1} at STP (standard temperature and pressure). It should be borne in mind that the He leak test is a rigorous test due to the small size of the He atom, and that a seal exhibiting a given He leak rate may have a much lower leak rate for larger atoms or molecules including water (moisture). It also has a leak rate around 2·7 times faster than that of standard air (Neyer, 2002). For typical hermetic seal applications a He leak rate of $< 10^{-8}$ cm^3s^{-1} would normally be regarded as a good seal.

Glass-to-metal seal systems fail hermeticity tests for a number of reasons, the most common of which include poor surface preparation prior to sealing, the presence of high residual stresses in the seal due to thermal expansion mismatch between the glass and the metal which causes the glass to fracture either during seal preparation or at some time afterwards, and mechanical damage to the seal component as a result of subsequent handling operations. Additional information on seal hermeticity may be found elsewhere (e.g. Ely, 2000).

6.10   Additional test methods

Many additional tests have been devised for assessing the suitability of specific components for their intended application. In the case of electronic package components, for example, a series of tests are suggested which include resistance to thermal shock and chemical environments, particularly moisture and salt atmospheres, in addition to temperature resistance and lead and electrical integrity.

7.   STRESSES IN SEALS

7.1   Generation and magnitude of bulk residual stresses

Unless the thermal expansion characteristics of a specific system over a given temperature range are identical, stresses will be generated in a glass- or glass-ceramic-to-metal seal or coating as a result of the influence of differential contraction during cooling from the sealing or coating temperature. As precise matching of thermal expansions will rarely if ever be possible in practice, considerable effort has gone into the design of suitable seal systems in order to alleviate or minimize the effects of residual tensile stresses and to ensure that any major residual stresses are compressive in nature, as described earlier in Chapter 2.

At temperatures close to the fabrication temperature, where the 'glass' is above T_g and in the supercooled liquid régime, rapid stress relaxation of the glass will occur so that the influence of differential contraction can normally be ignored. As the temperature is lowered, however, a point will be reached where stress relaxation of the glass becomes so slow that any differences in thermal contraction between the two materials can no longer be accommodated. A further decrease in temperature will then result in the creation, within the time-scales involved, of permanent stresses. If the contraction stresses arising within the brittle glass or glass-ceramic are tensile in nature, failure of this phase may occur either during cooling or, more seriously, at a later time when the component may be in service, due to the influence of static fatigue. It is generally accepted that residual tensile stresses should not exceed ≈ 10 MPa in order for a glass seal to remain hermetic over its service life time (Rawson, 1980).

The point at which stress relaxation can no longer accommodate contraction stresses is often referred to as the "setting" or "set" point of the glass, T_{set}. Like the glass transition temperature, T_g, T_{set} is not, of course, a fixed temperature, but occurs over a range of temperature; and similarly, it depends on such factors as the cooling rate of the glass and its prior thermal history. The set point may be regarded as the temperature below which the glass behaves as an ideal elastic solid. It has variably been taken as equal to either the strain point of the glass (viscosity $= 10^{13.5}$ Pa s) or its annealing point (viscosity $= 10^{12}$ Pa s), or alternatively, as the strain point plus some arbitrary value, e.g. +5 K, or the annealing point minus an arbitrary value, e.g. −15 K or −20 K.

Bulk residual stresses can be quantified analytically. Alternatively, finite stress analysis techniques can be employed, although in general, and as previously noted, these can be extremely time consuming and can also be open to misleading analysis, particularly if interfaces with many reaction products are involved. It is often better to resort to simple analytical techniques to estimate the influence of residual stresses on the bulk system, through knowledge of the material and processing

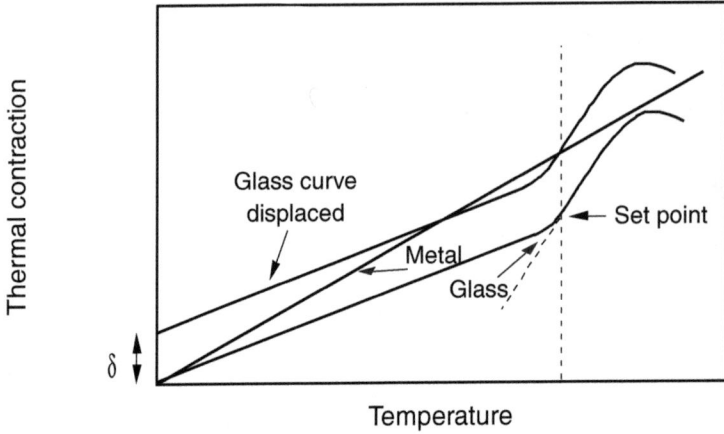

Figure 41: Thermal expansion curves for a metal and a glass illustrating one method that can be used for estimating the residual thermal stress in a seal or coating by measuring the resultant displacement of the glass curve

parameters, and use of intuition.

An estimate of the magnitude of the thermal strain due to differential contraction can be made by comparing the thermal expansion curves of the metal and the glass. The curve for the glass is then displaced, as shown in Figure 41, so that the set point coincides with the curve for the metal. The differential contraction, δ, can then be found at any given temperature from the relative displacement of the two curves at that point. As a rough guide (McLellan and Shand, 1984), δ should not exceed a value of $\approx 5 \times 10^{-4}$ $(\Delta L/L)$ in order to produce a satisfactory seal. From a knowledge of δ, it is possible to compute the corresponding residual stress. For example, for a thin glass coating on a metal substrate, the stress in the glass, σ_g, at a given temperature can be calculated to a first approximation using the expression (Rawson, 1980):-

$$\sigma_g = -E_g \delta / (1-v) \tag{7.1}$$

where E_g is Young's modulus of the glass, and v is its Poisson's ratio. If the glass has a lower thermal expansion than that of the metal, the coating will be in compression on cooling; but if the compressive stresses are too high, spalling of the

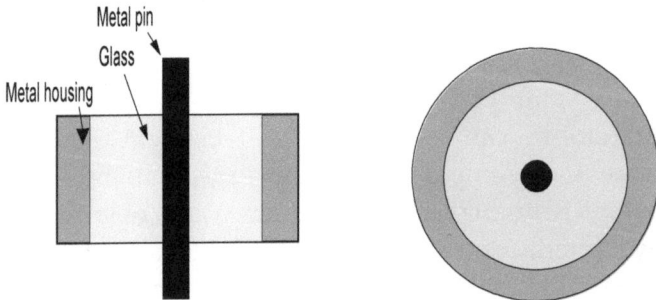

Figure 42: Schematic diagram of a simple cylinder seal

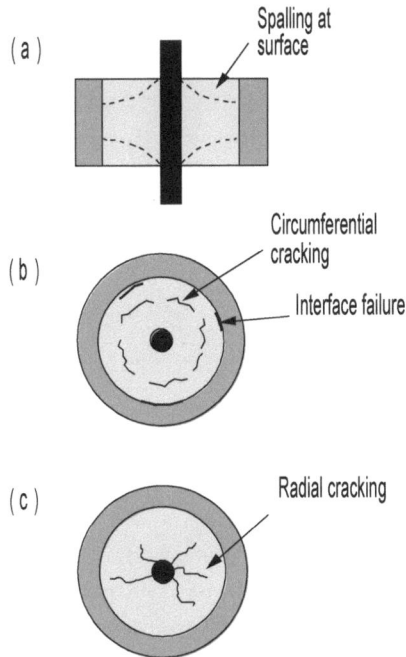

Figure 43: Effect of stresses associated with a simple cylinder seal due to thermal expansion mismatch

glass surface may occur. Conversely, if the glass has the higher expansion, the coating will be in tension, and if these stresses are of high enough magnitude cracking or crazing of the coating will occur.

As an illustration of the effect that residual stresses can have on a more complex metal–glass system, consider the example of a seal consisting of a central metal pin sealed into a cylindrical metal housing, as depicted in Figure 42. A number of scenarios are possible. For example, if the metal housing has a higher thermal expansion than that of the glass, the bulk glass will be in a state of radial compression on cooling, whilst the metal housing will be in a state of tension. Unless the metal is very thin this should present no serious problem for the integrity of the housing. If, however, the radial compression is sufficiently high, the glass surface near the metal may spall away under the influence of circumferential tension, as shown in Figure 43(a). In the case where the glass has a higher thermal expansion than the metal housing, on the other hand, the bulk glass will be in a state of radial tension, and failure may occur either at the metal–glass interface or at some distance within the glass, as illustrated in Figure 43(b). A similar case exists when the pin material has a higher expansion than that of the glass. The glass near to the pin will be in radial tension and either the metal–glass interface will fail, or circumferential cracks will develop in the glass near the pin. Finally, if the pin has a lower expansion than that of the glass, the glass will be in a state of circumferential tension, and cracks may form in the glass, as illustrated in Figure 43(c).

Approximate values for the stresses involved can be found using the following analysis, as described by Scherer(1986). This analysis assumes that plane strain conditions apply and also assumes, to a first approximation, that the elastic moduli and Poisson's ratios of the glass and the metal components are similar in magnitude, i.e. $E_{glass} \approx E_{metal}$, and $\nu_{glass} \approx \nu_{metal}$. The stresses in the glass due to thermal expansion mismatch between the pin and the glass (neglecting any effect from the outer metal cylinder), can be given by:-

$$\sigma'_{z(glass)} = -(r_1^2/r_2^2)(E/[1-\nu])\Delta\varepsilon' \tag{7.2}$$

$$\sigma'_{r(glass)} = 1/2(1-[r_2^2/r_1^2])\sigma'_{z(glass)} \tag{7.3}$$

$$\sigma'_{\theta(glass)} = 1/2(1+[r_2^2/r_1^2])\sigma'_{z(glass)} \tag{7.4}$$

Where, $\sigma_{z(glass)}$ is the axial stress, $\sigma'_{r(glass)}$ is the radial stress, $\sigma_{\theta}{}'_{(glass)}$ is the circumferential (hoop) stress, and $\Delta\varepsilon' = (\alpha_{glass}-\alpha_{metal\ pin})\Delta T$, with $\Delta T = T_{set}-T_{ambient}$.

Similarly, the stress in the glass due to expansion mismatch between the glass and the outer metal cylinder (neglecting any effect from the inner metal pin), can be given by:-

$$\sigma_z{}''_{(glass)} = (1-[r_2^2/r_3^2])(E/[1-\nu])\Delta\varepsilon'' \tag{7.5}$$

$$\sigma_r{}''_{(glass)} = \sigma_\theta{}''_{(glass)} = 1/2\sigma_z{}''_{(glass)} \tag{7.6}$$

where, $\Delta\varepsilon'' = (\alpha_{glass}-\alpha_{metal\ cylinder})\Delta T$.

Assuming superposition of stresses, the resultant stresses in the glass due to the influence of both the metal pin and the cylinder, can be given by the sums of the individual stresses:-

$$\sigma_{z(glass)} = \sigma_z{}'_{(glass)} + \sigma_z{}''_{(glass)} \tag{7.7}$$

$$\sigma_{r(glass)} = \sigma_r{}'_{(glass)} + \sigma_r{}''_{(glass)} \tag{7.8}$$

$$\sigma_{\theta(glass)} = \sigma_\theta{}'_{(glass)} + \sigma_\theta{}''_{(glass)} \tag{7.9}$$

Using equations [7.7], [7.8] and [7.9], an estimate of the stresses that are likely to occur in any given system can be made.

This description of the effect of stresses due to thermal expansion mismatch in glass–metal systems is, of course, greatly simplified, and many other factors may need to be taken into account when designing a suitable practical metal–glass component for a specific application. The above methods do, nevertheless, provide a useful initial screening method for down-selecting the most suitable material combinations. More comprehensive accounts of stresses in glass seals and coatings and methods for dealing with them can be found elsewhere (e.g. Chambers *et al*, 1989; Rulon, 1972; Varshneya, 1982; Scherer, 1986; Scherer and Rekhson, 1985; Lau *et al*, 1987).

The situation is complicated further in the case of glass-ceramic-to-metal seals and coatings due in part to the more complex thermal expansion behaviour of glass-ceramic materials, coupled with the fact that the overall behaviour may be very composition and process dependent. In general, less information is available on the creation of residual stresses in metal/glass-ceramic systems. Although in principle it is possible to match the thermal expansion characteristics of a glass-ceramic material to those of a specific metal or alloy, in practice this can rarely be achieved over the entire temperature range of interest. High expansion glass-ceramics (e.g. $\alpha > 14\times10^{-6}\,K^{-1}$) normally rely on the presence of a significant proportion of cristobalite, α-quartz or tridymite, all of which undergo phase inversions. With the exception of tridymite, these are also accompanied by relatively large volume changes, particularly in the case of cristobalite ($\approx 2\%$); this is illustrated by the thermal expansion curves shown in Figure 20. Such volume changes can lead to high residual stresses in a glass-ceramic system. The effect of these stresses can often be minimized by annealing, but only if there is sufficient residual glassy phase present with a low enough T_g. Unfortunately, determining a value for T_g of residual glass is not usually feasible employing standard DTA or DSC techniques, unless there is a high proportion of residual glass present. It is possible, however, to determine T_g of minor phases using Dynamic Mechanical Thermal Analysis, DMTA, and this technique has been successfully employed in devising a suitable heat-treatment schedule for the preparation of glass-ceramic-to-metal seals containing quartz and tridymite (Metcalfe *et al*, 1991). The presence of cristobalite is particularly problematic because the α–β cristobalite inversion occurs at relatively low temperatures, i.e. ≈ 150–$220°C$. It is therefore not generally practical to anneal out stresses associated with this transformation. For glass-ceramics containing cristobalite, this phase should be present as very small crystals, ideally $< 1\,\mu m$ in size, in order to minimize the detrimental effects of residual stresses caused by the α–β phase inversion. For glass-ceramics containing a sufficiently high proportion of residual glass, and which exhibit well-defined glass-like softening behaviour, it is possible to estimate the magnitude of residual stresses formed by sealing to a particular metal using the same method as employed for glass-to-metal seals, i.e. calculation of the thermal strain parameter δ, from the respective thermal expansion curves, or calculation of the stresses employing equations [7.7], [7.8] and [7.9] (for the case of glass-ceramics it must be assumed that a set temperature can be defined, below which the material behaves elastically).

7.2 Interfacial reactions

Reactions which occur in the interfacial region between diffusing metal ions and the constituents of the glass or glass-ceramic can play an extremely important role in determining the overall properties of the resultant metal-glass system. In addition, the presence of dissolved water in the glass can also affect seal quality due to reaction between the water and diffusing metal ions to form hydrogen gas. These influences are described in the following sections.

7.2.1   Reaction between the metal and glass or glass-ceramic

An additional complication in both glass and glass-ceramic-to-metal sealing can be caused by chemical reactions within the glass or glass-ceramic close to the metal interface which can alter the microstructure in this region, relative to the bulk material. This can have serious implications on the quality of the resulting seal or coating due to the fact that the thermal expansion characteristics of the interfacial zone may differ considerably from those of the bulk glass or glass-ceramic. Hence, an apparently thermally matched system may in fact exhibit high thermal expansion mismatch stresses within the interfacial zone.

Residual interfacial stresses can be estimated from a knowledge of the interfacial microstructures and the thermal expansion characteristics of individual phases and precipitates. Unfortunately, the characteristics of such phases and precipitates are often not known with any degree of accuracy. If property data are available, or at least can be estimated with reasonable accuracy, values for the radial and tangential stresses, σ_r and σ_t, respectively, associated with individual spherical particles may be made using an equation given by Selsing(1961), originally derived for particle reinforced ceramic matrix composites:-

$$-\sigma_r = 2\sigma_t = \Delta\alpha\Delta T/\{([1+v_m]/2E_m)+([1-2v_p]/E_p)\} \qquad [7.10]$$

Where, ΔT is the difference between the sealing temperature and ambient temperature; $\Delta\alpha$ is the thermal expansion mismatch between matrix and particle, with $\Delta\alpha = \alpha_m - \alpha_p$, v_m and v_p are the Poisson's ratios of the matrix and particle, respectively, and E_m and E_p are the Young's modulus of matrix and particle, respectively. For positive values of $\Delta\alpha$, microcracking is expected to occur within the glass-ceramic in the vicinity of the particle when the value of σ exceeds the tensile strength of the glass-ceramic.

Alternatively, an assessment of bulk interfacial stresses can be made experimentally employing strip seals, analogous to bimetallic strips, in which a thin glass-ceramic layer is sealed to a thin metal strip. It was shown by Walton and Sweo(1953), based on the bimetallic strip analysis of Timoshenko(1925), that the curvature of composite enamel/metal strips observed on cooling strips from the sealing temperature to ambient temperature can be used to compute the magnitude of the bulk stresses in the coating and the metal substrate. On cooling a composite enamel/metal strip, for which the metal and enamel exhibit different thermal expansion characteristics, to below the glass transition temperature of the enamel, where flow of the glass can no longer occur, stresses build up which distort the strip into the form of an arc, in much the same way as is noted when heating a bimetallic strip. The resultant stress can then be calculated from a knowledge of the deflection obtained, the dimensions of the strip seal, and the elastic moduli of the enamel and metals employed.

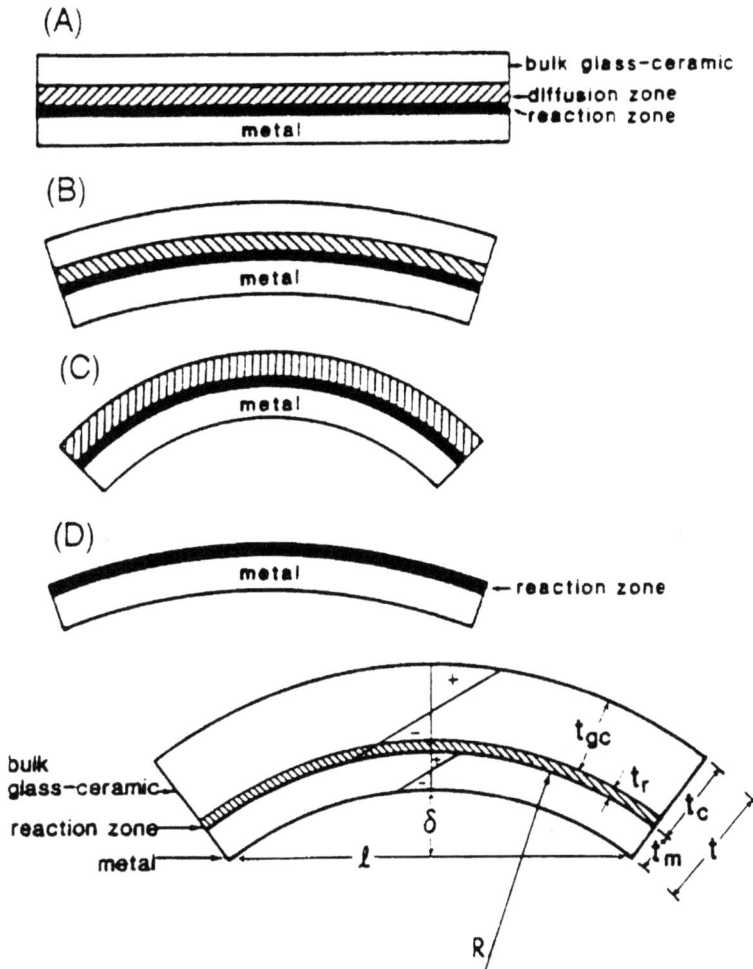

(A)

bulk glass-ceramic
diffusion zone
reaction zone
metal

(B)

metal

(C)

metal

(D)

metal
reaction zone

bulk
glass-ceramic
reaction zone
metal

t_{gc}
t_r
t_c
t_m
t
δ
ℓ
R

Figure 44: Bimetallic strip seal for determining the residual interfacial stresses in a glass-ceramic-to-metal seal (after Kunz and Loehman, 1987)

Kunz and Loehman(1987) extended this approach in order to calculate both the thermal expansion of the reaction zone in a glass-ceramic/metal system and the interfacial stresses developed due to this reaction zone. In their study, a lithium silicate glass-ceramic, the glass-ceramic having nominally the same thermal expansion as that of the alloy, was sealed to the Ni-based superalloy Inconel 718. Metal strips 2 cm wide by 10 cm long by 0·25 mm thick were employed in the investigation. After sealing they noted that the strip was distorted into an arc, as depicted in Figure 44, and microscopic analysis of the interfacial region between the glass-ceramic and the metal showed the presence of a reaction zone ≈ 25 µm thick adjacent to the metal substrate which was less crystalline than the bulk glass-ceramic and in which a strong crystal orientation effect was observed. Kunz and Loehman noted that the elastic modulus and thermal expansion of this region was likely to be quite different from that of the bulk glass-ceramic. Strip deflections were monitored after sealing and subsequently after removal of consecutive layers of glass-ceramic until only the reaction layer remained. The resultant stresses were calculated using the elastic analysis of Timoshenko(1925):-

$$\sigma_r = -1/2R[E_c t_c(1 + t_c/3t) + E_m(t_m^3/3tt_c)] \qquad [7.11]$$

$$\sigma_m = 1/2R[E_c(t_c^3/3tt_m) + E_m t_m(1 + t_m/3t)] \qquad [7.12]$$

$$\sigma_c = 1/2R[E_c t_c(1 - t_c/3t) - E_m(t_m^3/3tt_c)] \qquad [7.13]$$

$$\sigma'_m = 1/2R[E_c(t_c^3/3tt_m) + E_m t_m(t_m/3t - 1)] \qquad [7.14]$$

where σ_r is the compressive stress in the reaction zone, σ_m is the tensile stress in the metal, σ_c is the stress at the free surface of the glass-ceramic, σ'_m is the stress at the free surface of the metal, E_c is an effective composite Young's modulus for the glass-ceramic and reaction zone, t is the total strip thickness, t_m is the thickness of the metal, and t_c is the total thickness of glass-ceramic and reaction zone.

A value for the radius of curvature, R, is calculated knowing the strip deflection, δ, and strip length, ℓ, from:-

$$R = \ell^2/8\delta \quad \text{(valid only for } \delta < R) \qquad [7.15]$$

The effective composite modulus of the glass-ceramic and reaction zone, E_c, is approximated using the rule of mixtures:-

$$E_c = E_r(t_r/t) + E_{gc}(t_{gc}/t) \qquad [7.16]$$

where E_{gc} is the modulus of the glass-ceramic.

The strains at the interface in the reaction zone, ε_r, and the metal, ε_m, may be calculated from:-

$$\varepsilon_r = \sigma_r/E_c \qquad [7.17]$$

$$\varepsilon_m = \sigma_m/E_m \qquad [7.18]$$

The thermal expansion of the reaction zone may subsequently be found using the expression:-

$$\varepsilon_r + \alpha_r\Delta T = \varepsilon_m + \alpha_m\Delta T \qquad [7.19] \text{ or}$$

$$(\alpha_m - \alpha_r)\Delta T = \varepsilon_r - \varepsilon_m \qquad [7.20]$$

where ΔT is the temperature difference between the stress free and as-cooled samples. The temperature of the stress-free state may be found by heating the distorted strip until it becomes flat.

In the glass-ceramic/metal example reported by Kunz and Loehman (1987) it was calculated that the thermal expansion of the reaction zone was around 25%

lower than that of the bulk glass-ceramic and metal. This results in a significant compressive stress in the reaction zone, when previously this region was believed

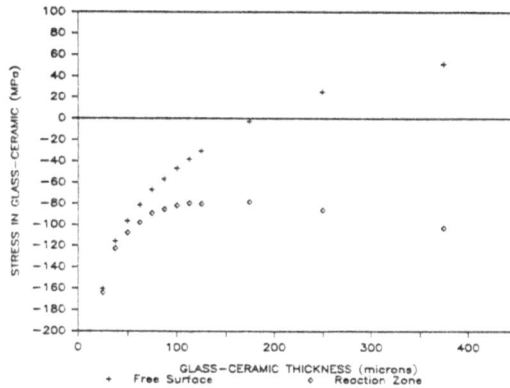

(a) Stress in reaction zone calculated from strip curvature with progressive removal of the glass-ceramic

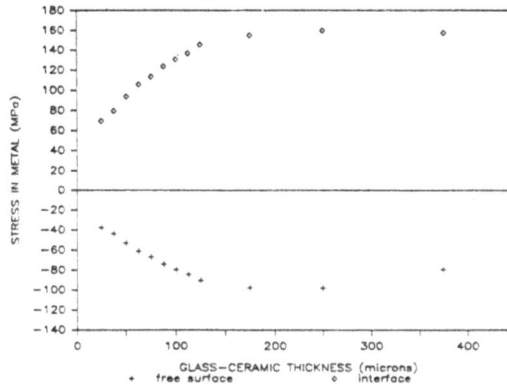

(b) Stress in metal calculated from strip curvature with progressive removal of the glass-ceramic

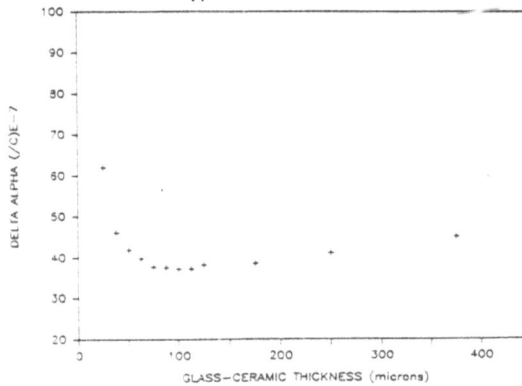

(c) Difference in thermal expansion coefficient between reaction zone and bulk glass-ceramic calculated using the strip-seal method

Figure 45: Stresses associated with a glass-ceramic-to-metal seal in the vicinity of the interface (after Kunz and Loehman, 1987)

to have been stress free. By progressively removing thin layers of glass-ceramic and monitoring the change in curvature obtained it was possible to build up a stress profile across the interfacial region, as shown in Figure 45.

It should be noted that values for residual stresses obtained by this method are effectively averaged out over the entire interface and therefore do not give any indication of the localised stresses associated directly with individual pre-cipitated phases or reaction products. Various simplifications and assumptions are also made in the analyses including, for example, the fact that no account is taken of end effects, viscoelasticity, temperature dependence, stress relaxation, etc. Kunz and Loehman (1987) concluded that detailed stress analysis, e.g. by finite element stress analysis, is required in order to determine more precisely the stress state produced in the reaction zone for more complex geometries, as would be found in most seal components (but as noted earlier, detailed analysis requires specific information on the interfacial products and precipitates, which may not be available). The strip seal method does, however, provide a simple test for determining comparative values for the stress associated with different glass-ceramic/metal systems, and does therefore provide a useful screening method for the selection of the most suitable combinations.

7.2.2   Influence of water on interfacial reactions

Normal glass can contain significant quantities of dissolved water, typically 0.02-0.06%. It was noted by Haws and Kramer et al(1985) that this can significantly affect seal quality through reaction of the water with diffusing metallic species from the substrate, in particular Cr, to yield hydrogen gas:-

$$Cr + 3/2H_2O \rightarrow 1/2Cr_2O_3 + 3/2H_2\uparrow; (\Delta G° = -129 \text{ kJ/mole}) \qquad [7.21]$$

It was found that seals prepared using starting glass with higher dissolved water contents contained more bubbles. It was noted that a number of methods could be used to minimize the influence of water. These included melting under dry conditions; for example, by bubbling dry gas through the melt; using dry starting constituents or materials low in water, e.g. Li_3PO_4 as a source of P_2O_5 in prefer-ence to $NH_4H_2PO_4$; or by use of specific additives that would react preferentially with Cr during seal manufacture, e.g. CuO:-

$$Cr + 3CuO \rightarrow 1/2Cr_2O_3 + 3/2Cu_2O; (\Delta G° = -381 \text{ kJ/mole}) \qquad [7.22]$$

In addition to Cr, reaction of water with many other metals commonly found in commercial alloys is also thermodynamically favourable; for example:-

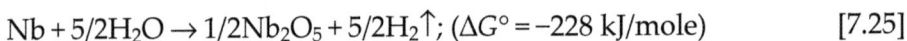

$$Al + 3/2H_2O \rightarrow 1/2Al_2O_3 + 3/2H_2\uparrow; (\Delta G° = -367 \text{ kJ/mole}) \qquad [7.23]$$

$$Ti + 2H_2O \rightarrow TiO_2 + 2H_2\uparrow; (\Delta G° = -354 \text{ kJ/mole}) \qquad [7.24]$$

$$Nb + 5/2H_2O \rightarrow 1/2Nb_2O_5 + 5/2H_2\uparrow; (\Delta G° = -228 \text{ kJ/mole}) \qquad [7.25]$$

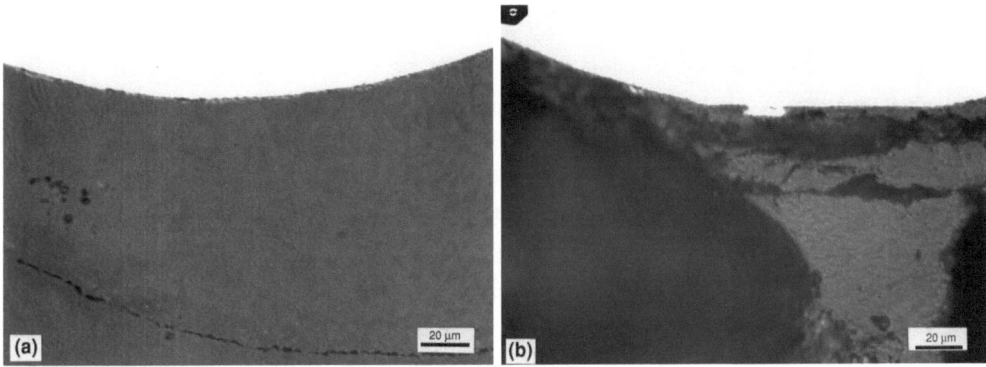

Figure 46: Micrographs of seals between lithium zinc silicate glass-ceramic and Hastelloy C276 alloy prepared under (a) 'dry' and (b) 'wet' conditions – note extensive bubbling (AWE)

Other examples are summarized in Table 12.

Glass-ceramic-to-metal seals have also been manufactured using a lithium zinc silicate starting glass prepared under 'wet' and 'dry' conditions (wet glass by bubbling argon through a water bath and then through the melt, and dry glass by bubbling dry argon through the melt) in addition to normal ambient atmosphere melting conditions (Sambrook *et al*, 1993). Glasses were sealed to a variety of alloys including Nimonic 90, Hastelloy C276 and stainless steel. It was observed that the seals prepared using the wet glass generally contained numerous large bubbles, whereas the glasses prepared under dry or normal conditions generally contained only small and/or very few bubbles, as shown in Figure 46. It was observed, however, that the seals to stainless steel were far less susceptible to bubble formation, even for the wet glass, and whereas seals to the Ni-based alloys using the wet glass failed a He leak test all the seals to stainless steel passed this test. This suggests that some systems are far more susceptible than others to bubble formation due to the presence of water in the glass.

It should be noted, of course, that seals and coatings containing bubbles are not only weakened mechanically, but their lifetime behaviour will also be affected through the generation of enhanced fracture paths. In the case of electrical components, the electrical integrity of the item may also be affected.

7.3   Ageing mechanisms and factors influencing lifetime behaviour

Many factors need to be taken into consideration in the successful design and manufacture of high quality glass- and glass-ceramic-to-metal seals and coatings if an adequate lifetime behaviour is to be achieved. In order for a glass or glass-ceramic to form a mechanically strong, adherent and if necessary hermetic seal to a metal or alloy, for example, a number of specific criteria must be met. In particular, it is important that a chemical bond be formed at the interface between

the glass or glass-ceramic and the metal. In addition, the thermal expansion characteristics of the glass or glass-ceramic must be matched as closely as possible
to those of the metal or alloy in question in order to prevent the formation of
undesirable tensile stresses in the seal or coating after cooling from the fabrication temperature, or during subsequent thermal cycling. This requires a thorough
understanding of the relevant glass/metal or glass-ceramic/metal interactions in
order that steps can be taken to avoid or at least minimize reactions within the
interfacial region that may lead to localized modifications of the glass-ceramic
microstructure. Such reactions can, due to the creation of a mismatch in thermal
expansion between phases present, lead to the formation of undesirable highly
localized internal stresses, the presence of which can initiate failure of a seal or
coating either during manufacture or, more seriously, at some later stage due to
the influence of static fatigue.

7.3.1   Thermal expansion mismatch stresses

As noted earlier, unless the thermal expansion characteristics of a given glass- or
glass-ceramic-to-metal seal or coating system are identical, stresses will be generated due to differential contraction during cooling from the fabrication temperature. As precise matching of thermal expansions will rarely if ever be possible in
practice, considerable effort has gone into the design of suitable seal systems in
order to alleviate or minimize the effects of residual tensile stresses and to ensure
that any major residual stresses are compressive in nature (e.g. Partridge, 1949;
Varshneya, 1982; Beauchamp and Burchett, 1991; Tomsia et al, 1991).

If the thermal expansion mismatch stresses are high enough, failure of a seal
will normally occur by debonding at the interface or by fracture through the
glass or glass-ceramic either during manufacture or shortly afterwards. However,
even if the residual stresses are not high enough to initiate cracking in the short
term, they may be high enough to do so in the longer term due to the influence of
static fatigue (defined as the growth of pre-existing flaws under the influence of
a static stress lower than the ultimate instantaneous failure stress, and which can
ultimately lead to failure). Failure of the metallic components very rarely occurs,
because stresses can normally be accommodated by plastic flow of the metal.

To remain hermetic, a seal must exhibit strong bonding at the interface. It
is recognized that for any seal system residual stresses must be kept as low as
possible and must be predominantly compressive in nature. This requires that
not only are the thermal expansion characteristics of the metal and bulk glass or
glass-ceramic closely matched, but also that interfacial reactions are limited and
do not lead to the formation of reaction products and precipitated phases with
thermal expansion or elastic characteristics that differ significantly from those
of the bulk phases. The formation of such phases may lead to unacceptably high
residual thermal stresses within the interfacial reaction zone. In addition, in the
case of glass-ceramics, reaction between diffusing metallic species and the nucleating agent employed for the particular glass-ceramic must also be avoided.

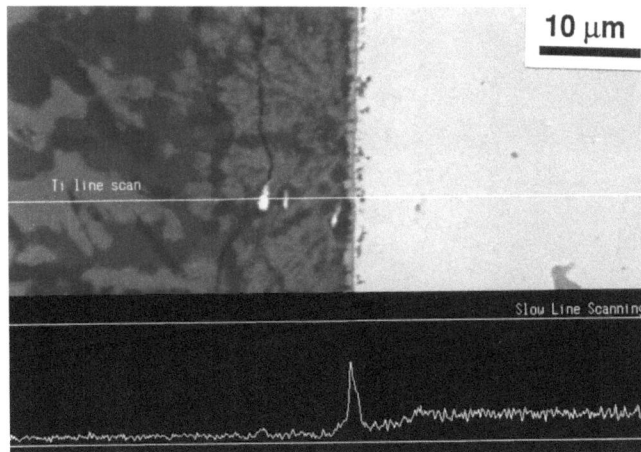

Figure 47: Micrograph showing cracking due to the influence of residual thermal expansion mismatch stresses in the vicinity of a precipitate in the interfacial zone of a glass-ceramic-to-metal seal
(after Donald, 1992)

If reaction does occur, the effectiveness of the nucleating agent may be impaired, and this may result in the formation of a coarse microstructure in the interfacial region, or a region with a high proportion of residual glass. In either case, the thermal expansion characteristics in the interfacial region will differ from those of the bulk glass-ceramic and this will again lead to the generation of highly localized residual stresses, detrimental to the lifetime behaviour of the component.

Any mismatch in thermal expansion between the bulk glass or glass-ceramic and the metal is expected to play a significant contribution to the lifetime behaviour through the influence of static fatigue. In addition, potential ageing mechanisms will also be activated through the effect of any highly localized residual thermal stresses caused by interfacial reactions. Figure 47 shows cracking around a precipitate in the interfacial zone due to the presence of residual stresses.

Bulk residual stresses can be quantified analytically (e.g. Scherer, 1986) as covered in Section 7.1. Alternatively, finite stress analysis techniques can be employed although, as noted earlier, in general these can be extremely time consuming and can also be open to misleading analysis, particularly if interfaces with many reaction products are involved. As noted earlier, it is often better to resort to simple analytical techniques to estimate the influence of residual stresses on the bulk system through knowledge of the material and processing parameters, and use of intuition.

Residual interfacial stresses can be estimated from a knowledge of the interfacial microstructure and the thermal expansion characteristics of individual phases and precipitates (e.g. Wissuchek, 1998). Unfortunately, the characteristics of such phases and precipitates are often not known with any degree of accuracy. As pointed out earlier, stresses can also be measured directly using the bi-metallic

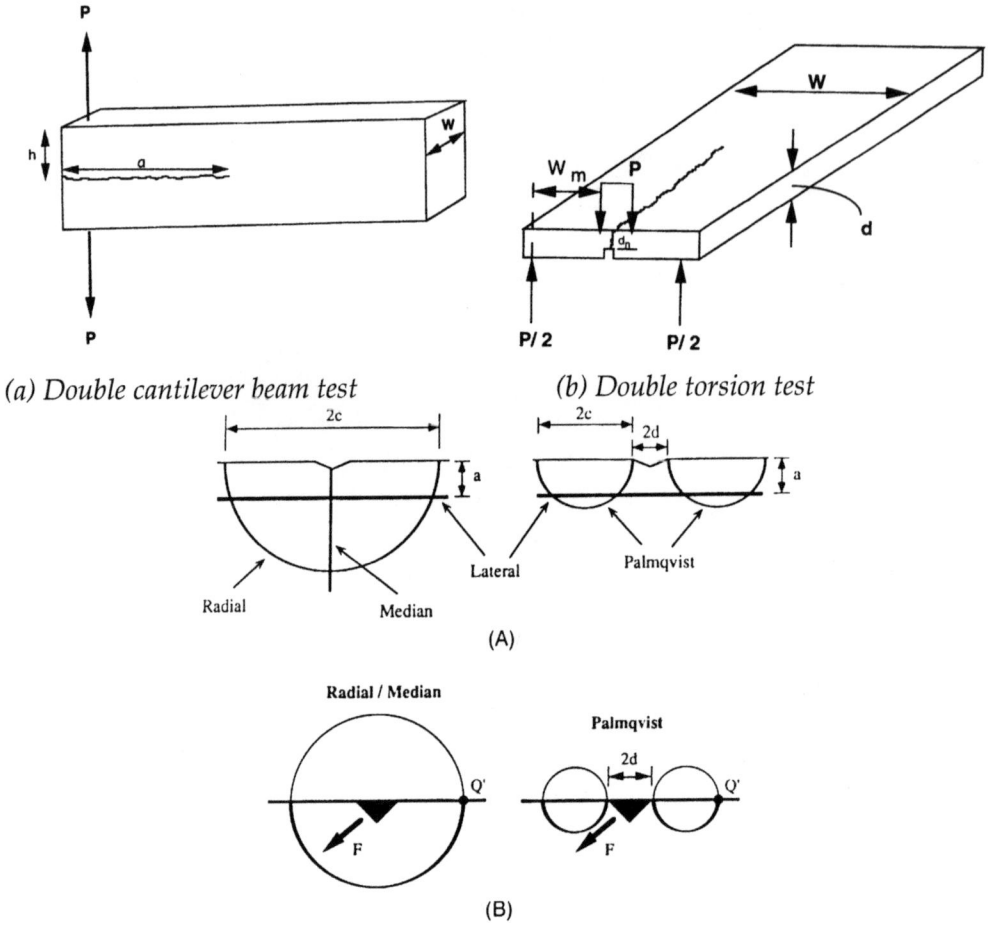

(a) Double cantilever beam test   (b) Double torsion test

(A)

(B)

(c) Indentation test showing system of cracks around indentation in glass (after Smith and Scattergood, 1992)

Figure 48: Schematic diagram of specimen geometries used for fracture mechanics measurements (after Breder and Wereszczak, 1998)

strip method, applied successfully to enamelled and glass-ceramic systems. This analysis allows stresses due to the interfacial region to be quantified, although it does not provide direct information on the highly localized stresses that may be associated with individual precipitated phases and which may be particularly detrimental to the lifetime behaviour of glass-metal systems.

7.3.2   Experimental fracture mechanics

Some information may be sought from fracture mechanics data for the glasses, glass-ceramics and ceramics employed in seals or coatings. The fracture mechanics parameters for the bulk materials can be determined experimentally using strain-energy release methods. These involve introducing a large crack into a material and determining the resistance to the propagation of this crack under different stressing conditions and environments. Methods that can be used include the double cantilever beam technique, the double torsion technique, the notched beam test, and indentation methods, as illustrated in Figure 48. Historically, the double cantilever beam method was the first specimen type to gain popularity in the testing of brittle materials. For this particular geometry it can be shown that:-

$$K_I = 3 \cdot 45 \{ Fc(a + 0 \cdot 7h/c) \} / bh^{3/2} \qquad [7.26]$$

where F is the applied force, b is the beam width, h is the beam height, a is the beam length, and c is the crack length.

To obtain meaningful results using this method the growth of the crack must be accompanied by a decreasing value of F in order to keep K constant. This may be achieved in practice by allowing the crack to propagate into an ever increasing specimen width, although experimentally this is complex. An alternative approach is to employ a double torsion geometry in which the critical force is independent of the crack length. For the double torsion geometry it can be shown that:-

$$K_I = Fb' \{ 3(a + v)/bd^3 d_n \}^{1/2} \qquad [7.27]$$

where b' is the distance from the edge of the beam to the notch, d_n is the distance from the top of the specimen to a groove which guides the propagating crack, and v is the crack velocity.

The crack velocity can be measured using a travelling microscope, as described by Weiderhorn and Bolz (1970). Alternatively, the relaxation method of Evans (1972) may be used in which the specimen is loaded quickly until the crack begins to propagate. Loading is then stopped and the crack allowed to extend under conditions of constant displacement. The velocity is given by the relationship:-

$$v = -P_i c_i (dp/dt)/p^2 \qquad [7.28]$$

where P_i is the load at the onset of relaxation, c_i is the initial crack length, and dp/dt is the gradient of the relaxation curve at a given value of the load P.

A further method, which is applicable for crack growth rates $> 10^{-4} ms^{-1}$, employs modulation of the fracture surface by sound waves (Richter, 1983). This method produces a pattern of striations on the fracture surface, the spacing of which depends on the frequency of the sound waves employed and the magnitude of the crack velocity.

For these methods, the crack should ideally be "atomically" sharp at its tip. This may be achieved by introducing a notch into the sample surface by means of a diamond saw, followed by initiation of a suitably sharp crack from the root of this notch. This may be accomplished by means of a suitable loading arrangement that initiates a sharp crack but prevents catastrophic crack growth and subsequent propagation.

Plots of K_I, determined using any of these methods, versus crack velocity can be built up, from which the critical stress intensity factor, K_{Ic}, can be found. A representative curve is shown in Figure 49. There may be a threshold value of

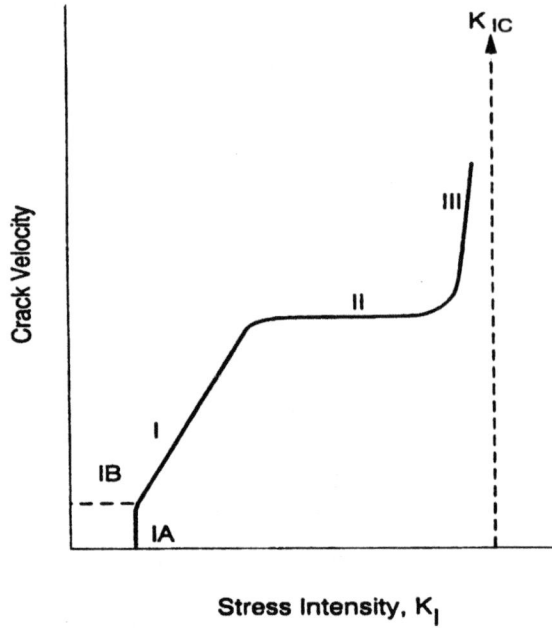

Figure 49: Idealized representation of stress intensity factor, K_I, as a function of crack velocity illustrating the different regimes of crack growth (after Freiman, 1998)

the stress intensity factor, K_{Io}, below which there is no observable crack growth. As illustrated in Figure 49, three distinct regions are often apparent in the crack velocity versus K behaviour.

In region I it can be shown that:-

$$v = \alpha_1 K_I^n \qquad\qquad\qquad [7.29]$$

where α_1 and n are constants.

In region II, on the other hand, it can be shown that:-

$$v = \alpha_2 \qquad\qquad\qquad [7.30]$$

where α_2 is a constant.

In the case of glasses, Region III is of less practical interest, with the origins still under debate. This region culminates in near instantaneous fracture at a critical value for the stress intensity factor, K_{Ic}.

Although, historically, the fracture toughness, K_c, of materials is determined in bending on notched samples using the double cantilever beam method, it has been shown that results in reasonable agreement with the bending method can be achieved using the relatively simple technique of indentation fracture (as reviewed for example by Shetty *et al*, 1985). This method employs a standard Vickers (or equivalent) indentation hardness tester. Above a critical indenter load it is observed that a system of cracks is generated around an indentation,

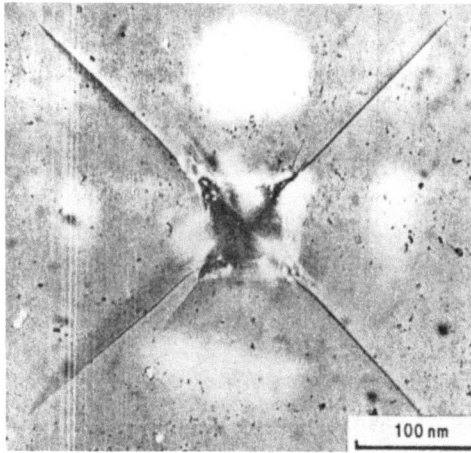

Figure 50: System of cracks produced in the vicinity of a Vickers indentation test (after Cho et al, 1990)

as illustrated in Figure 50. The following relationship is then employed:-

$$K_c = \S(E/H)^{1/2}(P/c^{3/2})$$ [7.31]

where $\S$ is a dimensionless constant which depends on the indenter geometry (0·016 determined empirically for a standard Vickers indenter), E is Young's modulus, H is the hardness, P is the load, and c is the radius of the radial crack emanating from the indentation (c should have a value around three times that of the radius of the plastic zone and this can be achieved by use of a sufficiently high load).

Alternatively, (Miyata and Jinno, 1982; Matthewson and Field, 1980):-

$$kK_c = P/(\pi^{3/2} \tan\Psi\, c^{3/2})$$ [7.32]

where k is a correction factor dependent on frictional and surface effects and also on hardness and Young's modulus, and Ψ is the half angle of the indenter. For a standard Vickers indenter, $2\Psi = 136°$; therefore, $1/(\pi^{3/2}\tan\Psi) = 0\cdot07256$, and:-

$$K_c = (0\cdot07256P/c^{3/2})/k$$ [7.33]

The relative fracture toughness, K_c', may be given by:-

$$K_c' = 0\cdot07256P/c^{3/2}$$ [7.34]

In the absence of data for E, H or k, this is a useful and very simple relationship which can be used for estimating comparative fracture toughness for similar types of material. For more accurate values of fracture toughness it has been shown empirically that for many glasses $k = 0\cdot64$ (Miyata and Jinno, 1982).

Further information on fracture mechanics techniques can be found in many sources (e.g. Bar-On, 1991; Perez, 2004).

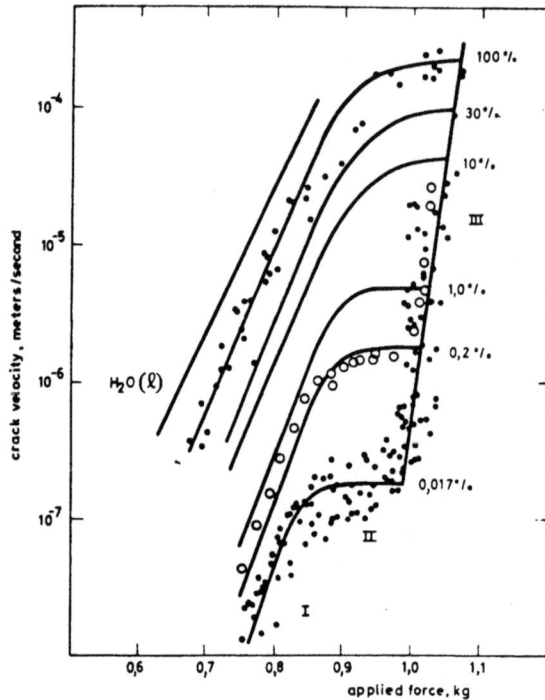

Figure 51: Characteristic K_I versus crack velocity diagram for a glass as a function of atmosphere relative humidity
(after Weiderhorn, 1967)

7.3.3   Static fatigue

Static fatigue can be somewhat likened to stress-corrosion of crystalline metallic systems for which, in the presence of a corrosive environment, existing surface defects grow under the influence of an external stress below the normal failure stress. For oxide glasses, the corrosive factor is usually water present in the environment. Of particular importance during static fatigue is the relationship between the stress intensity factor and the crack velocity.

The effect of static fatigue on lifetime behaviour is well documented for the case of bulk oxide glasses and ceramics (e.g. Davidge, 1979; Kelly and MacMillan, 1986; Gy, 2003). Unfortunately, although some data do exist for ceramic/metal systems (e.g. Huh and Kobayashi, 1997; Shaw, 1998; Cai and Bao, 1998; White, 1998), very little information is available for glass/metal or glass-ceramic/metal combinations, although the influence of static fatigue may be very important in seal and coating environments. It is clear that more studies are required in this area.

For the case of bulk glasses, Weiderhorn (1967), in his classic paper, reported the measurement of crack velocities in glass as a function of K in double cantilever beam specimens held in a nitrogen atmosphere containing varying quantities of water vapour. He showed that the overall crack growth could indeed be divided into three distinct regions and that these were displaced towards higher crack velocities with higher atmospheric water contents, as shown in Figure 51. In region I crack velocities are governed by the rate of stress corrosion by water at the crack tip, i.e. the rate of crack growth is reaction-rate controlled and the crack velocity is proportional to K_I. The higher the concentration of water, the greater is this effect, and the higher is the resulting crack velocity for a given value of K_I. Region II is independent of K_I and is governed by the rate of diffusion of the corroding species, e.g. water, to the crack tip. Region III, on the other hand, is thermally activated. This region is intrinsic to the material and does not depend on the corroding species. The critical value of stress intensity, K_{Ic}, where fracture is virtually instantaneous, occurs at the end of region III, whilst there is a threshold value, K_{Io}, below which no crack growth with time occurs. The stress corrosion governed crack propagation regime corresponding to region I has been explained by Charles and Hillig (1962) on the basis of crack sharpening and blunting behaviour. At stresses greater than some threshold value the crack tip radius decreases due to stress corrosion effects. Below this threshold value, on the other hand, the crack tip radius increases, thereby blunting the crack tip and leading to an increase in strength of the material. Although the presence of water is the main environmental factor controlling the time dependent behaviour of the strength of glass, it has been shown that other environments may also affect the performance (Michalske and Freman, 1983; Michalske and Bunker, 1987). In particular, environments similar to that of water with a molecular diameter of <0·5 nm and with a molecular structure containing a lone-pair electron orbital across from a proton donor site have the greatest effect on crack growth. Such environments are found, for example, with ammonia, hydrazine and formamide.

A number of recent reviews have been published in the area of static fatigue (e.g. Gy, 2003; White, 1998; Breder and Wereszczak, 1998) and these should be consulted for detailed information on this subject.

7.3.4 Weibull statistics and life prediction

Failure of a brittle material is dependent on the fracture toughness of the material which is itself a function of the susceptibility of the material to subcritical crack growth. Failure prediction and life assessment of real materials is a complex subject but can be assessed, to a greater or lesser extent, employing Weibull statistics. If a large number of nominally identical samples are tested under a uniform uniaxial tensile stress, Weibull statistics, based on the weakest link hypothesis, show that the probability, P_f, of failure at a given stress, σ, may be given by:-

$$P_f = 1 - \exp\{-[(\sigma - \sigma_u)/\sigma_o]^m)V\} \qquad [7.35]$$

where σ_u is a threshold stress below which the failure probability is zero (this must usually be set equal to zero unless there are physical reasons to expect an upper limit to the size of a flaw), m is the Weibull modulus (this is a measure of the variability of the quantity $\sigma_f - \sigma_u$, where σ_f is the fracture stress, σ_o is a quantity which does not correspond to any readily determined physical quantity but can be related to the arithmetic mean, σ_f, of the values of failure for all specimens in the sample, and V is the volume of the sample.

Equation [7.35] may be rewritten as:-

$$\ln\ln(1/P) = \ln V + m\ln(\sigma-\sigma_u)-m\ln\sigma_o \qquad [7.36]$$

The survival probability can then be plotted as a function of the fracture stress, as shown in Figure 52. The slope of the graph will be equal to $-m$. The value of m gives an indication of the variability in strength, the higher the value of m the more consistent is the material.

Calculation of the Weibull modulus for a given material and test conditions enables predictions to be made about the survival probability at particular levels of stress. This is of considerable interest in general engineering design, but only relates to the samples as tested experimentally. In particular, the larger the size of the sample, the greater can be the maximum possible size for a flaw. Hence, the larger the specimen, the lower its strength, the greater the variability in its strength, and the lower its survival probability at a given stress. It can be shown quantitatively that the volume of the sample affects the properties in the following manner:-

$$\sigma_{v1}/\sigma_{v2} = (V_2/V_1)^{1/m} \qquad [7.37]$$

If there is a batch-to-batch variation in Weibull modulus an effective modulus for all the batches may be obtained by pooling the data and using the normalised Weibull equation. This treats the distribution of strengths related to the mean strength, σ, of each batch as follows (taking σ_u equal to zero):-

$$P_f(\sigma/\sigma_o) = 1-\exp\{-\{\sigma/\sigma_o(1/m!)\}^{1/m}\} \qquad [7.38]$$

Having evaluated the mean strength under the design conditions the safe

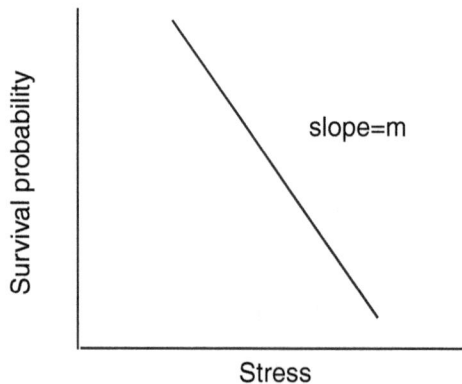

Figure 52: Survival probability as a function of the applied stress. The slope is equal to the Weibull modulus, m, of the material (the value of m gives an indication of the variability in strength – the higher the value of m, the more consistent is the material, i.e. the less variability in strength)

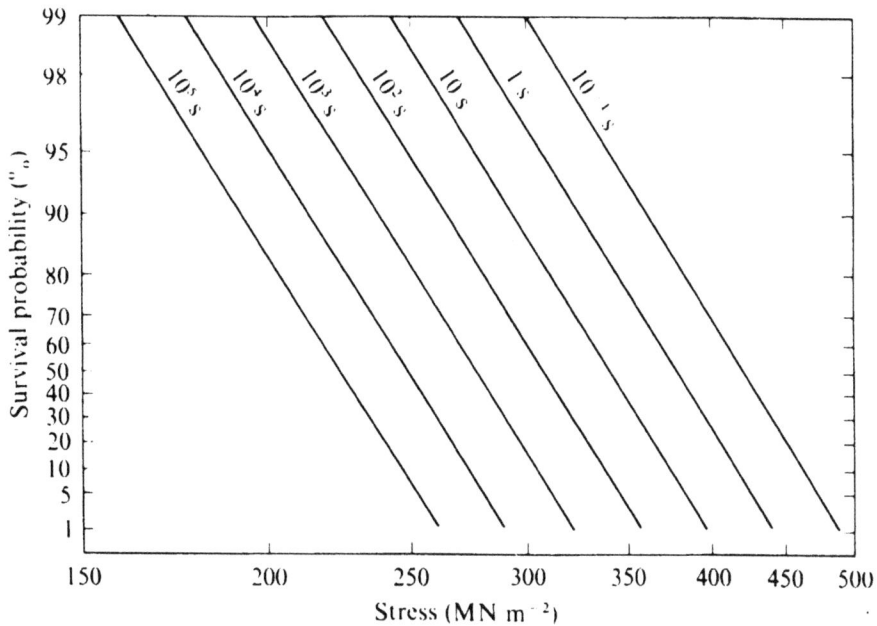

Figure 53: Typical strength–probability–time, SPT, diagram constructed using data obtained from Weibull statistics and the susceptibility of the material to static fatigue. This gives the survival probability as a function of fracture stress at different strain rates and provides an indication of the strain rate dependence of strength from which an estimate of the susceptibility of the material to subcritical crack growth, i.e. static fatigue, can be made – the lower the strain rate dependence, the lower the susceptibility (after Davidge et al, 1973)

working stress, σ_s, for an acceptable failure probability, P, may be calculated:-

$$\sigma_s = \sigma\{[-\log(1-P)]^{1/m}/(1/m!)\} \qquad [7.39]$$

Unfortunately, failure probabilities and strength predictions are complicated further by the time dependence of strength at a given stress, i.e. the influence of static fatigue. It is necessary, therefore, to determine the survival probability after a given time of service under stress. This life assessment can be carried out by combining the statistical and time dependent properties of the material. For materials which exhibit region I behaviour, for example, it can be shown that:-

$$\sigma_{t2} = \sigma_{t1}(t_1/t_2)^{1/n} \qquad [7.40]$$

where σ_{t1} is the mean strength in a short term test, i.e. at $t = t_1$, and σ_{t2} is the mean strength in a longer term test, $t = t_2$. The parameter "n" determines the susceptibility of the material to subcritical crack growth; the higher the value of n, the less susceptible the material is.

This relationship, i.e. equation [7.40], combined with Weibull statistics allows the construction of a strength/probability/time diagram for any given material and set of environmental conditions. Experimentally, the mean strength at a given strain rate, ε_1, is found, and this is repeated for different strain rates, as depicted in Figure 53. The value of n may then be computed on the basis of equation [7.40] and a knowledge that :-

$$n + 1 = \log(\varepsilon_1/\varepsilon_2)/\log(\sigma_{\varepsilon 1}/\sigma_{\varepsilon 1}) \qquad [7.41]$$

It should be noted that Weibull statistics are subject to a number of limitations. In particular, the data are subject to very large errors if the sample size is too small. In general, a minimum of 30 samples is required and, ideally, 100 samples should be chosen. For determination of the slow crack growth exponent, n, it is necessary, in order to obtain meaningful results, to employ strain rates at least three orders of magnitude apart. This, coupled with the requirement for at least 30 samples per test, makes the gathering of reliable and intelligible Weibull statistics an extremely time consuming and laborious operation. There is often a tendency to use too few data. This can lead to serious misinterpretation of the statistics, and consequently to misinterpretation of the properties and behaviour of a given material in a practical situation. The translation of Weibull data gathered in simple test geometries in uniaxial tension to the multiaxial stresses encountered in real situations is also difficult, as is the extrapolation of data to other sample shapes and sizes. Although ideas can be formulated as to the general behaviour of real materials in practical situations, any conclusions that are drawn should be treated with caution and regarded as reliable to a factor of perhaps two or three only.

Weibull statistics may be applied to the brittle materials used in sealing and coating applications to give an indication of the likelihood of failure of a bulk material under a given set of conditions. Great caution should, however, be taken in applying the data obtained to the scenario of a composite seal or coating, where the influence of the other materials making up the system may play a very significant role. This is another area worthy of more research effort.

7.3.5   Thermal and dynamic fatigue

In addition to the effects of static fatigue, lifetime behaviour may also be influenced by thermal or dynamic fatigue brought about by thermal or mechanical cycling during service (see for example Roebben et al, 1996; Breder and Wereszczak, 1998, for information on bulk ceramics). Unfortunately, this is not well documented in the case of glass or glass-ceramic/metal systems, although again some data are available for the more general case of ceramic/metal combinations (e.g. Selverian and Kang, 1992).

More recently, Kruzic et al(2001) have monitored the time dependent properties of alumina–aluminium interfaces under both static and cyclic loading conditions. The samples employed consisted of 5 to 100 μm thick aluminium layers bonded between polycrystalline or single crystal alumina. It was observed that static loading in moist conditions led to the propagation of interfacial cracks into the alumina, whilst under cyclic loading conditions crack growth occurred predominantly by interfacial debonding. In both cases, the properties, including fatigue behaviour, were noted to improve with decreasing aluminium layer thickness.

Some thermal cycling studies have been undertaken in the case of thermal barrier coatings. For example, Vaben *et al*(2004) examined the influence of thermal cycling on the properties of yttria-stabilized zirconia coatings prepared by plasma spraying and found that resistance to thermal cycling increased with increasing horizontal crack density, and Zhu *et al*(2004) monitored thermal cycling resistance in similar coatings using a laser to heat the surface. Xiong *et al*(2005) have monitored the behaviour of functionally graded thermal barrier coatings under cyclic fatigue conditions, finding that thin coatings failed by surface spallation whilst thicker coatings were susceptible to interface cracking. A finite element based fracture mechanics analysis of a thermal barrier coating in which the effects of thermal fatigue on pre-existing microcracks within the coating has been carried out by Liu *et al*(2004).

With a relative scarcity of data, it is clear that the area of thermal and dynamic fatigue is yet another area where more studies are required.

7.4   Component failure and ageing studies

Many different types of test have been used for life assessment studies of materials (see, for example, Nelson, 1990). Accelerated ageing tests are employed in order to gather data in realistic timescales. From the data obtained, predictions are then made as to the likely lifetime of a given material or component under specified service conditions. Mathematical models are available for accelerated life tests under constant stress. The Arrhenius life relationship is widely used to model material or component life as a function of temperature, and the life of some materials and components in temperature accelerated tests may be described by an Arrhenius–Weibull model. The inverse power relationship, on the other hand, is often used to model material or component life as a function of an accelerating stress.

Although extensive literature exists on the ageing and lifetime behaviour of bulk glass and ceramic materials (e.g. Cook, 1999) there are relatively few data specifically on the measured ageing characteristics of glass-, glass-ceramic, or ceramic-to-metal seal systems. Details of some of these studies are given below.

In one study carried out at AWE, both real time and accelerated ageing tests have been carried out on a wide range of systems. These included lithium zinc silicate and lithium aluminosilicate glass-ceramics bonded to stainless steels and Ni-based superalloys, with selected samples in the form of multi-pin connectors remaining leak tight after more than 10 years ageing at 50°C and 50% RH.

Progressive deterioration of glass-to-metal seals in diode packages employed in space flight hardware has been noted by McPhilmy *et al*(1989), this deterioration

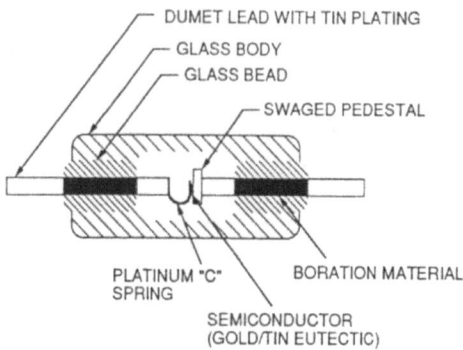

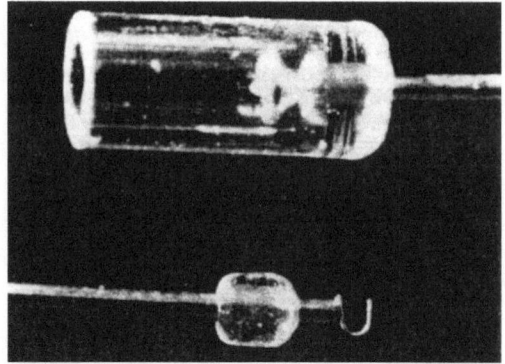

Figure 54: Schematic illustration of a diode envelope and glass-beaded lead with Pt spring prior to being fused into the glass tubing used in construction of the diode (after McPhilmy et al, 1989)

eventually causing the diode to open circuit. The cause of the deterioration was traced to chemical attack of the seal due to plating process chemicals used during the manufacturing process becoming entrapped in the seal. The diodes under investigation consisted of Dumet alloy wire leads sealed into a glass tube (see Figure 54). Such seals are prepared by first oxidizing the wires, coating them with sodium tetraborate glass, followed by sealing them into the glass tube. The diode assembly is subsequently subjected to an acid bath finishing operation to clean the exposed wires, which are then electroplated with tin. Accelerated testing of diode units showed that if the oxide remaining on the wires after the sealing and plating process was below a certain critical amount, which was dependent on the precise manufacturing conditions, premature failure of the component resulted.

It has also been observed that glass-to-metal seal components can exhibit variable leak rates under certain pressure conditions, and that such a fault can also lead to premature failure in service. A recent example of such a failure has been reported by Clarke and DerMarderosian (1999) for a glass-to-metal seal component used in a hybrid electronic packaging system. This component passed a standard helium leak test at $< 10^{-9}$ cm^{-3}s^{-1}, yet failed a residual gas analysis test when it was found to contain > 5000 ppm of moisture. The moisture content was also found to increase after 1000 hours ageing at 125°C. In an attempt to find the cause of this fault seals were tested using an arrangement that allowed pressurization of the seals both externally and internally whilst monitoring the opposite side of the seal for helium. It was noted that some seals were pressure sensitive to leak rate, suggesting the presence of pressure sensitive flaws. In the case of the failed package, it was considered that some leakage of moisture and air occurred during assembly of the package components due to the presence of a pressure sensitive flaw in the seal, but subsequently the glass-to-metal seal remained leak tight as determined by a standard helium leak test. It was believed that this accounted for the initial moisture content determined by the residual gas analysis test. The moisture content subsequently increased further during ageing due to reaction between residual oxygen with hydrogen outgassed by the package base metal. In another example, an investigation has been carried out by Cannon *et al*(1991) aimed at assessing the influence of crack growth characteristics and extension rates on the mechanical behaviour of glass/metal and ceramic/metal systems under conditions of both static and cyclic loading. Crack growth studies were performed on Cu/glass and alumina/Al–Mg alloy interfaces using the double cantilever beam technique. Moisture enhanced subcritical crack growth was noted for both systems and it was observed that cyclic loading increased interfacial crack growth rates substantially. It was stressed that this may have implications for the lifetime of microelectronic packages and related components. In a further example, Okabe *et al*(2002) and Takahashi *et al*(2002) have studied the reliability of strength and the effect on lifetime behaviour under neutron irradiation of metal–ceramic joints used in neutron detectors. These detectors employ a silicon nitride-to-stainless steel seal bonded using a Cu–Ti alloy braze. It was observed that under neutron irradiation swelling of the ceramic and work hardening of the metal interlayer occurred, this leading to delayed failure of the seal. A fracture mechanics analysis of the materials and geometry of the system predicted a 41 year lifetime for this component. It was noted that the properties of the seal could be improved through optimization of the interlayer thickness, which was determined to be of the order of 0·2 mm.

The time dependent failure of alumina/aluminium interfaces in the form of alumina layers bonded via a thin aluminium interface 5–100 μm thick has also been studied under both static and cyclic loading conditions (Kruzic *et al*, 2001). Crack growth rates were measured experimentally under various conditions. It was noted that under conditions of cyclic loading crack growth occurred predominantly by interfacial debonding, although for thicker interfaces cracks could deviate into the ceramic where they then propagated more rapidly. Thin interfaces exhibited higher fatigue thresholds, although for higher loads the failure mode could change to crack growth within the ceramic phase, this leading to lower crack growth resistance. Similar crack growth behaviour was noted under static conditions in a humid environment.

Other studies include analyses of glass-to-metal failures in integrated circuit components by Zhiting *et al*(2003), failure of metal/ceramic interlayers under cyclic and static loading conditions (Kruzic *et al*, 2004), failure of silicon-to-metal and borosilicate glass seals in microfluidic devices (Shim *et al*, 2003), and fatigue failure of turbine ceramic thermal barrier coatings (Mutoh *et al*, 2003).

It is very clear that there is a lack of systematic data on the lifetime behaviour of seal components employing glass, glass-ceramic, ceramic and metal systems, and on the failure mechanisms responsible. This is an important area ripe for further study, the ultimate aim being to be able to predict with more confidence a realistic lifetime for components under service conditions.

8. SEALING OF GLASSES AND GLASS-CERAMICS TO SPECIFIC METALS AND ALLOYS

In the following sections a comprehensive review is given of studies aimed at determining the factors responsible for providing satisfactory bonding to a number of specific metals and alloy systems.

8.1 Bonding to mild and low-alloy steels and iron

A considerable body of work has been reported on the bonding of glasses to Fe-based alloys, much of this being directly related to the early technological importance of porcelain enamels, and in particular the enamelling of cast irons and related materials. The work on enamels is very adequately covered elsewhere (e.g. Andrews, 1961; Wratil, 1984; Maskell and White, 1986).

More recently, Yang *et al*(2003) have studied in some detail the influence on the interfacial microstructure of enamel coatings on hot-rolled steel with and without the presence of a NiO enamel undercoat. Hot-rolled steel differs from low-carbon steel in having a higher carbon content and this can lead to problems associated with fishscaling due to the formation of hydrogen gas bubbles at the interface. The precise reason why the higher carbon contents acerbate this effect is not fully understood, although it seems to be related to the ability of lower carbon steels to retain higher hydrogen contents in solution at lower temperatures. It is known that the presence of a NiO undercoat can alleviate fishscaling. In their study, Yang *et al* observed that in the absence of an undercoat, bonding to the substrate was weak. On the other hand, the presence of a NiO undercoat promoted strong bonding, not unexpected in view of the well established adherence promoting abilities of NiO. Perhaps surprisingly, however, large bubbles were evident in the NiO undercoated samples, whilst fewer and smaller bubbles appeared in the samples that did not have an undercoat. The presence of large bubbles in the NiO samples and their relative absence in the samples without an undercoat appears contradictory as it would be expected that the samples with larger and more extensive bubbling would undergo fishscaling whilst the samples with fewer bubbles would not. This apparent contradiction was explained on the basis that the larger bubbles present in the samples with an undercoat contained gases at a lower pressure than necessary for the occurrence of fishscaling. Contrary to expectation large bubbles may therefore be beneficial at least in the special case of fishscaling by reducing the overall pressure within the individual bubbles.

Glass-ceramic enamels have been applied to metals for a number of years

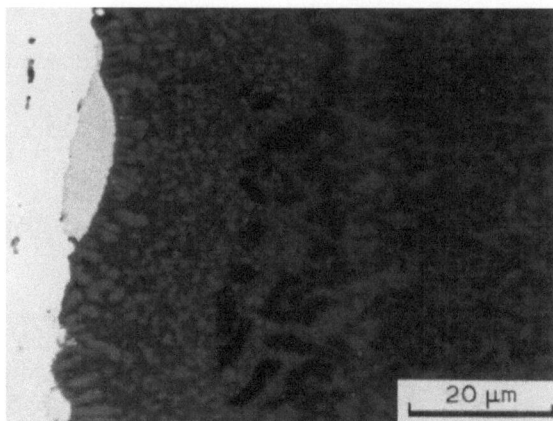

Figure 55: Coating of lithium silicate glass-ceramic on pre-oxidized mild steel showing the presence of a dendritic phase within the glass-ceramic (after Sturgeon et al, 1986)

(Penkov and Gutzow, 1991). These enamels contain a significant amount of crystalline phase and take advantage of the improvement in properties associated with these glass-ceramic materials. They normally contain a proportion of such crystalline phases as quartz, tridymite, cristobalite, perovskite, wollastonite, celsian and forsterite, and more recently sanbornite, $BaO.2SiO_2$.

Work on sealing of glasses to Fe–Ni and Fe–Ni–Co alloys, e.g. Nilo-K, for electrical applications has also been extensively reported, with the early work fully covered by Partridge (1949), this forming an important and comprehensive part of his monograph.

More recently, a number of papers have reported on the bonding of glasses or glass-ceramics to mild and low alloy steels (Sturgeon *et al*, 1986; Nicholas, 1986). In the work by Sturgeon *et al*(1986) a low alloy mild steel was coated with a lithium aluminosilicate glass-ceramic containing small additions of K_2O, ZnO and P_2O_5. The metal substrate was pre-oxidized prior to coating to yield a layer of $FeO + Fe_2O_3$ approximately 15 μm thick on the mild steel. During subsequent firing, it was observed that rapid diffusion of Fe^{2+} into the coating occurred, resulting in the formation of a dendritic lithium iron silicate phase (possibly $Li_2FeO.8ZnO.2SiO_4$) within the interfacial reaction zone. This zone extended > 40 μm into the coating, as shown in Figure 55. Diffusion of this nature can lead to the formation of a composition gradient and a corresponding variation in thermal expansion coefficient within the interfacial region, as noted in Figure 56.

Nicholas(1986) examined the interaction of sodium disilicate glass with Fecralloy (an Fe–6Cr–4Al–0·25Y alloy) in the temperature range 900 to 1100°C under vacuum using the sessile drop technique. Both unoxidized and pre-oxidized samples (with a layer of Al_2O_3 2 to 6 μm thick) were employed. Not unexpectedly, severe de-gassing of the glass was observed during these experiments (glass prepared under conditions of ambient pressure contain dissolved gases which, if the glass is subsequently remelted under reduced pressure will give

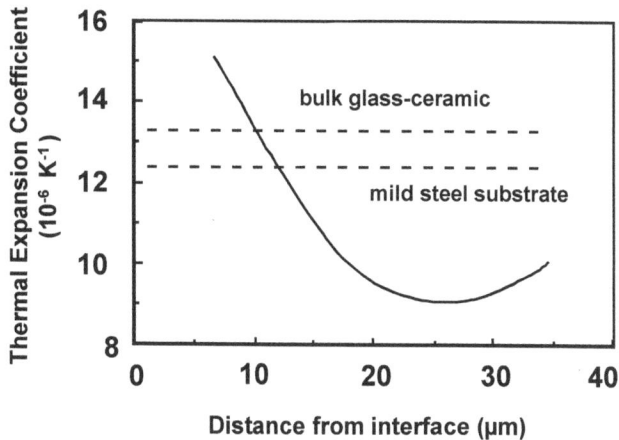

Figure 56: Variation in thermal expansion coefficient within the interfacial region for a seal between mild steel and a glass-ceramic due to diffusion of Fe^{2+} from the metal substrate into the glass-ceramic (after Holland, 1995)

rise to the evolution of bubbles within the glass). Using unoxidized material, stable contact angles of 38° to 47° were obtained, depending on the temperature, whilst with pre-oxidized samples excellent wetting with contact angles of 2° to 6° were achieved. It was noted that the predominant redox reaction when using unoxidized metal was:-

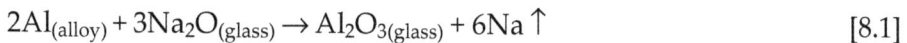

$$2Al_{(alloy)} + 3Na_2O_{(glass)} \rightarrow Al_2O_{3(glass)} + 6Na \uparrow \qquad [8.1]$$

For the pre-oxidized samples, on the other hand, wetting and bonding was achieved via the Al_2O_3 oxide film. Due to the sealing conditions employed, i.e. under vacuum, it is not clear to what extent bubbling of the glass used for sealing to the unoxidized metal is due to the evolution of Na vapour and whether or not seals contained more bubbles as a result of this additional source of gas bubbles.

Reasonably refractory glass-ceramic coatings (stable to 750–800°C) have been reported for the protection of cast iron tubes used in the melting of Al alloys (Penkov and Gutzow, 1991). Two different families of glass-ceramics were investigated in this work. One, based on the K_2O–BaO–Cr_2O_3–NiO–FeO–SiO_2 system, was nucleated with MoO_3 and contained barium disilicate as the major crystalline phase. Alternative glass-ceramics were prepared in the Li_2O–Na_2O–ZnO–P_2O_5–NiO–SiO_2 system. These glasses were also nucleated using MoO_3, but contained cristobalite as the major crystalline phase.

8.2   Bonding to high alloy and stainless steels

The formation of high quality, durable seals to stainless and related high alloy steels initially proved difficult, due in part to the relatively high thermal expansion of the majority of these alloys (typically ≈ 17 to 22×10^{-6} K^{-1} for conventional stainless steels). Nevertheless, there have been a number of reports on the bonding of glasses and glass-ceramics to a variety of alloy steels, with varying degrees of success (Metcalfe *et al*, 1991; Sturgeon *et al*, 1986; Birkbeck *et al*, 1987; Cassidy and Fagin, 1987; Moddeman *et al*, 1989a, 1989b; Cassidy and Moddeman, 1989).

Birkbeck *et al*(1987) and Cassidy and Fagin(1987), for example, prepared glass-ceramic seals to a precipitation hardenable Al-containing stainless steel with a thermal expansion coefficient $\approx 17\times10^{-6}$ K^{-1}. Although historically, and in practice, metal parts are normally pre-oxidized prior to any sealing or coating operation, a separate pre-oxidation schedule is not always necessary, nor indeed desirable. In this particular study, both unoxidized and pre-oxidized metal samples were sealed to a lithium silicate glass-ceramic containing small additions of K_2O, Al_2O_3, B_2O_3 and P_2O_5 using a standard heat-treatment schedule. The best seals were obtained using the pre-oxidized metal, but only if a vacuum annealing stage was carried out prior to sealing. If the vacuum treatment was not carried out, a porous interface was obtained during sealing. This effect was believed to be due to the presence of water adsorbed onto the Al_2O_3 layer. Sealing was also attempted between the glass-ceramic and a 304L stainless steel substrate. In this instance, however, successful sealing could not be achieved, possibly due to the higher thermal expansion of this particular stainless steel relative to the Al-containing alloy. Subsequently, Cassidy and Moddeman(1989) and Moddeman *et al*(1989a, 1989b) reported the successful bonding of a similar glass-ceramic to 304L stainless steel. In this study, significant differences in the microstructure of the interfacial zone were noted, relative to that of the bulk glass-ceramic, depending on whether or not the alloy samples were pre-oxidized prior to sealing. In the case of unoxidized samples, a reaction zone containing Fe and Cr phosphide precipitates was obtained extending ≈30 μm into the glass-ceramic with a further diffusion zone containing Cr extending ≈200 μm into the glass-ceramic. In the reaction zone the crystalline structure of the glass-ceramic was disrupted due to the removal of the P_2O_5 nucleating agent through reaction with Fe and Cr, with a high proportion of glassy phase noted in this region. When using pre-oxidized samples, on the other hand, only a very narrow diffusion zone was apparent, and there was an absence of phosphide precipitates. In addition, the glass-ceramic retained its fine crystalline microstructure up to the interface. In this instance, the oxidized layer acts as a diffusion barrier, preventing excessive diffusion of metal ions into the glass-ceramic.

Knorovsky *et al*(1991) have also prepared seals to 304L stainless steel using a similar lithium silicate glass-ceramic to that employed in the previous study. The heat-treatment schedule was chosen to give a mixture of lithium metasilicate, lithium disilicate and tridymite, with the desired thermal expansion characteristics. Unexpectedly, despite the apparently favourable thermal expansion behaviour, it was found that seals failed during cooling. The reason for the failures was traced to excessive diffusion of Cr from the metal immediately adjacent to the interface into the glass-ceramic. This depletion of Cr (from $\approx12\%$ Cr within

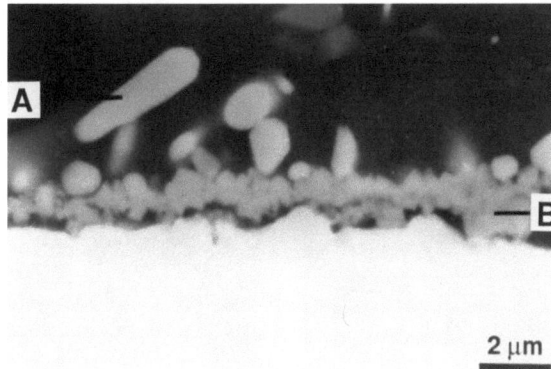

*Figure 57: Micrograph of a lithium zinc silicate glass-ceramic-to-stainless steel
interface showing the presence within this region of iron phosphide precipitates (A)
and a semi-continuous chromium silicide based precipitated phase (B)
(after Donald, 1993)*

the bulk of the metal to ≈5% at the interface with the glass-ceramic) resulted in the partial transformation of the fcc structure of the austenitic steel into a bcc phase immediately adjacent to the interface. Not only is this particular phase transformation accompanied by a large volume contraction (≈2%) but, in addition, the bulk material possesses a lower thermal expansion coefficient. Hence, the thermal expansion of the metal close to the interface was no longer matched to that of the glass-ceramic, and this resulted in failure at the interface. A number of solutions to this problem were suggested, including plating the surface of the stainless steel with Cr or Ni to act as a barrier against the diffusion of Cr from the bulk metal, or alternatively pre-oxidation.

Metcalfe *et al*(1991) and Donald(1993) have studied the sealing of a lithium zinc silicate glass-ceramic containing a high proportion of ZnO to 321 stainless steel. A standard sealing and heat-treatment schedule of 5 minutes at 950°C followed by 60 minutes at 465 or 585°C and 60 minutes at 800 to 1000°C was employed. Sealing was carried out in argon using unoxidized metal samples. After sealing, a relatively thin interfacial reaction zone was apparent, extending ≈4 to 5 μm into the bulk glass-ceramic. This zone consisted of a semi-continuous precipitate of a phase which EDS revealed to contain Cr and Si adjacent to the interface (possibly a complex mixture of CrO or Cr_2O_3 with Cr_3Si_2 or $LiCr(SiO_3)_2$), whilst acicular crystals rich in Fe and P (an iron phosphide phase) were found within the glass-ceramic close to the interface. Due to the partial removal of the P_2O_5 nucleating agent from the interfacial region, this zone contained a high proportion of glassy phase which extended ≈5–6 μm into the bulk glass-ceramic. A micrograph is shown in Figure 57 which clearly shows the presence of precipitated phases within the interfacial zone. It was noted that pre-oxidation of the substrate minimised the problem associated with the formation of these undesirable precipitated phases, as shown in Figure 58. The presence of a discrete oxide layer helps to

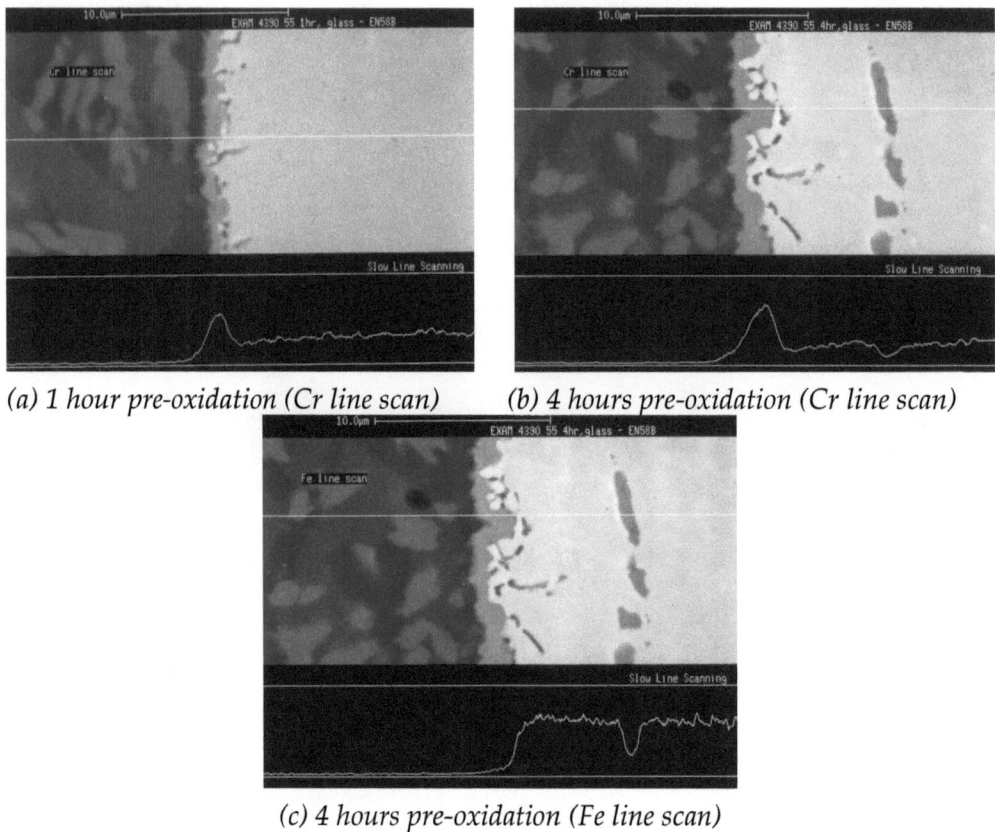

(a) 1 hour pre-oxidation (Cr line scan)   (b) 4 hours pre-oxidation (Cr line scan)

(c) 4 hours pre-oxidation (Fe line scan)

Figure 58: Micrographs of a lithium zinc silicate glass-ceramic bonded to pre-oxidized stainless steel. EDS line scans for Cr and Fe

minimise the diffusion of the Fe and Cr from the metal in the timescales involved. Figure 58(a) shows elemental line scans across the interface and shows the oxide layer to be rich in Cr with little diffusion of Cr across the oxide/glass-ceramic interface and into the bulk glass-ceramic. Similarly, diffusion of Fe across the interface is low. As noted in Figure 58(b) increasing the pre-oxidation time to 4 hours results in a thicker oxide layer. Also evident is enhanced grain boundary corrosion at this longer oxidation time, possibly caused by reaction of Zn from the glass-ceramic with the metal substrate. Diffusion of Cr into the bulk glass-ceramics remains low. The influence of a number of transition metal oxide additions on the sealing behaviour of the glass-ceramics employed for bonding to unoxidized stainless steel was also investigated, with additions including CuO, NiO, Cr_2O_3, WO_3 and MoO_3. It was noted that CuO resulted in the formation of a thinner reaction zone, relative to the seals which did not contain this addition, and there was a marked absence of phosphide phases in this zone. Bonding was achieved via a very thin (≈ 0.2 to 0.5 μm) layer which EDS revealed to be rich in Cr. Seals made with the glass-ceramic containing NiO also exhibited a narrow reaction zone rich in Cr, but differed in the presence of small (≈ 1–2 μm) particles

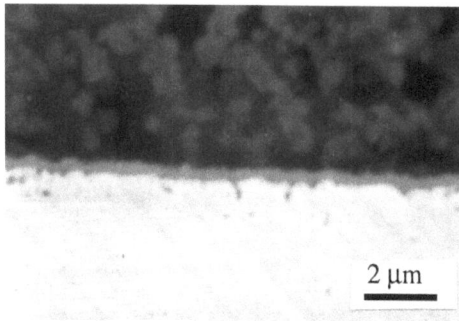

(a) CuO

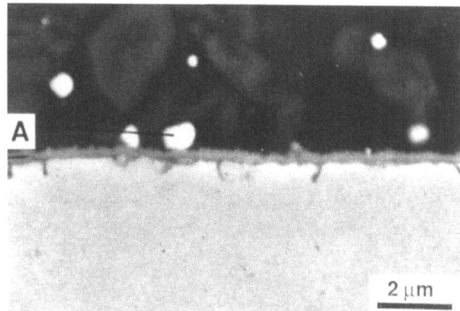

(b) NiO A = precipitate rich in Ni

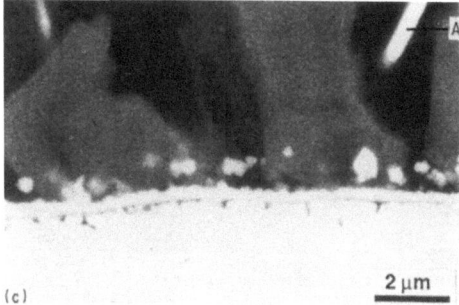

(c) MoO₃
A = precipitate rich in Mo and Si;
possibly Mo₃Si

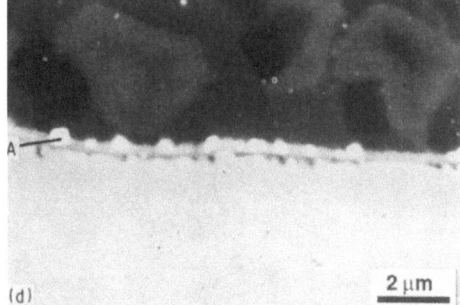

(d) WO₃
A = precipitate rich in W and Si;
possibly W₃Si

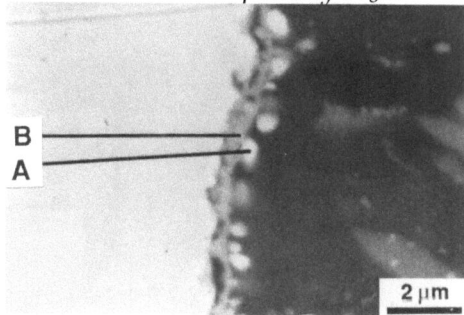

(e) Cr₂O₃
A = precipitates rich in Fe and P; possibly FeP
B = layer rich in Cr; most likely Cr₂O₃

Figure 59: Micrograph of a lithium zinc silicate glass-ceramic-to-stainless steel
interface for glass-ceramics containing a variety of transition metal oxides
(after Donald, 1993)

rich in Ni dispersed both along the interface and within the glass-ceramic close
to the interface. Seals made with the WO_3-containing glass-ceramic were very
similar to those containing NiO, exhibiting a very thin interfacial layer rich in
Cr, and a dispersion of particles along the interface which were rich in both W
and Si (possibly WSi_2 or more likely W with some Si dissolved in it). The seals
containing MoO_3 also exhibited a thin interfacial layer but this was very rich in

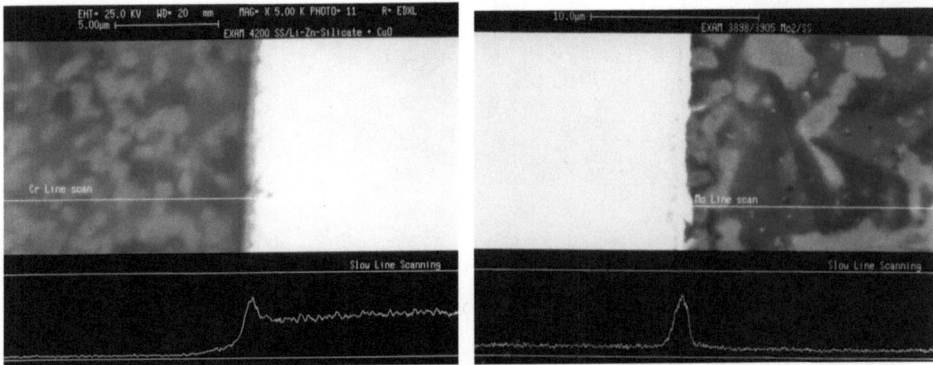

(a) CuO addition, Cr line scan

(b) MoO₃, Mo line scan

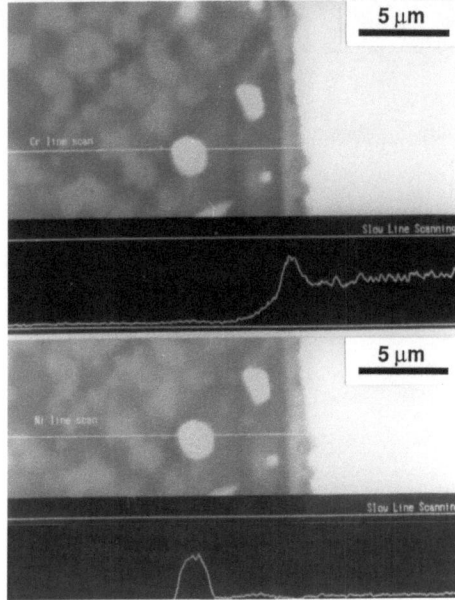

(c) NiO, Cr and Ni line scans

Figure 60: Micrograph of a lithium zinc silicate glass-ceramic-to-stainless steel interface for glass-ceramics containing a variety of transition metal oxides. EDS line scans for Cr, Mo and Ni
(after Donald, 1993)

Mo, rather than Cr. Particles rich in both Mo and Si (possibly $MoSi_2$ or a Mo–Si solid solution) were dispersed along the interface and within the glass-ceramic close to the interface. The TMO additions appear to influence the diffusion of Fe and Cr into the bulk glass-ceramic, and to hinder the formation of phosphide and related phases. In this respect, they are beneficial by preventing the depletion of the P_2O_5 nucleating species from the interfacial zone. Micrographs are shown in Figure 59. Some of the corresponding EDS line scans are shown in Figure 60. This extended study clearly emphasises the influence of either a pre-oxidation stage or the use of specific additives to the starting glass in promoting the formation

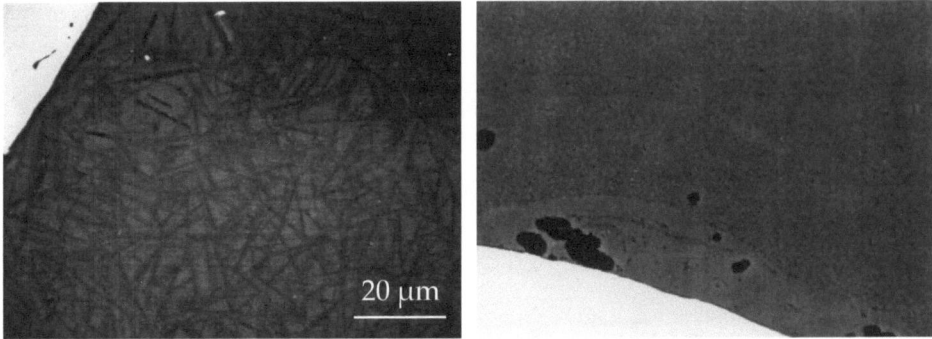

(a) Nilo 48 alloy sealed to AS8
glass-ceramic

(b) Nilo 48 alloy sealed to AS7
glass-ceramic

(c) AISI 304 stainless steel sealed to AS7 glass-ceramic
Figure 61: Micrographs of lithium aluminosilicate glass-ceramics sealed to
iron-based alloys

of acceptable interfaces through control of the interfacial chemistry and diffusion of metallic species.

A lithium aluminosilicate glass-ceramic, LAS, has also been employed at AWE for bonding to stainless steel, as noted in Figure 61, which shows the interfaces of seals to Nilo 48 alloy and AISI 304 stainless steel. In these instances, a very much 'cleaner' interface is obtained relative to the lithium zinc silicate glass-ceramic. This suggests that the LAS glass-ceramics are significantly less reactive than the lithium zinc silicate compositions.

Mantel(2000) investigated sealing of a glass to titanium-stabilized ferritic stainless steel. In this study, it was observed that a double oxide layer is formed with the outer layer composed of a spinel phase and TiO_2 which dissolves in the glass during sealing, whilst the innermost layer next to the metal is rich in Cr_2O_3. Although the presence of Cr_2O_3 at the interface promotes bonding it was noted that counteracting this beneficial effect, Si may diffuse from the steel into the interfacial region and lead to a reduction in seal strength.

In the case of Fe–Cr alloys Sturgeon et al(1986) have examined the coating behaviour of an Fe–17% Cr alloy (with a moderately low $\alpha \approx 12 \times 10^{-6}$ K^{-1}) with a lithium aluminosilicate glass-ceramic nucleated with P_2O_5. The thermal expansion

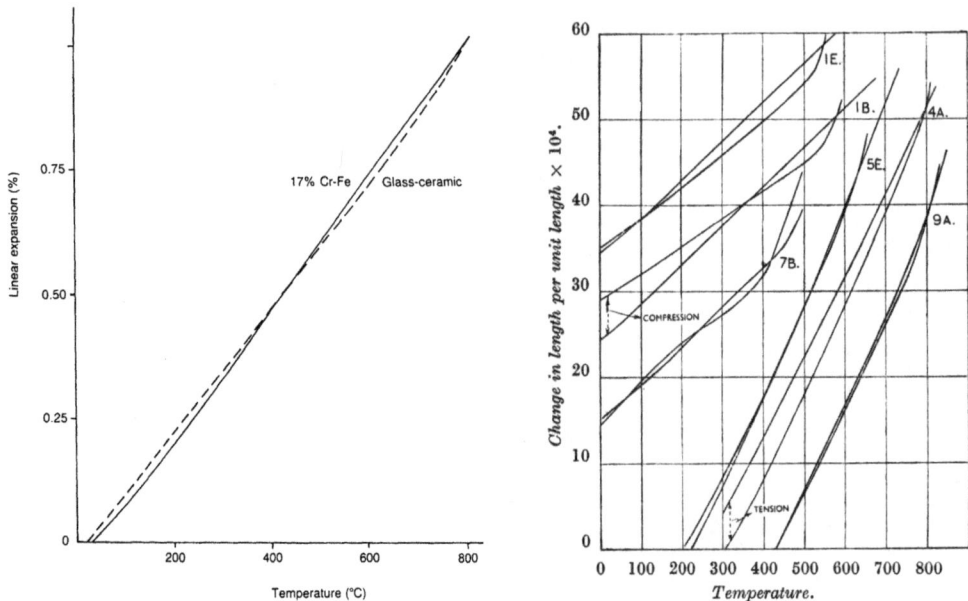

(a) 17%Cr–Fe alloy and a glass-ceramic   (b) A selection of metals and alloys and
(after Partridge, 1990)   matched glasses (after Partridge, 1949)

Figure 62: Thermal expansion curves for a selection of metals and alloys with expansions
matched to those of glasses and glass-ceramics

curve for a 17%Cr–Fe alloy compared with a glass-ceramic developed for sealing to this alloy is shown in Figure 62(a). The metal substrate was pre-oxidized prior to coating to yield a layer of $Cr_2O_3 \approx 1$ μm thick. It was noted that strong bonding to the substrate was achieved via this Cr_2O_3 layer, which remained substantially unaltered during the coating process. Only limited diffusion of Cr^{3+} into the coating was observed, this extending to a depth of the order of 4 μm. The coating of an Fe–17% Cr alloy with a glass-ceramic containing a dispersion of SiC particles has been reported by Hyde and Partridge(1990), this study showing that mechanically stronger coatings can be obtained through the particle reinforcement mechanism. The thermal expansion of the glass-ceramic matrix was matched to that of the substrate in this instance, the major crystalline phases present being quartz and lithium disilicate.

In addition to the use of 'conventional' silicate compositions for sealing to stainless steel, limited work has been reported on the use of high expansion phosphate-based glasses. For example, Chambers et al(1989) employed phosphate-based glasses containing additions of Na_2O, K_2O, BaO and Al_2O_3 for sealing to 304L stainless steel. This work did indicate, however, that although the expansion of the glass could be matched reasonably well to yield mildly compressive seals, failure of some seals occurred. The failures were explained on the basis of tensile stresses generated in the glass due to viscoelastic relaxation during cooling. These stresses were not predicted using conventional elastic theory.

High thermal expansion phosphate glasses based on the K_2O–Al_2O_3–(Fe_2O_3)–P_2O_5 system, and reported to exhibit reasonable chemical durability, have also been suggested as potential candidates for sealing to stainless steel (Peng and Day, 1991a, 1991b). The composition of the overall most suitable glass was found to be $15Na_2O$–$15K_2O$–$10BaO$–$10Al_2O_3$–$50P_2O_5$, with a softening temperature of 405°C, expansion coefficient $19 \cdot 2 \times 10^{-6}$ K^{-1} and a durability in water at 30°C of < 10 mg $m^{-2}day^{-1}$.

8.3   Bonding to other Fe-based alloys

A number of alternative Fe-based alloys have been used extensively in the electrical and electronics industries for the preparation of glass-to-metal seal devices. In particular, relatively low expansion binary Fe–Ni and ternary Fe–Ni–Co alloys have been used for many years (originally as a substitute for Mo and W) because their thermal expansion characteristics match those of a number of common "hard" borosilicate-type glasses (Partridge, 1949; Price, 1984; Kramer *et al*, 1985; McMillan and Hodgson, 1966). These alloys include the Nilo series, e.g. Nilo-K. Expansion curves of some Fe–Ni alloys are given in Figure 6(a). A comparison of the expansion behaviour of some of these alloys with glasses developed for sealing purposes is given in Figure 62.

8.4   Bonding to Ni-based alloys

There have been a large number of studies aimed at the preparation of glass and glass-ceramic seals and coatings to Ni-based alloys. Some of the most important of these studies are described below.

Sealing of glass to Ni–Cr based alloys used in dental applications has been studied by Fairhurst *et al*(1985). It was noted that strong metal oxide adhesion to the alloys was necessary in order to form a satisfactory product. Kelsey *et al*(1981) reported the interaction of a barium silicate glass with Inconel X-750, a Ni–Cr–Fe alloy containing 2·5% Ti and minor additions of Al and Nb. Glass was allowed to react with the metal at 1100°C in various partial pressures of oxygen, and the resulting interfaces were examined using scanning electron microscopy. It was observed that pre-oxidation of the metal in a partial pressure of oxygen of 10–14 Pa led to the formation of a $NiTiO_3$–Cr_2O_3 solid solution oxide layer which is strongly bonded to the metal substrate. It was noted that the glass bonded well to this oxide layer. Pre-oxidation of the metal at higher oxygen pressures, on the other hand, resulted in the formation of a poorly adherent layer containing TiO_2 and $NiCr_2O_4$ which produced non-hermetic bonds to the glass.

Tomsia and Pask(1986) reported the sealing of a potassium aluminosilicate glass to two 80Ni–20Cr alloys, one prepared from high purity (99·99%) metals, the other a commercially available alloy containing 1·5% Si together with traces of Al, Fe and Mn. The alloys were pre-oxidized prior to sealing. It was noted that the high purity alloy formed a multi-layer oxide scale, whilst the commercial alloy produced a single mixed scale with the outer portion richer in Cr_2O_3. Subsequent sealing produced good quality coatings in each case. The seal to the pure alloy was, however, mechanically weaker with fracture occurring through the oxide layer. This difference in behaviour was explained by the presence of a high degree of porosity in the oxide layer of the commercial alloy, this allowing glass to reach the metal substrate and react directly with it. Saturation of the glass with Cr_2O_3 was maintained through dissolution of oxide from the scale. This study illustrates the large effect that relatively small amounts of impurities may sometimes have on the resultant properties of glass-metal systems.

McCollister and Reed(1983) employed a lithium silicate glass-ceramic containing small additions of Al_2O_3, Li_2O, K_2O and B_2O_3 together with P_2O_5 nucleating agent to prepare hermetic seals to the Ni-based superalloys Inconel 625 and 718, and Hastelloy C276. Sealing was carried out using a three-step heat-treatment schedule consisting of a high temperature stage at 1000°C to melt the glass, a low temperature hold at 650°C to initiate nucleation, and an intermediate temperature hold at 800 to 850°C to crystallize the glass. In order to achieve the correct thermal expansion for sealing to the Ni-based alloys, the glass-ceramic produced using this treatment contained a high proportion of cristobalite (up to ≈30%). Further work by Headley and Loehman(1984) showed that crystallization of the glass occurred by epitaxial growth of lithium metasilicate, lithium disilicate and cristobalite onto relatively large (0·1 to 1·0 μm) Li_3PO_4 nuclei which formed in the glass mainly during the high temperature sealing stage (rather than in the classical nucleation stage). An example is shown in Figure 14(b). This work indicated that the use of a conventional low temperature nucleation stage could in fact be superfluous to the production of the microstructure required for this particular application.

In order to analyse the detailed interfacial chemistry of this type of glass-ceramic seal to Inconel 718 alloy, Haws et al (1985) studied seals that had been prepared under prolonged heating at the sealing temperature (i.e. 66 hours at 1000°C). The resultant seals were studied by scanning electron microscopy, wavelength dispersive x-ray analysis and x-ray photoelectron spectroscopy, XPS. Many particles were observed in the glass-ceramic which were identified as metal phosphides of stoichiometry close to M_2P, where M = Cr, Ni, Nb and Fe. It was noted that the chromium phosphide particles extended well into the bulk glass-ceramic, whilst the other metal phosphides only occurred very close to the interface between the metal and the glass-ceramic. It was also found that many metallic species, including Ni, Nb, Ti, Cr and Fe, were dissolved in the residual glassy phase, glass being present near to the interface due to the depletion of the P_2O_5 nucleating species in this region. The oxidation states of these metals were determined using XPS. It was observed that niobium was present as Nb^{5+}, titanium as Ti^{4+}, Fe and Cr were in mixed oxidation states, and Ni was apparently present in the zero-valent state. Further evidence for the presence of zero-valent metallic species in glass-ceramic seals was reported by Moddeman *et al*(1985). In addition to the XPS data, direct evidence for the presence of zero-valent Ni has also been found using high resolution TEM where small particles ≈100–200 nm in size, identified as metallic Ni by electron diffraction, were observed within the glass-ceramic.

Studies by Loehman and co-workers using the same glass-ceramic/metal system as that employed by Moddeman (Loehman, 1987, 1989; Headley *et al*, 1986; Loehman *et al*, 1986; Watkins and Loehman, 1986) indicated that during the sealing operation the glass reacts directly with the Inconel 718 by a redox mechanism, presumably after having dissolved the thin (20 to 200 nm) layer of oxide which is present on its surface. It was confirmed that reaction occurs in the interfacial zone between Cr, which diffuses into the glass from the metal, and the lithium phosphate nuclei which are present in the crystallizing glass, to form chromium phosphide. The following reactions were postulated (Watkins and Loehman, 1986):-

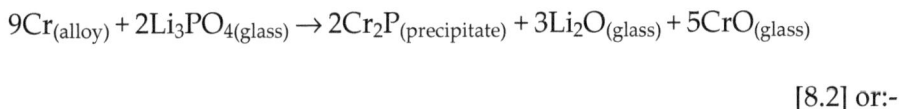

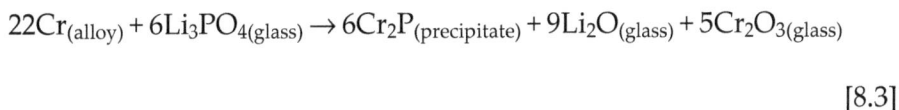

$$9Cr_{(alloy)} + 2Li_3PO_{4(glass)} \rightarrow 2Cr_2P_{(precipitate)} + 3Li_2O_{(glass)} + 5CrO_{(glass)}$$

$$[8.2] \text{ or:-}$$

$$22Cr_{(alloy)} + 6Li_3PO_{4(glass)} \rightarrow 6Cr_2P_{(precipitate)} + 9Li_2O_{(glass)} + 5Cr_2O_{3(glass)}$$

$$[8.3]$$

This was later revised by Loehman(1987) to:-

$$59Cr_{(alloy)} + 14Li_3PO_{4(glass)}$$
$$\rightarrow 2Cr_{12}P_{7(precipitate)} + 21Li_2O_{(glass)} + 35CrO_{(glass)} \qquad [8.4] \text{ or:-}$$

$$142Cr_{(alloy)} + 42Li_3PO_{4(glass)}$$
$$\rightarrow 6Cr_{12}P_{7(precipitate)} + 63Li_2O_{(glass)} + 35Cr_2O_{3(glass)} \qquad [8.5]$$

These reactions destroy the nuclei with the consequence that during subsequent crystallization of the glass this depletion of the nucleating species results in the formation of a coarse crystalline microstructure with properties, including thermal expansion characteristics, which differ markedly from that of the bulk glass-ceramic. In particular, thermal expansion is lowered (Notis, 1962) and this has a detrimental effect on seal quality and life. It has also been noted (Headley *et al*, 1986) that reaction can occur at the alloy grain boundaries in contact with the glass during sealing to form an intermetallic silicide phase, $M_6Ni_{16}Si_7$, where M may be a mixture of the alloy constituents, e.g. Fe, Mo and Nb.

In the case of Hastelloy C276 a different set of redox reactions occur (Loehman *et al*, 1986; Loehman, 1987). This alloy contains a significant proportion of Mo (15 to 17%) which, together with Cr, diffuses into the glass during sealing. This results in the formation of precipitates of molybdenum phosphide and lithium chromium silicate according to the reaction:-

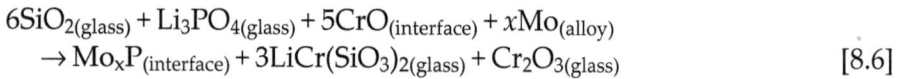

$$6SiO_{2(glass)} + Li_3PO_{4(glass)} + 5CrO_{(interface)} + xMo_{(alloy)}$$
$$\rightarrow Mo_xP_{(interface)} + 3LiCr(SiO_3)_{2(glass)} + Cr_2O_{3(glass)} \qquad [8.6]$$

The lithium phosphate nuclei are similarly destroyed by this mechanism, again leading to the formation of a coarse interfacial zone with substantially different properties from that of the bulk glass-ceramic.

In addition to the possibility of many reactions taking place between diffusing metal species and the glass or glass-ceramic constituents during the formation of glass- and glass-ceramic-to-metal seals and coatings, bubbles are also often formed in the interfacial zone. The formation of bubbles has normally been attributed to reaction of the glass with the metal or with surface contaminants during sealing to give gaseous reaction products, or to outgassing of either the metal or the glass. One common mechanism involves reaction of the glass with carbides present in the metal to give CO or CO_2. This effect can be eliminated by de-carburizing the metal surface prior to sealing or coating (Notis, 1962). It has also been noted in the case of some of the Ni-based superalloys bonded to a lithium silicate glass-ceramic that reaction can occur between metal species diffusing into the glass from the metal substrate and water present in the starting glass (Craven et al, 1986; Haws et al, 1985). In a number of cases large bubbles were observed in the glass-ceramic after sealing. Leading on from these observations it was pointed out that thermodynamic data suggest that hydrogen gas formation may in fact be responsible for the occurrence of bubbles in a very wide range of seal systems. The metals, M, interact with this absorbed water to form hydrogen gas according to the reaction:-

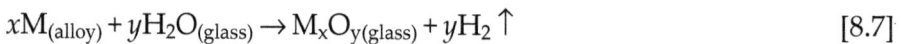

$$xM_{(alloy)} + yH_2O_{(glass)} \rightarrow M_xO_{y(glass)} + yH_2 \uparrow \qquad [8.7]$$

Thermodynamic data for specific reactions have been calculated and are summarized in Table 12. The data are computed from the information given in the publication by Barin et al(1973). It is clear from this Table that reaction of water with many metallic species, e.g. Fe, Mo, Mn, Cr, Nb, Ta, Ti, Y, Al, etc, is thermodynamically favourable, whilst reaction with other metals, e.g. Ni, Co and Cu, is not. It is noteworthy that reaction with Cr, which is present in high concentrations in most superalloy systems, is particularly favourable thermodynamically through the reaction:-

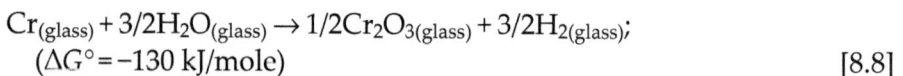

$$Cr_{(glass)} + 3/2H_2O_{(glass)} \rightarrow 1/2Cr_2O_{3(glass)} + 3/2H_{2(glass)};$$
$$(\Delta G^\circ = -130 \text{ kJ/mole}) \qquad [8.8]$$

In order to confirm the hypothesis that reaction with dissolved water can lead to hydrogen gas bubble formation, Haws et al(1985) prepared seals to Inconel 718 using glass that had been melted under standard atmospheric conditions and these were compared with seals made from glass that had been prepared by bubbling dry nitrogen through the melt. Glass-ceramic seals fabricated from the glass prepared under ambient conditions were found to contain a greater degree of porosity than those fabricated employing the alternatively prepared glass. In addition, it was observed that seals prepared at increased pressure, by hot isostatic pressing at 170 MPa, were relatively bubble free. It is likely that this over-pressure is high enough to suppress the formation of pores by keeping any hydrogen formed in solution in the glass. In addition to lending clear support to the detrimental influence of water on bubble formation in seals this work also showed that the crystallization behaviour of the glass at increased pressure differed substantially from the behaviour under ambient conditions. Sealing at ambient pressure gave a mixture of lithium metasilicate, lithium disilicate and cristobalite, whilst sealing at increased pressure gave lithium metasilicate and quartz. This study was extended by Craven et al(1986) by employing glasses that had been melted and cast under a variety of different atmospheric conditions so that they contained different proportions of dissolved water, the approximate amounts of water being determined by infra-red analysis. Seals prepared using glass that had been prepared under high relative humidity atmospheric conditions were observed to contain bubbles. The gas within these bubbles was identified as hydrogen by gas chromatographic analysis. Seals prepared employing glass that had been melted and cast under dry room conditions, on the other hand, were relatively free from bubbles. On the basis of thermodynamic data, it was proposed that as an alternative to preparing sealing glass under dry conditions (which is not always practical) it may be feasible to incorporate into the glass batch additions of suitable additives that will react preferentially with Cr, so eliminating or minimising the Cr–water reaction. Such additives include various metal oxides, e.g. CuO (Table 5) which promotes the reaction:-

$$Cr + 3CuO \rightarrow 1/2Cr_2O_3 + 3/2Cu_2O; (\Delta G° = -381 \text{ kJ/mole}) \quad\quad [8.9]$$

This reaction has a ΔG value of -380 kJmol^{-1}, compared to -130 kJmol^{-1} for the Cr–water reaction. Chromium should therefore react preferentially with the CuO instead of the dissolved water in the glass. This behaviour was confirmed experimentally by adding 1 wt% CuO to the glass batch prior to melting followed by casting under normal atmospheric conditions to yield a "wet" glass. Sealing with this glass led to the production of bubble-free, high quality seals with Inconel 718. It was also demonstrated experimentally that doping the glass with products of equation [8.7], e.g. 1 wt% Cr_2O_3 in the case of Cr-containing metals (eqn. [8.8]), could shift the equilibrium to the left and so reduce hydrogen bubble formation.

The coating characteristics of a lithium silicate glass-ceramic containing small additions of K_2O, ZnO and P_2O_5 to a Ni–Co–Cr–Mo alloy, Nimonic 263, have been examined in considerable detail by Holland and co-workers (Hong and Holland, 1989a, 1989b; Holland et al, 1990; Hong, 1991). The glass-ceramic was applied to both unoxidized and pre-oxidized metal surfaces in the form of a dispersion of glass particles ≈10 μm in size in an organic binder employing a screen print-

(a) Unoxidized substrate showing a coarse glass-ceramic microstructure

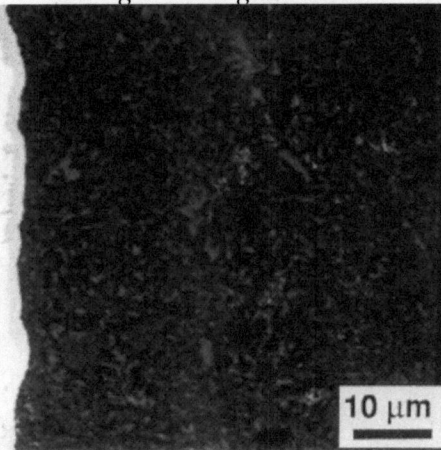

*(b) Pre-oxidized substrate showing an adherent oxide interfacial region and a fine
grain size glass-ceramic microstructure*
*Figure 63: Glass-ceramic coating on a nickel-based superalloy illustrating the
influence of an oxide layer formed by pre-oxidation of the metal substrate on the seal
quality (after Hong and Holland, 1989)*

ing technique. The coatings were then fired either in nitrogen or air at various
temperatures in the range 800 to 1100°C. An optimum temperature of 980°C was
subsequently chosen, this providing adequate wetting of the metal substrate by
the glass without the occurrence of excessive chemical reaction. It was found that
totally different reaction behaviour was observed, dependent on whether or not
the substrate had been pre-oxidized. Examples are shown in Figure 63.

For a substrate that had not been subjected to a pre-oxidation treatment, the
following reactions were suggested. Firstly, Cr at the alloy interface reacts with
ZnO in the glass to form a CrO layer at the interface; the reduced metallic Zn

then alloys with the substrate:-

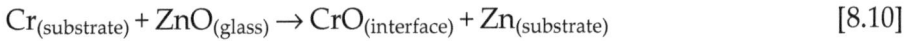

$$Cr_{(substrate)} + ZnO_{(glass)} \rightarrow CrO_{(interface)} + Zn_{(substrate)} \qquad [8.10]$$

(It should be noted that metallic zinc could be present as vapour at the coating temperature suggesting that in the case noted here rapid alloying of the Zn vapour with the substrate occurs to minimize loss of Zn from the system). So long as the CrO layer remains, the interface will be saturated with the substrate oxide and a strong chemical bond should exist. As firing continues, however, the rate of reaction [8.10] will decrease due to a build up of reaction products. This is also coupled with a fast diffusion rate for Cr^{2+}. Under these conditions, the CrO will dissolve rapidly into the glass and the CrO layer will disappear. Consequently, the glass is then again brought into direct contact with the substrate; however, further oxidation of the substrate through reduction of ZnO in the glass is now restricted due to the depletion of Zn^{2+} in the interfacial region. Under the driving force of chemical equilibrium, further redox reactions will take place between Cr^{2+} which has diffused into the glass from the substrate and the P_2O_5 nucleating agent present in the glass. The following reactions were suggested:-

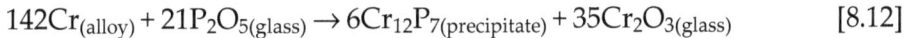

$$59Cr_{(alloy)} + 7P_2O_{5(glass)} \rightarrow 2Cr_{12}P_{7(precipitate)} + 35CrO_{(glass)} \qquad [8.11] \text{ or}$$

$$142Cr_{(alloy)} + 21P_2O_{5(glass)} \rightarrow 6Cr_{12}P_{7(precipitate)} + 35Cr_2O_{3(glass)} \qquad [8.12]$$

These reactions result in the formation of acicular precipitates within the interfacial region. They are highly undesirable because they remove the nucleating agent from the interfacial zone. During subsequent crystallization of the glass this depletion of P_2O_5 results in the formation of a coarse crystalline microstructure at the interface with properties, including thermal expansion, that may differ markedly from that of the bulk glass-ceramic. Redox reactions also occur at the coating surface leading to the formation of $LiCr(SiO_3)_2$ crystals through the following sequence of reactions:-

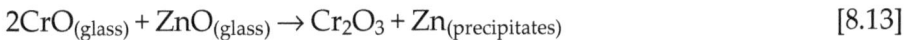

$$2CrO_{(glass)} + ZnO_{(glass)} \rightarrow Cr_2O_3 + Zn_{(precipitates)} \qquad [8.13]$$

$$Cr_2O_{3(glass)} + Li_2O_{(glass)} + 4SiO_2 \rightarrow 2LiCr(SiO_3)_{2(precipitates)} \qquad [8.14]$$

This effect is also undesirable, leading to physical wrinkling of the coating surface due to the relatively low thermal expansion ($\alpha \approx 5\cdot6\times10^{-6} \text{ K}^{-1}$) of the $LiCr(SiO_3)_2$ phase. Reactions associated with the coating of a pre-oxidized substrate, on the other hand, were found to be very different in behaviour. Suitable pre-oxidation leads to the formation of a 3 to 4 μm thick adherent, non-porous layer of Cr_2O_3. This layer prevents the glass from reacting directly with the metal. Consequently, all the redox reactions are eliminated, or at least significantly reduced. Saturation of Cr_2O_3 in the glass adjacent to the interface can easily be maintained (the diffusion of Cr^{3+} is significantly slower than that of Cr^{2+}), so creating the desired bonding conditions. Prolonged firing can, however, lead to the saturation limit

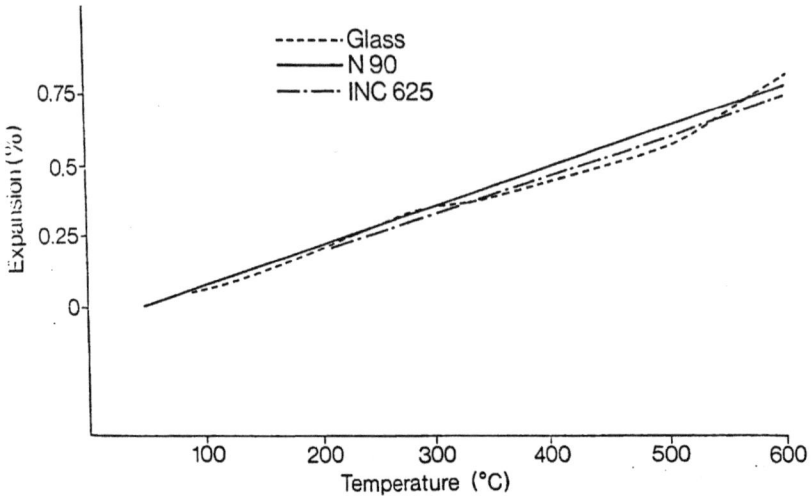

Figure 64: Experimental thermal expansion curves for a lithium zinc silicate glass-ceramic matched to Nimonic 90 and Inconel 625 alloys

of Cr_2O_3 in the glass being exceeded at the interface, and this can subsequently result in the undesirable formation of $LiCr(SiO_3)_2$ at the interface between the glass and the substrate oxide.

The bonding characteristics of a lithium zinc silicate glass-ceramic containing a high proportion of ZnO to Ni-based superalloys including Nimonic 90, Inconel 625 and Hastelloy C276 have also studied in detail by Metcalfe and Donald(1990). Superimposed thermal expansion curves are shown in Figure 64 for the materials investigated, illustrating the good match between the glass-ceramic and the metals. All sealing was carried out in argon using a standard heat-treatment cycle of 5 minutes at 950°C followed by 60 minutes at 465 or 585°C and 60 minutes at 800 to 1000°C, using both unoxidized and pre-oxidized metals. In the case of the unoxidized alloys, high quality seals were obtained which were relatively free from porosity. In general, a narrow interfacial reaction zone was clearly visible, although this normally extended less than 1 μm into the glass-ceramic. In the case of sealing to Nimonic 90 the interfacial region consisted of a semi-continuous precipitate of a phase which EDS showed to be rich in Cr, Ti and Si. It is possible that this layer is a complex mixture of CrO, Cr_2O_3 and TiO_2, together with Cr_3Si_2 or $LiCr(SiO_3)_2$. Unlike other Ni-based alloy/glass-ceramic systems, no evidence was found for the presence of transition metal phosphide phases in the interfacial zone. This suggests that the nucleating agent, P_2O_5, has not been removed by reaction with diffusing metal species for this particular system. It was noted that reduced metallic Zn had been formed at the metal interface, particularly at grain boundaries, most likely as a consequence of reaction [8.10] or [8.13]. In addition, small cubic crystals of a phase identified as $ZnCr_2O_4$ were dispersed throughout the bulk glass-ceramic, suggesting rapid diffusion of Cr from the metal substrate

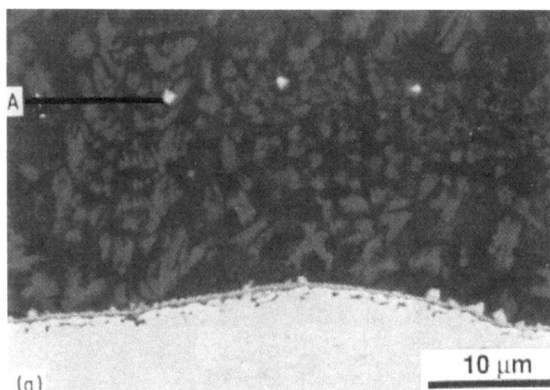

(a) *Unoxidized substrate (the bright particles marked 'A' are $ZnCr_2O_4$ precipitates within the bulk glass-ceramic)*

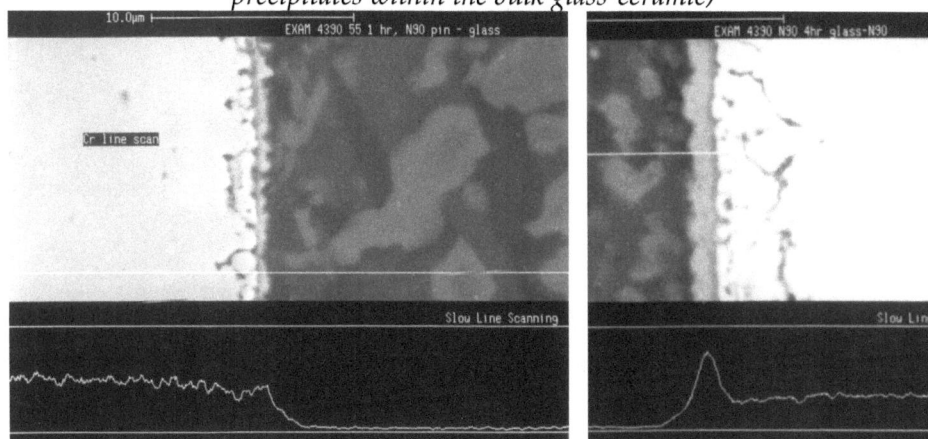

(b) *Substrate pre-oxidized for 1 hour in air prior to sealing (note the well defined oxide interface)*

(c) *Substrate pre-oxidized for 4 hours in air prior to sealing (note the thicker oxide interface which is richer in Cr than that of the corresponding seal to metal pre-oxidized for 1 hour)*

Figure 65: *SEM micrographs of lithium zinc silicate glass-ceramic seals to the Ni-based superalloy Nimonic 90*

into the glass-ceramic, as illustrated in Figure 65(a). Similar microstructures are found when using pre-oxidized metal, although in this case there is an absence of crystals in the bulk glass-ceramic, as shown in Figure 65(b,c), suggesting that diffusion of Cr is lower relative to unoxidized samples. Note that the oxide layer is thicker for the sample pre-oxidized for 4 hours and that the concentration of Cr as shown by the EDS line scan would appear to be higher in this layer.

When sealing this lithium zinc silicate glass-ceramic to Inconel 625 a layered interfacial reaction zone was noted similar to that of the un-oxidized Nimonic 90 alloy, as shown in Figure 66. In this case, EDS revealed the concentration of Ni and

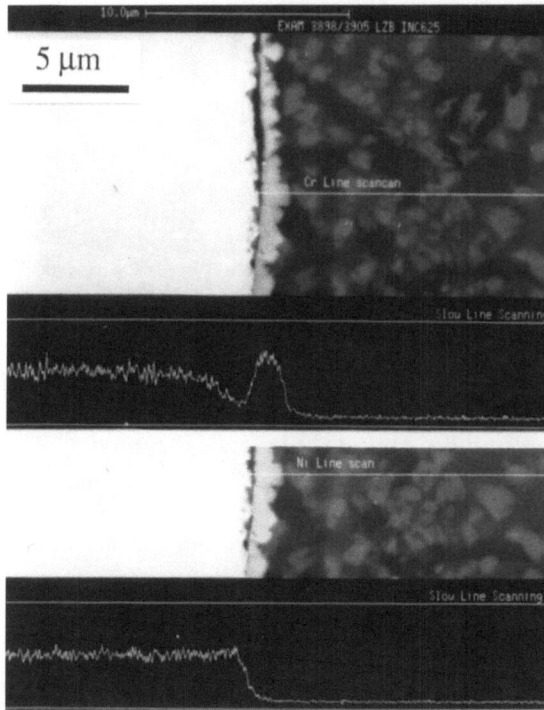

Figure 66: SEM micrographs of a lithium zinc silicate glass-ceramic sealed to the Ni-based superalloy Inconel 625

Si to be low in this zone whilst the concentration of Cr was similar to that of the bulk alloy. This suggests that this zone consists mainly of CrO and/or Cr_2O_3, as is also the case for the Nimonic 90 alloy. Similarly, there was no evidence for the formation of transition metal phosphide phases. Seals to Hastelloy C276 are shown in Figure 67. Unoxidized samples show the presence of a very thin layer rich in Cr and Mo. This gives rise to an excellent interface with few reaction products and with the glass-ceramic retaining a fine microstructure right up to the interface. The other line scans shown, for Ni, Si and Zn, highlight the effectiveness of EDS for defining the elements present in specific phases in the glass-ceramic and the inter-diffusion of elements between the glass-ceramic and the metal substrate. In contrast to the unoxidized sample, pre-oxidized samples, on the other hand, gave rise to a very complex and disordered interface consisting of precipitated phases rich in Ni together with interfacial layers rich in Cr and Mo, as shown in Figure 68. A thin Mo-rich layer is noted at the metal boundary and there are two discrete layers rich in Cr. Inter-dispersed between these layers and just present into the bulk glass-ceramic are Ni-rich precipitates. This microstructure highlights the very complex reactions that can occur in the interfacial region of some systems. As noted earlier, this is not a desirable situation, the introduction of complex phases into the interfacial zone introducing stresses which may affect the lifetime behaviour of the seal.

(a) Cr line scan

(b) Ni line scan

(c) Mo line scan

(d) Si line scan

(e) Zn line scan

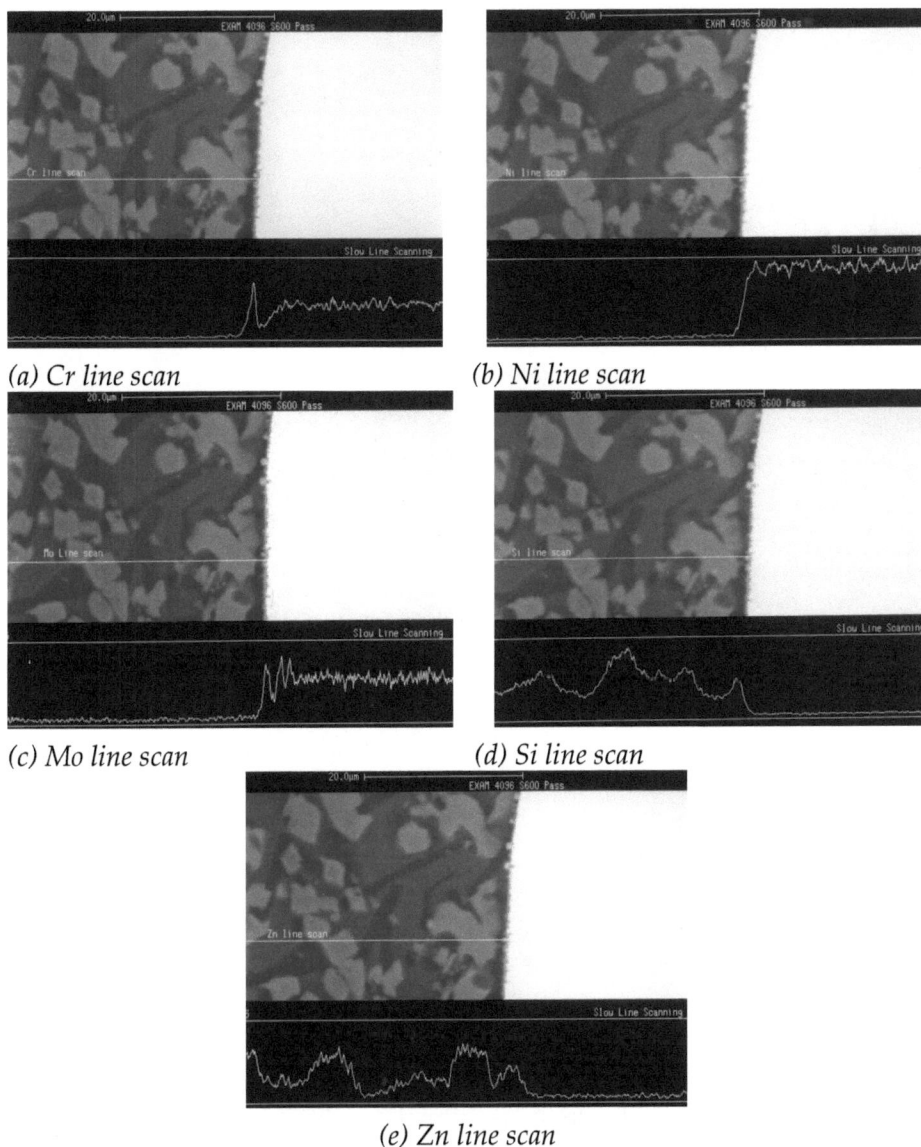

Figure 67: SEM micrographs of lithium zinc silicate glass-ceramic-to-Hastelloy C276 alloy interfaces. EDS line scans for Cr, Ni, Mo, Si and Zn

The work described above was extended to include the effect of a number of transition metal oxide additions including CuO, NiO and MoO_3, added to the glass batch at the 2 mole% level, on the sealing behaviour of the lithium zinc silicate glass-ceramic to Inconel 625 and Nimonic 90 alloys. Micrographs are shown in Figure 69 which show layered interfacial regions. In the case of the CuO addition, Figure 69(a), sealing to Nimonic 90 resulted in the formation of a much coarser interfacial zone than when this oxide was absent, consisting of four or five discrete layers with a total thickness of the order of 3 to 4 μm. EDS analysis showed that this region was rich in Ni and Cr, with smaller additions of Co, Ti and Si. It was

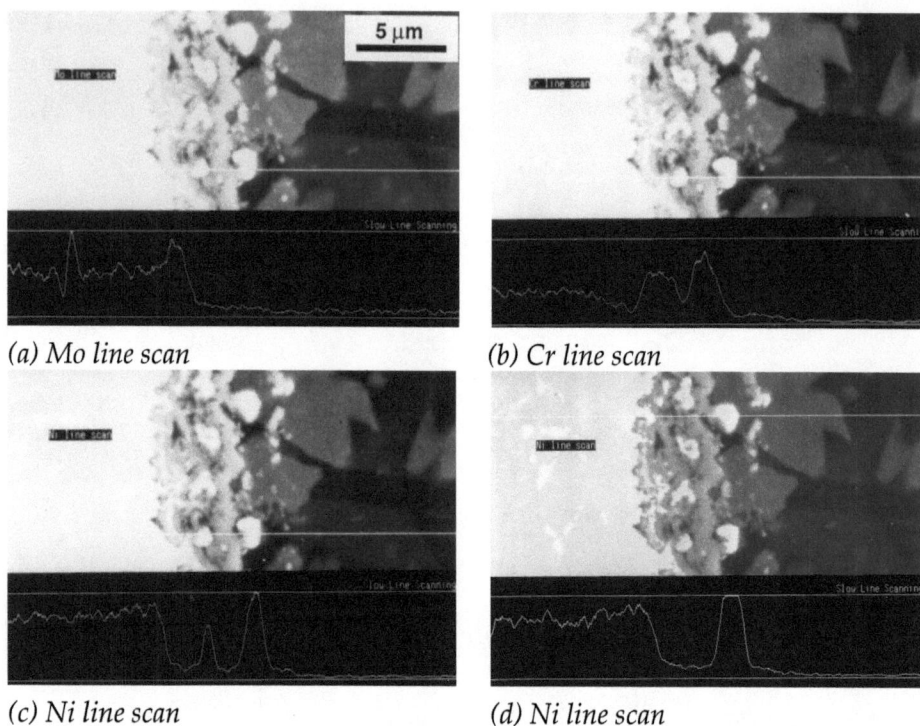

(a) Mo line scan

(b) Cr line scan

(c) Ni line scan

(d) Ni line scan

Figure 68: SEM micrographs of a lithium zinc silicate glass-ceramic sealed to pre-oxidized Hastelloy C276 alloy. EDS line scans for Mo, Cr and Ni

noted that the outermost interfacial layer was richer in Cr, whilst the inner layer was richer in Ni. The microstructure of the glass-ceramic at the interface was very fine, and there was no evidence for the formation of transition metal phosphide phases. Similar effects were noted for the addition of NiO. Seals to Inconel 625 were also similar in appearance. Addition of MoO_3 to an Inconel 625 seal, Figure 69(b,c), similarly resulted in the formation of a layered interfacial zone consisting of a coarse granular inner layer and a thinner continuous outer layer. The inner zone was found to be rich in Ni, Mo and Cr, whilst the outer region was rich in Mo and Cr with some Si. In addition, small discrete precipitates, also rich in Mo and Cr with some Si (possibly a mixture of $MoSi_2$ and Cr_3Si_2), were observed within the glass-ceramic close to the interface. The formation of such complex interfacial structures again emphasises the importance of glass composition, and the effect that relatively small additions can have on such structures.

Very recently, Bengisu *et al*(2004) have examined the bonding characteristics of lithium silicate based glasses to a range of Ni-based superalloys, including Inconel 600 and 718, and Hastelloy C276, with particular reference to the long term stability of seals in the temperature range 700–900°C. It was noted that small additions of ZnO to the glasses improved the bonding characteristics. It was also observed that seals to all the alloys investigated were stable under prolonged heating at temperatures < 800°C, but failed when heated to 900°C. The reason-

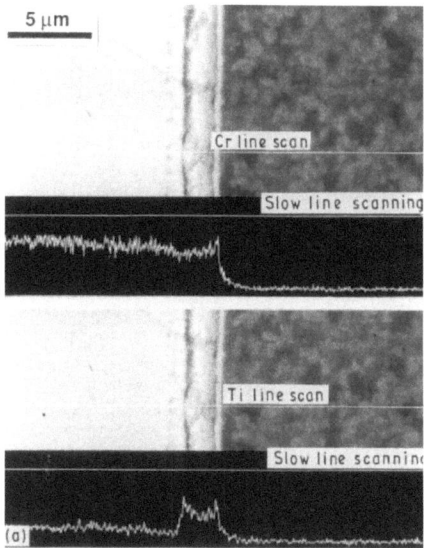

(a) Seal to Nimonic 90 alloy for glass-ceramic containing 2 mol% CuO addition

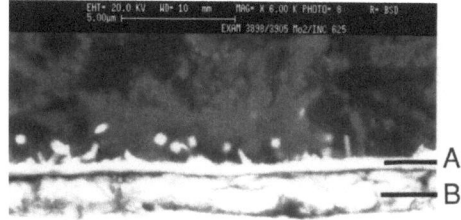

(b) Seal to Inconel 625 alloy for glass-ceramic containing 2 mol% MoO₃ addition. Note the layered interface. Zone 'A' is rich in Ni, Cr and Mo; zone 'B' in Mo, Cr Zn and Si

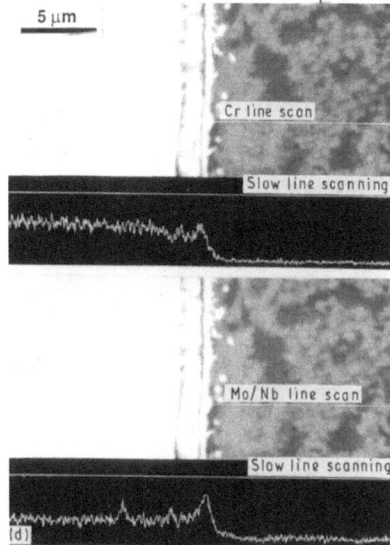

(c) As (b); line scans for Cr and Mo

Figure 69: SEM micrographs of lithium zinc silicate glass-ceramic seals to a variety of Ni-based superalloys, the starting glasses containing a number of transition metal oxide additions

ably good stability of these systems up to 800°C suggests that they may be very useful for some fuel cell applications.

8.5   Bonding to platinum, copper, silver and gold

Coating (enamelling) of Cu, Ag and Au has been practised for many years, with particular reference to the preparation of enamelled jewellery and related regalia. These enamels are usually based on relatively low softening temperature alkali lead silicate compositions. Some typical enamel compositions are given in Table 9. Further details of the early work in this area can be found elsewhere (e.g. Andrews, 1961).

Copper and less recently platinum have also been widely utilized in the production of electrical feed-through seals, with glass normally being employed as the sealing media. In addition, seals have been produced to silver and gold for electrical applications. Less work has been reported on the bonding of these metals to glass-ceramic materials, although McMillan and Hodgson (1966) showed that a lithium zinc silicate glass nucleated with P_2O_5 and exhibiting a thermal expansion coefficient of 17.4×10^{-6} K^{-1} over the range 20–400°C could be successfully bonded to copper. Thermal expansion curves for Cu and a glass-ceramic are shown in Figure 70. More recently, Risbud et al(1986) have prepared Cu/cordierite glass-ceramic systems where it was observed that diffusion of Cu^+ into the glass-ceramic led to the formation of small (≈100–200 nm) metallic precipitates of copper within the interfacial zone.

The bonding behaviour between Cu and a crystallizable refractory lithium aluminosilicate glass has been reported by Donald(1977). In this work, the metal was sealed to the glass by fusing the metal in contact with the glass at 1150°C under an inert atmosphere. The molten metal took several minutes to wet and

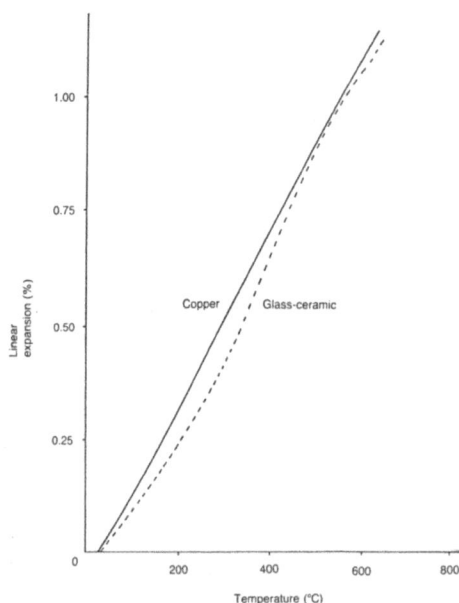

Figure 70: Thermal expansion curves for Cu and a glass-ceramic
(after Partridge, 1990)

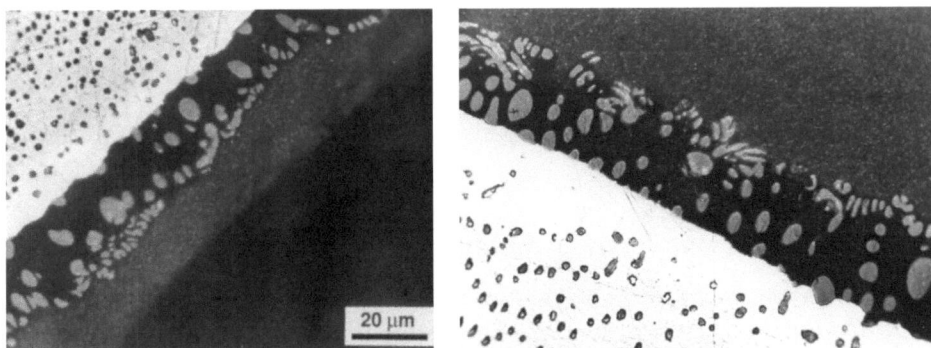

Figure 71: Glass-to-copper seal interface showing presence of oxide precipitates and dendrites (after Donald, 1977)

spread over the glass surface. It was noted that mutual dissolution occurred at the interface to give strong bonding via a Cu–CuO eutectic layer. A transitionary zone ≈40 μm thick was observed which contained many precipitates of copper oxide. Bonded systems were also produced by mutual fusion of both metal and glass at 1400°C. In this instance, there was no eutectic layer and bonding was achieved due to mutual solubility at the interface, with the precipitation of oxide particles within both the glass and the copper. Examples are shown in Figure 71. Bonding was also achieved to a Cu–Ti alloy by melting the alloy in contact with the glass, with the metal almost instantaneously wetting and spreading across the glass surface to produce a strongly bonded system due to the reactive nature of Ti metal.

8.6 Bonding to chromium

Very little work has been reported on the sealing of glass or glass-ceramic materials to unalloyed Cr, due in part to the very high chemical reactivity of this metal. In one study, Tomsia *et al*(1985) examined the bonding and associated chemical reactions between a model sodium disilicate glass and Cr, with the aim of attempting to gain a clearer understanding of the role played by Cr in the formation of seals to practical Cr-containing alloys, including stainless steels. A number of reactions were identified, depending on the partial pressure of oxygen in the sealing environment and the overall thickness of the glassy layer, including:-

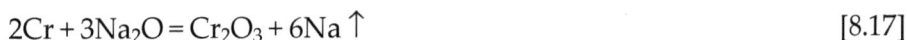

$$Cr + Na_2O = CrO + 2Na \uparrow \qquad\qquad [8.15]$$

$$2CrO + Na_2O = Cr_2O_3 + 2Na \uparrow \qquad\qquad [8.16]$$

$$2Cr + 3Na_2O = Cr_2O_3 + 6Na \uparrow \qquad\qquad [8.17]$$

It was noted that the formation of Cr^{2+} (CrO) may result in the formation of a mechanically weak interface, whilst formation of Cr^{3+} (Cr_2O_3) is more likely to result in the formation of a strong bond between the glass and Cr. This difference in bonding behaviour was related to the lower solubility of Cr_2O_3 in the glass, resulting in saturation of this oxide within the interfacial zone being readily achieved during sealing.

8.7   Bonding to molybdenum and tungsten

Glass-to-metal seals employing Mo and W have been used extensively in electrical applications (e.g. Partridge, 1949; Smithells, 1952). More recently, Leichtfried *et al*(1998) have examined the sealing characteristics of glass to oxide dispersion strengthened Mo, ODS-Mo. It was noted that sealing of quartz glass to ODS-Mo gave rise to stronger seals relative to pure Mo and also led to improved lifetime behaviour of halogen lamps employing Mo seals.

Limited work has also been reported on the sealing of glass-ceramics to molybdenum. In one study, Nash *et al*(1983) successfully sealed a zinc aluminosilicate glass-ceramic to unoxidized Mo. Sealing was initially carried out employing a conventional three-stage heat-treatment schedule consisting of a high temperature stage to melt the glass, followed by a nucleation stage of 2 hours at 650°C and a crystallization stage of 2 hours at 850°C. It was subsequently noted, however, that crystal nucleation *and* crystal growth occurred during the initial heating and final cooling stages up to and down from the sealing temperature. Conventional nucleation and crystallization stages were therefore deemed unnecessary. The final heat-treatment consisted of a single high temperature cycle to 1040°C over 30 minutes. This was followed by a hold at this temperature before cooling down to ambient temperature. Thermal expansion curves for Mo and a glass-ceramic are shown in Figure 72.

Thorp *et al*(1991) and Holland *et al*(1990) have also reported the coating of Mo employing a magnesium aluminosilicate glass-ceramic. Screen printing was used to apply a coating of around 100 μm in thickness, which was then fired in a nitrogen atmosphere. It was noted that a well-defined interfacial region was apparently absent as monitored by SEM/EDX analysis, but by measuring the permittivity of the coating as a function of thickness it was proposed that some diffusion of Mo^{4+} into the coating did occur.

8.8   Bonding to titanium and its alloys

The service temperature of Ti alloys is limited to around 590°C in air due to the formation of an oxide scale with a brittle sub-surface layer which can initiate surface cracking. In addition, two allotropic forms of Ti exist, α-Ti with an hcp crystal structure and β-Ti which is bcc. The α phase exhibits greater toughness and improved fatigue performance, whilst the β phase offers superior creep behaviour. The α to β transformation occurs at 882°C, and is accompanied by a volume change. This can hinder coating of this metal, unless a low coating temperature

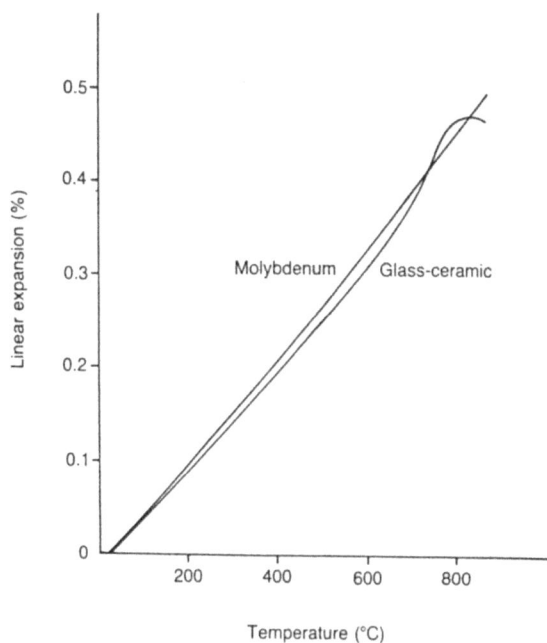

Figure 72: Thermal expansion curves for Mo and a glass-ceramic (after Partridge, 1990)

is employed, or a phase-stabilized Ti alloy is used. Until relatively recently, only limited work has been reported on the bonding of glasses and glass-ceramics to Ti and its alloys, due in part to the problems associated with this volume change, and also in part due to the high reactivity of this metal.

In general, conventional silicate sealing glasses cannot be readily employed to prepare high quality seals or coatings to Ti and its alloys due to the occurrence of severe interfacial reactions. These reactions lead to porous structures and mechanically weak interfaces due to the evolution of gaseous reaction products and the formation of brittle and/or poorly adherent silicide phases. For example, Passerone et al(1977) noted that molten lithium silicate glasses react with Ti at temperatures in the range 1100 to 1400°C under reducing conditions to form a number of intermediate phases including Ti_2O_3 and Ti_5Si_3. In addition, Si diffuses into the Ti substrate embrittling the metal. At the higher temperatures extensive solution (up to 15 to 20 at%) of Si in Ti occurs, and this leads to the formation of a liquid phase. Based on thermodynamic data for the free energy of reaction of Ti with various oxides, Sitnikov et al(1974) predicted that introduction of certain oxides into a lithium silicate glass should reduce the corrosion of Ti by the molten glass. Their predictions were confirmed experimentally, with observation that addition of Li_2O, BeO, CaO, SrO, BaO and CeO_2 did indeed retard the corrosion of Ti by the molten glasses. Similarly, substitution of oxides in the glass by less reactive oxides also retarded corrosion; for example, replacement of Na_2O by B_2O_3, MnO or ZnO.

Brow and Watkins (1987) examined the potential of non-silicate boroaluminate glasses for sealing to pure Ti and a β-phase stabilized Ti alloy. The performance of these glasses was compared to that of a commercial sodium silicate sealing glass. Bonding to this glass at temperatures in the range 760 to 950°C in argon was, surprisingly, found to produce good quality hermetic seals with only a thin reaction zone ≈1 μm thick. The presence of a titanium silicide phase was confirmed in this zone, the most likely reaction being:-

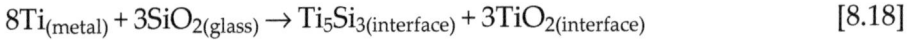

$$8Ti_{(metal)} + 3SiO_{2(glass)} \rightarrow Ti_5Si_{3(interface)} + 3TiO_{2(interface)} \qquad [8.18]$$

The formation of reasonable quality seals to a silicate glass was believed to be a consequence of the relatively low sealing temperatures employed in this work. A number of boroaluminate glass seals were also investigated, these glasses containing the alkaline earth oxides CaO (these glasses are known as "CABAL" glasses), SrO (SRBAL glasses) or BaO (BABAL glasses). Sealing was carried out in the temperature range 670 to 745°C. High quality hermetic seals were obtained with no obvious interfacial reaction products visible in sectioned seals, although the presence of Ti was detected in the glass ≈15 μm from the interface. There was some evidence from XPS of a mixed TiB_2/TiO_2 interfacial region, possibly through the reaction:-

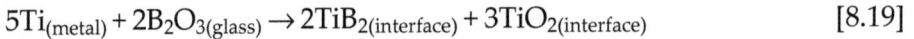

$$5Ti_{(metal)} + 2B_2O_{3(glass)} \rightarrow 2TiB_{2(interface)} + 3TiO_{2(interface)} \qquad [8.19]$$

Seal strengths were monitored by measuring the load required to fracture a simple seal configuration. It was noted that the load required to fracture a seal with the boroaluminate glass was approximately 50% greater than that required to fracture the equivalent silicate seal.

Brow *et al*(1997) have reported the development of alkaline earth lanthanum borate glasses and their subsequent sealing to titanium and titanium alloys at temperatures in the range 700–800°C. Suitable compositions were given as (wt%):- 40–70% B_2O_3, 5–30 BaO and 5–20% La_2O_3 with one or more of Al_2O_3 (0–20%), CaO (0–12%), Li_2O (0–8%), Na_2O (0–8%), SiO_2 (0–8%) and TiO_2 (0–15%). Thermal expansions of these glasses are in the range $8 \cdot 7$–$10 \cdot 3 \times 10^{-6}$ K^{-1}, with leach rates in water at 70°C reported to be around 100 mg m^{-2}day^{-1}. Sealing was carried out in nitrogen at temperatures in the range 700–800°C, about 150–200°C higher than T_g, to provide cylinder seals comprising a Ti or Ti alloy housing with a central metal pin of Mo, Pt or Ni–48Fe alloy. Helium leak rates for seals produced using these glasses were reported to be $< 10^{-9}$cm^{3}s^{-1}.

Hong and Holland(1989a, 1989b) and Hong(1991) have reported the coating of Ti with a lithium aluminosilicate glass-ceramic. Coating of unoxidized Ti in an inert atmosphere of argon at 970°C was observed to give a very porous coating. This was believed to be due to reaction of the glass with Ti under these specific sealing conditions to give Ti_5Si_3 together with gaseous oxygen:-

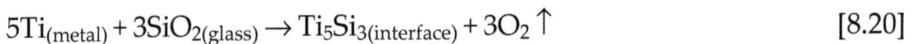

$$5Ti_{(metal)} + 3SiO_{2(glass)} \rightarrow Ti_5Si_{3(interface)} + 3O_2 \uparrow \qquad [8.20]$$

In addition, it was noted that a number of further reactions are possible (in particular between Ti and the P_2O_5 nucleating agent), including:-

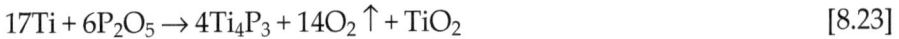

$$16Ti + 6P_2O_5 \rightarrow 4Ti_4P_3 + 15O_2 \uparrow \qquad\qquad [8.21]$$

$$9Ti + 3P_2O_5 \rightarrow 2Ti_4P_3 + 7O_2 \uparrow + TiO \qquad\qquad [8.22]$$

$$17Ti + 6P_2O_5 \rightarrow 4Ti_4P_3 + 14O_2 \uparrow + TiO_2 \qquad\qquad [8.23]$$

Coating at lower temperatures, down to 900°C, was unsuccessful because rapid crystallization of the glass powder occurred, and this prevented adequate wetting of the substrate, in addition to inhibiting sintering of the powder. It was subsequently observed that use of suitably pre-oxidized Ti can eliminate the reactions that evolve gaseous oxygen. Due to the fact that TiO_2 dissolves readily in the glass during coating it was found necessary, however, to pre-oxidize the metal to give a relatively thick (≈ 8 to $10\ \mu m$) oxide layer. Unfortunately, although the resultant coatings were non-porous, they were also mechanically weak, due to the brittle nature of the oxide layer and its poor adhesion to the substrate. It was further noted that non-porous coatings can also be formed by coating unoxidized Ti in air. In the presence of air, reaction [8.20] is replaced by reaction [8.18]; however, it was found that the brittle behaviour of Ti_5Si_3 again led to mechanically weak coatings. In addition to pre-oxidation, the effect of nitriding the surface of Ti on the coating chemistry was also examined. A thin layer of TiN was formed by heating the metal in nitrogen at 900 to 980°C. Unfortunately, on coating, a porous microstructure again resulted despite the presence of this layer. This was believed to be due to the reaction:-

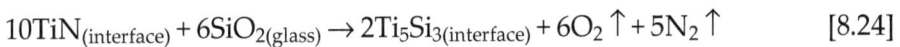

$$10TiN_{(interface)} + 6SiO_{2(glass)} \rightarrow 2Ti_5Si_{3(interface)} + 6O_2 \uparrow + 5N_2 \uparrow \qquad [8.24]$$

In addition to the lithium aluminosilicate system alternative aluminosilicate compositions were investigated by Hong(1991) for sealing to Ti. It was found that glasses from the calcium aluminosilicate system containing 10 wt% TiO_2 gave particularly good coating results. It was suggested that for this system reaction [8.18] again occurs in preference to [8.20]. Unfortunately, the crystallization characteristics of this system are poor. The overall coating behaviour of these silicate systems led Holland et al(1990) and Hong(1991) to examine the use of traditional adhesion promoting oxide additives on the coating behaviour of Ti. This work showed that addition of CoO, for example, can lead to the formation of a strong, non-porous interface through the reaction:-

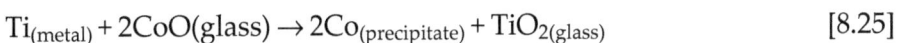

$$Ti_{(metal)} + 2CoO(glass) \rightarrow 2Co_{(precipitate)} + TiO_{2(glass)} \qquad\qquad [8.25]$$

This reaction occurs because the free energy of formation of CoO has a less negative value than that of TiO_2. Reaction [8.25] therefore occurs in preference to reaction [8.20]. Precipitates of metallic Co are formed in the glass-ceramic near to the interface and improve the bond strength further by mechanical keying effects.

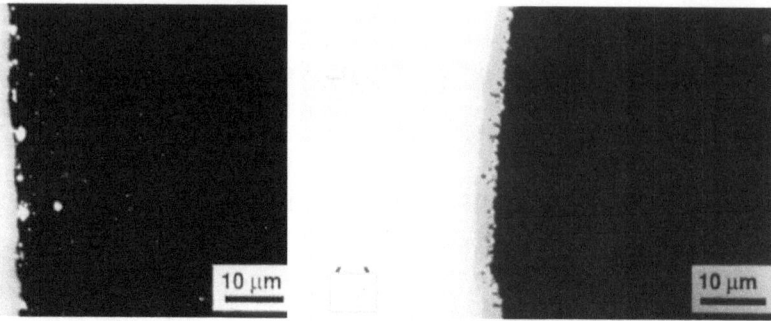

(a) With CoO addition   (b) Without CoO addition

Figure 73: SEM micrographs of a glass-ceramic coating on titanium, for a glass-ceramic with and without the addition of CoO (after Hong and Holland, 1989)

It was noted that failure of these coatings was always within the glass-ceramic, rather than at the interface, in contrast to the other coatings on Ti. Micrographs showing coatings applied to Ti are given in Figure 73, for glasses both with (a) and without (b) the addition of CoO.

Coatings have also been successfully applied to Ti by Donald *et al*(1995) employing a calcium borosilicate, CBS, glass and a similar glass doped with rare earth metal oxide additions. Examples are shown in Figure 74. The interface between Ti and a CBS glass of composition (mol%) $41.7CaO–27.6B_2O_3–30.7SiO_2$ designed to give a thermal expansion coefficient of $8.7\times10^{-6}\,K^{-1}$ is shown in Figure 74(a). Some surface crystallization along the interface is evident for this system. The interface between Ti and the same CBS glass doped with 9.6 mol% Er_2O_3 is shown in Figure 74(b). In this instance, the glass has partially crystallized to give a dispersion of crystals rich in Er_2O_3 within the glass matrix. This partially crystallized glass exhibits a slightly higher thermal expansion of $9.6\times10^{-6}\,K^{-1}$, which is a reasonable match to Ti but which may introduce small tensile stresses into the coating, although in general the coating was noted to remain intact.

The bonding characteristics of Ti to an SnO-containing enamel coating prepared under vacuum at 800°C have been investigated by Lui *et al*(2004) employing SEM and EDS analysis. It was believed that reaction of SnO and SiO_2 with Ti to give TiO_2 at the interface occurred, and this resulted in good bonding.

More recently, there has been great interest in the coating of Ti and Ti alloys, driven by the use of this metal for biomedical implant applications. For example, coatings of a calcium phosphate based glass-ceramic have been successfully applied to titanium alloys by Kasuga *et al*(2001a, 2001b). In this work, a glass-ceramic of composition $60CaO–30P_2O_5–3TiO_2–7Na_2O$ was coated onto a new β-titanium alloy of composition Ti–29Nb–13Ta–4.6Zr developed for biomedical applications. Samples were prepared by coating Ti alloy discs by dipping in a slurry of glass particles in methanol. After drying, the coated discs were heated in air at 800°C for 1 hour to yield strongly bonded glass-ceramic coatings exhibiting tensile bond strengths of the order of 20 MPa, a value significantly higher than that

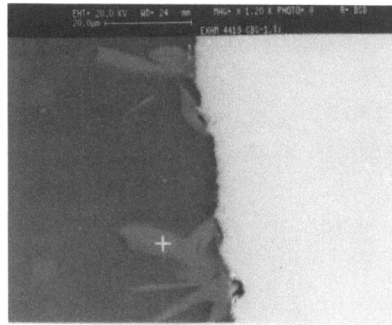

(a) Calcium borosilicate glass showing the presence of some surface crystallization at the interface.

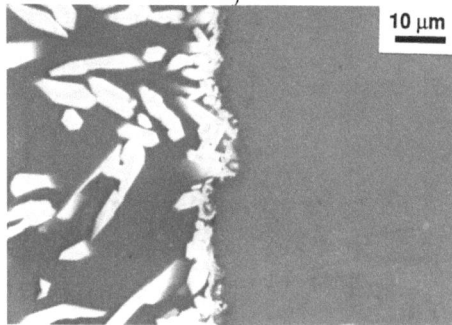

(b) Calcium borosilicate glass-ceramic doped with Er_2O_3
Figure 74: SEM micrographs of glass-ceramic coatings on Ti
(after Donald et al, 1995)

achieved by plasma spraying hydroxyapatite coatings onto titanium. (Heating at temperatures < 800°C, on the other hand, resulted only in very weakly bonded systems). During the process, the glass was noted to crystallize to predominantly β-$Ca_3(PO_4)_2$. Microscopic examination of the coatings revealed the presence of extensive porosity, although apparently this did not seriously effect the bonding strength. A large difference in thermal expansion behaviour between the glass-ceramic coating and the titanium substrate was noted, i.e. 18×10^{-6} K^{-1} for the glass-ceramic compared with 9×10^{-6} K^{-1} for the metal. It was suggested that good bonding was achieved, without coating failure despite the large thermal expansion difference, due to the presence of a thin transitional layer 4 μm thick which acted to relieve the stresses between the coating and the substrate.

Bioactive glass coatings have also been applied to titanium implant alloys by firing a slurry of glass powder in isopropyl alcohol applied to the titanium surface (Bloyer, 1999). Good bonding was achieved via a very thin reaction layer of Ti_5Si_3, 50–100 nm thick, at the interface between the glass and the metal. Thin bioactive glass-ceramic films have also been deposited on titanium by RF magnetron sputtering (Mardare *et al*, 2003). Phases present included enstatite and forsterite in addition to calcium magnesium phosphate. The strength of the glass-ceramic/Ti bond determined in pull-off testing was around 40 MPa. Bioactive coatings of a fluorapatite-mullite glass-ceramic have also been successfully applied to implant alloys by electrophoretic deposition (Bibby *et al*, 2004).

Glass coatings 25–100 μm thick based on the sodium calcium borosilicate system containing smaller additions of K_2O, MgO and P_2O_5 have been applied to titanium implant alloys using an enamelling technique (Gomez-Vega *et al*, 1999). A suspension of glass powder in ethanol was applied to the metal surface. After drying, samples were fired in air or nitrogen at temperatures in the range 700–860°C. Optimum conditions in terms of coating quality and adhesion were found to be 800–840°C for 1–15 minutes in nitrogen. It was noted that during firing in air or nitrogen an oxide or nitride layer formed on the substrate surface. On further heating, this layer dissolves into the coating. It was found to be important to stop the process before all of this layer dissolved and before direct reaction of the glass with the substrate took place. Glass coatings from the calcium borosilicate system have also been applied to titanium by Brichi *et al*(2002) using enamelling techniques. In addition, coatings were prepared with glasses containing a 0·1–0·2% dispersion of titanium particles 20 μm in size, aimed at improving both the strength of the coating and its adherence to the substrate, with some success noted.

Additional information on coating of biomedical implant materials is give in Section 10.11.

Titanium has been bonded to Macor® machinable glass-ceramic by active metal brazing using a 64Ag–34·5Cu–1·5Ti filler alloy (Guedes *et al*, 2001). Bonding was carried out using a brazing alloy foil between the parts to be joined, which were clamped and heated under vacuum at temperatures in the range 850–930°C for 10 to 30 minutes. A multilayered interface 50 to 150 μm thick was obtained, which consisted of diffusion, solid solution and reaction layers. Optimum bonding conditions were found to be 850°C for 10 minutes, which gave a joint with an acceptable interfacial shear strength of 68 MPa compared with 80 MPa for that of the bulk glass-ceramic.

8.9   Bonding to aluminium and its alloys

Coating (enamelling) of Al and Al-alloys has been practised for many years (e.g. Andrews, 1961). Due to the relatively low melting temperature of Al (660°C), coating glasses have been confined to low softening temperature lead silicate, barium borosilicate and alkali phosphate compositions. Some typical compositions are given in Table 9. Further details of these systems are given by Andrews(1961).

More recently, work on coating of a silicon carbide reinforced Al-matrix composite with a lead borosilicate glass for high power electronic support structures has been reported by Ison *et al*(2000). In this study the bonding characteristics of a glass of composition $61·05PbO–26·80B_2O_3–12·15SiO_2$ (mol%) with thermal expansion matched to that of the Al composite were monitored using a sessile drop method. It was noted that bonding to pure Al in air at 530°C gave rise to a good interface with a minimal diffusion zone only 1–2 μm in width. Bonding to the composite resulted in some porosity together with some Pb-rich precipitates, most likely due to the formation of CO, CO_2 and Pb through the following thermodynamically favourable reactions:-

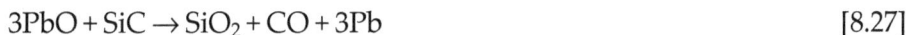

$$4PbO + SiC \rightarrow SiO_2 + CO_2 + 4Pb \qquad [8.26]$$

$$3PbO + SiC \rightarrow SiO_2 + CO + 3Pb \qquad [8.27]$$

In contrast, very little work has been reported on the *sealing* of glasses or glass-ceramics to aluminium or its alloys. This is due to a combination of factors that include the relatively low melting temperature of Al, its high thermal expansion ($\alpha \approx 25\times10^{-6}\ K^{-1}$) and its high chemical reactivity. It has nevertheless been shown that certain low melting point phosphate-based glasses and glass-ceramics are suitable for sealing to Al alloys (Wilder, 1980; Wilder *et al*, 1982; Peng and Day, 1991a, 1991b; Brow *et al*, 1999, 2000). For example, Wilder(1980) successfully sealed sodium barium phosphate and sodium calcium phosphate glasses and glass-ceramics to aluminium. It was noted that for glasses of composition (50–x) Na_2O–xRO–$50P_2O_5$ (R = Ca or Ba) the thermal expansion increased from 16·0 to 20·8×$10^{-6}\ K^{-1}$ for addition of BaO between the range 10–40 mol% and from 12·6 to 18·4×$10^{-6}\ K^{-1}$ in the case of CaO. The chemical durability of glass of composition $40Na_2O$–$10BaO$–$5Al_2O_3$–$45P_2O_5$ in water at 25°C was observed to be quite reasonable, exhibiting a leach rate of around 100 mg m^{-2}day^{-1}. Heat-treatment of these glasses was carried out successfully to yield glass-ceramic materials with improved mechanical properties whilst maintaining high thermal expansion. Depending on the precise composition, the crystal phases formed in the crystallized glass included $NaPO_3$, $NaBa(PO_3)_3$, $Ba_3(PO_4)_2$, $NaCa(PO_3)_3$ and $Na_4Ca(PO_3)_6$. Unfortunately, no durability data were provided for these glass-ceramics, but for glass-ceramics containing $NaPO_3$ the durability is expected to be poor, this crystalline phase being quite soluble in water.

Brow and Tallant(1997) and Kilgo *et al*(1999, 2000) have developed a series of alkali phosphate based glasses for sealing to aluminium with potential electronic packaging applications. Suitable compositions were reported to contain 10–25% Na_2O, 10–25% K_2O, 4–15% Al_2O_3 and 35–50% P_2O_5. Additional constituents included B_2O_3 at < 10%, and BaO, PbO, CaO or MgO at < 12% total. Thermal expansions were in the range 16·0–21·0×$10^{-6}\ K^{-1}$. Leach rates for these glasses in water at 70°C were generally very poor at > 1 gm^{-2}day^{-1} with one exception, a glass containing the highest amount of PbO of those studied (9%) for which a remarkably low leach rate of 10 mg m^{-2}day^{-1} was recorded. Various seals were prepared, including cylinder seals in which a central Cu/Be alloy pin was sealed into an aluminium housing with a diameter of 6·4 mm and wall thickness 1·3 mm using a cast glass preform with a centrally drilled hole to accommodate the pin. Sealing was carried out in a nitrogen atmosphere at 500°C.

A relatively high melting point lithium silicate glass has also been successfully sealed to Al and a number of its alloys employing a novel injection moulding technique (Kramer *et al*, 1987) shown in Figure 29. Although the glass employed had a higher working temperature than the melting point of Al, careful control over the injection moulding parameters was shown to lead to the formation of high quality, hermetic seals. The technique is made possible by the fact that the injection mould acts as an efficient heat sink, thereby preventing the molten glass from heating the Al component to beyond its melting point.

8.10   Bonding to tantalum

Limited work has been reported on the coating or sealing of glasses to Ta or its alloys. Practical bonding to Ta is rendered particularly difficult because the metal undergoes embrittlement at temperatures greater than about 300°C in air. Conventional pre-oxidation treatments are therefore not feasible without seriously degrading the metal.

Reasonably successful seals to Ta have nevertheless been produced by Donald *et al*(1994) employing lithium magnesium aluminosilicate, LMAS, and calcium borosilicate, CBS, glasses and glass-ceramics, with and without addition of Ta_2O_5. Examples are shown in Figure 75. For the LMAS coatings, extensive precipitation of a Ta-rich phase was observed at the interface, as noted in Figure 75(a). If the LMAS glass was doped with Ta_2O_5 (the Ta_2O_5 added as an 'adherence promoter' in order to promote saturation of the glass with the substrate metal oxide in the interfacial region) a thick, ≈20 μm, continuous oxide layer was noted separating the substrate from the glass. Although bonding between this oxide layer and the glass is good, bonding between the oxide and the Ta substrate is relatively weak and partial separation of the oxide from the substrate may occur, as noted in Figure 75(b) where the oxide layer has just started to detach from the substrate. On the other hand, excellent bonding is achieved for an LMAS glass containing both Ta_2O_5 and P_2O_5, as observed in Figure 75(e). The interface between Ta and a calcium borosilicate, CBS, glass is also relatively weak and the glass easily becomes detached from the substrate. If, however, the glass is doped with Ta_2O_5 this yields a well-bonded system, as seen in Figure 75(f), although in this instance extensive surface crystallization of the glass occurs, with the formation of long crystals aligned perpendicular to the substrate surface. The influence of Ta_2O_5 and P_2O_5 on the bonding behaviour in these systems does illustrate yet again the powerful influence that appropriate additives can have on the resulting interfacial chemistry and hence seal quality and integrity.

8.11   Bonding to zinc

Very little work has been reported on the bonding of glasses to zinc. In one study, Chen *et al*(1997) examined the reactions likely to take place between similar systems including AsSe glasses with Zn. It was noted that a thin layer of As_2Zn_3

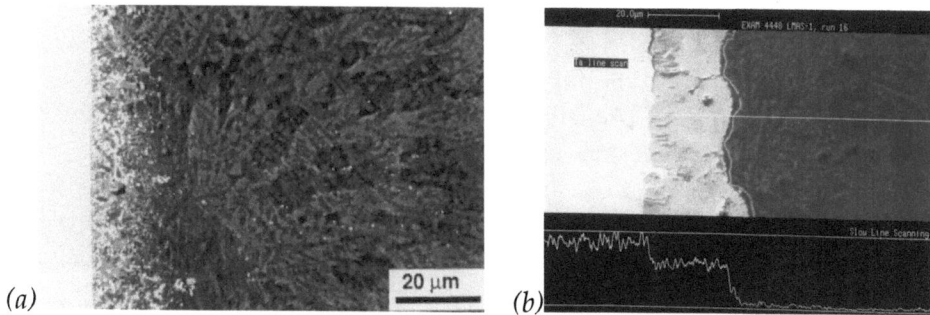

(a) Lithium magnesium aluminosilicate glass-ceramic
(b) Lithium magnesium aluminosilicate glass-ceramic doped with Ta$_2$O$_5$. Note
the thick oxide interfacial layer. EDS line scan for Ta is shown

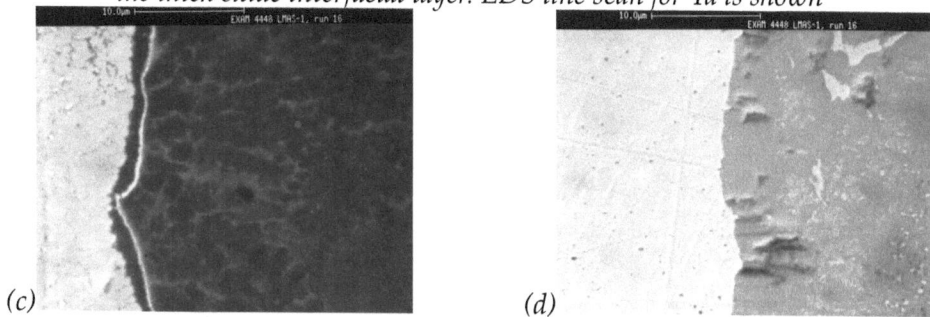

(c) Lithium magnesium aluminosilicate glass-ceramic doped with Ta$_2$O$_5$. Higher
magnification of oxide interfacial region. Note the thin well-defined diffusion
layer on the glass-ceramic side of the interface
(d) Lithium magnesium aluminosilicate glass-ceramic doped with Ta$_2$O$_5$. Higher
magnification of oxide interfacial region. Note the absence of precipitated phases
or reaction products at the interface

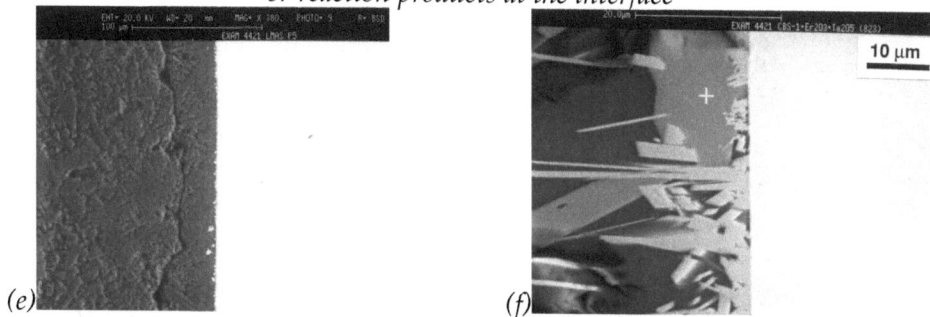

(e) Lithium magnesium aluminosilicate glass-ceramic doped with Ta$_2$O$_5$ and
containing additional P$_2$O$_5$. Note the small Ta-rich precipitates near the interface
(f) Calcium borosilicate glass doped with Er$_2$O$_3$ and Ta$_2$O$_5$
Figure 75: SEM micrographs of glass-ceramic coatings on Ta
(after Donald et al, 1995)

formed adjacent to the zinc substrate with ZnSe adjacent to the glass phase. The
total thickness of this dual reaction layer was < 10 μm and it was thought that
this would not to be detrimental to the properties of optical systems which might
use this glass as a solder.

8.12 Bonding to niobium

There are few reports on the bonding of glass to niobium or its alloys. In one study (Sedmale *et al*, 1999) barium aluminium phosphate glasses containing 12–20% silica and with thermal expansion coefficients in the range from 6.5 to 7.8×10^{-6} K^{-1} were used to coat niobium in an attempt to improve the oxidation resistance of this metal. Coatings were applied in the form of a suspension of glass particles in a water/isopropanol mixture and subsequently fired at 1050–1150°C under vacuum. It was noted that the interfacial region was partially crystalline and contained a dispersion of NbP and Al$_9$Nb phases.

8.13 Overall conclusions to be drawn from bonding studies

Reference to the many studies reported on the sealing of specific glasses and glass-ceramics to a very wide variety of metals and alloys allows a number of important observations and conclusions to be drawn. Of particular importance is the effect on the sealing characteristics and interfacial chemistry of such factors as the precise compositions of the glasses and metals selected for sealing, the state of the metal surface, the heat-treatment schedule adopted, and the sealing atmosphere employed.

It is clear, for example, that the state of the metal surface can play a crucial role in determining the quality of the resultant seal or coating. In some instances, pre-oxidation of the metal is essential in order to prevent or minimise excessive diffusion of metal alloying elements into the glass or glass-ceramic interfacial zone, whilst in other instances it is better to rely on redox reactions between the glass and metal or on the influence of residual oxygen in the sealing atmosphere to provide the bonding conditions necessary. In the case of pre-oxidation, different oxidation conditions can produce quite different oxide compositions and characteristics, including the degree of adherence to the metal substrate, overall porosity, and possible occurrence of composition gradients, with differences between oxide present at the metal–oxide interface and at the oxide-atmosphere interface. In the case of Cr-containing alloys, it is highly desirable to have an oxide layer present which consists predominantly of Cr$_2$O$_3$ rather than CrO. If CrO is present, rapid diffusion of Cr^{2+} into the glass may occur and this may alter the characteristics of the glass in the interfacial zone or may react with glass constituents to produce undesirable products; for example metal phosphide or silicide phases. The influence of additives to the glass in eliminating or at least minimizing undesirable reactions has been stressed; for example, use of transition metal oxide additions which react preferentially with diffusing Cr or Fe to yield less harmful products, or which prevent the formation of hydrogen caused by reaction of Cr with water present in the glass. In addition, the fact that some systems are apparently more prone to reaction of diffusing metal species with water requires further investigation.

The presence of impurities or small amounts of metal alloying elements can also influence the sealing behaviour, having either a detrimental or a beneficial effect depending on the specific system under consideration. The presence of carbon or carbides can be particularly problematic. It has also been noted that the presence of water dissolved in the glass can have a particularly undesirable influence on the sealing behaviour. It is thermodynamically favourable for many diffusing metal species to react with dissolved water to form hydrogen gas which may lead to severe bubbling within the interface.

Another important variable is the heat-treatment schedule employed to produce a seal or coating. In the case of a glass-ceramic system, a three-stage heat-treatment cycle is conventionally employed consisting of a high temperature stage to melt the glass, a lower temperature stage to nucleate the glass and a final higher temperature stage to bulk crystallize the glass. It has been observed in many instances, however, that nucleation and crystallization may occur effectively and efficiently during the heating and cooling stages to and from the sealing temperature, making a more complex three-stage treatment unnecessary.

Overall, when determining the optimum sealing conditions for any given system, it is clear that each individual system must be treated according to its own merits and particular characteristics and, in addition, to obtain a high-quality seal good house-keeping practices must be rigorously adhered to at all times.

9. CERAMIC-TO-METAL, GLASS-TO-GLASS, GLASS-TO-CERAMIC, AND CERAMIC-TO-CERAMIC SEALS AND COATINGS

When reviewed by Partridge in 1949 there was only a very modest degree of interest in alternatives to glass-to-metal systems. The main emphasis in this area was in fact on glass-to-glass systems used in the manufacture of graded glass-to-metal seals; for example, seals involving bonding of low expansion fused silica or borosilicate glass to higher expansion metals. In this instance, a gradual change in thermal expansion between the glass and the metal is achieved through the joining of a succession of glasses with different thermal expansion characteristics, thereby reducing residual stresses in the individual glassy layers to safe levels. The graded seal method is still often employed for sealing wires into glass envelopes. In this instance, the wires are first coated with an intermediate glass of suitable thermal expansion coefficient before sealing this assembly into the envelope, which may have quite different thermal expansion characteristics to those of the wire.

Glass-to-ceramic seals were reported by Partridge(1949) for a limited range of ceramic materials including china clay, sillimanite, pyrophylite, mullite, zircon, thoria, alumina, beryllia and magnesia. Partridge noted that, as a rough guide, the thermal expansion of the ceramic should ideally be equal to that of the glass from ambient temperature to the upper annealing point of the glass. He placed ceramics into two classes, those suitable for forming seals with hard, low expansion glasses, and those for sealing with soft, high expansion glasses. China clay, sillimanite and zircon fell into the first class, whilst ceramics based on steatite and magnesia fell into the second. In the case of commercial porcelains, it was stressed that these possess very different thermal expansion characteristics, depending on the source. Today, the range of commercially available ceramics has, of course, expanded considerably.

Ceramic-to-metal seals were also reported by Partridge in which the ceramic component would be treated to apply an adherent metal coating, usually of silver, followed by joining of the metal and ceramic parts by soldering. A metallic coating of silver was applied to the ceramic surface first by painting the surface with a suspension of silver particles, platinum chloride and silver oxide in an organic medium, often lavender oil. This was followed by heating at low temperature under oxidizing conditions to drive off the organic material. The temperature was subsequently increased until a mirror-like coating of silver appeared, of thickness around 30 μm. This process was the forerunner of the Mo–Mn metallization process, until recently employed extensively and almost exclusively in the manufacture of ceramic-to-metal components. Ceramic-to-metal and ceramic-to-ceramic bonding is also an extremely important topic in relation to metal and ceramic matrix composites (e.g. Donald, 1995), although this is outside the scope of this monograph. Some of the more recent specific developments in the areas of ceramic-to-metal, glass-to-glass, glass-to-ceramic and ceramic-to-ceramic seals and coatings are briefly reviewed in the following sections.

9.1   Ceramic-to-metal seals

There is a wealth of information available on the joining of ceramics to metals. As this topic is beyond the main scope of this monograph only a brief overview is given here. Additional and more comprehensive details may be found elsewhere (e.g. Bondley, 1947; Bender, 1954; Palmour, 1955; Van Houten, 1959; Kohl, 1960, 1964; Pattee, 1978; Erz and Hennicke, 1984; Nicholas and Mortimer, 1985; Moorehead, 1987; Bates *et al*, 1990; Hey, 1990; Courbiere, 1991; Akselsen, 1992; Selverian *et al*, 1992; Selverian and Kang, 1992; Santella, 1992; Mizuhara, 2000). Traditionally, ceramic-to-metal components are used when a glass-to-metal system is inadequate in terms of some specific property requirement, which may include temperature capability or electrical or mechanical behaviour; for example high temperature ceramic vacuum tubes, microwave and radar tubes, structural materials in engines and turbines, aerospace components, etc.

Ceramics generally exhibit higher melting temperatures than the most common metals to which bonding is desirable. In addition, pure molten metals do not as a rule wet ceramics and therefore, unlike glass-to-metal systems, fusion processes cannot normally be employed to bond a ceramic directly to a metal. It is therefore generally necessary to apply techniques that allow wetting of both ceramic and metal parts by a suitable braze filler metal. This may be accomplished by indirect brazing, where the ceramic surfaces are metallized prior to brazing with conventional filler metals, or by direct brazing, where the filler braze contains active metals such as titanium or zirconium that will react directly with the ceramic.

Many different techniques for joining ceramics to metals have been studied over the years, with much of the early work during the 1930s and 1940s carried out in Germany and the United States prior to and during World War II. Traditionally, bonding of ceramics to metals has been achieved by brazing the parts together after applying to the ceramic surface a coating which the braze will wet and adhere to. An early method used by the Telefunken company in Germany for the manufacture of microwave tubes consisted of applying a fine suspension of W, Ta, Mo, Re, Cr or Fe powders in a nitrocellulose lacquer to the ceramic surface by brushing or spraying. The ceramic was then fired in a hydrogen atmosphere at 1300 to 1500°C to bond the metal to the ceramic. A thin film

Metallizing conditions	Resulting microstructure	Effects
Temperature too low Short time Excessive humidity or Dry atmosphere	 Ceramic Molybdenum Glass	Metallizing friable No bonding Difficult to plate Metallizing structure inconsistent
Optimal	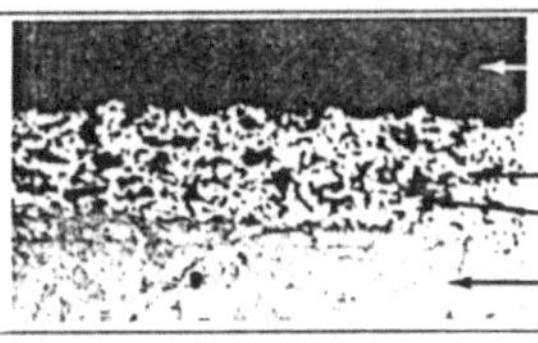 Ceramic Molybdenum (grey) Glass (black) Braze	Excellent adherence Bright molybdenum Easy plating Homogenous/coherent structure
Temperature too high Long time Excessive humidity or Dry atmosphere	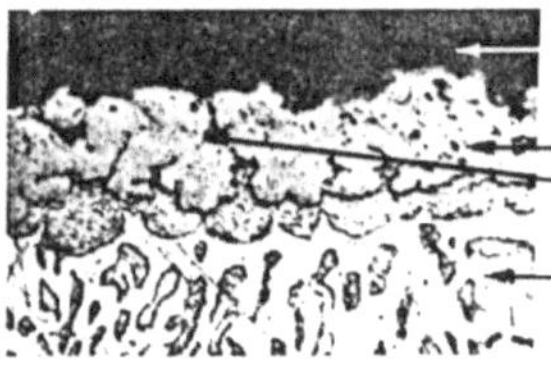 Ceramic Molybdenum (grey) Glass (black) Braze	Poor adherence Large Mo grains No glassy phase or an excess of glass on surface Braze penetration Leaks Plating difficulties

Figure 76: Schematic representation of the Mo/Mn process for bonding ceramics to metals by conventional brazing (after Hey, 1990)

of Ni or Cu was subsequently applied to this coating by electroplating and the metallized ceramic joined to the metal parts by conventional brazing. The addition of MnO, TiO_2, BaO or CaO to the metal powder mixture prior to applying to the ceramic surface and firing was observed to improve the adherence of the metallized layer. This reported process was a modification of much earlier processes involving the use of metal powders. Nolte and Spurek(1954) subsequently modified the Telefunken process by adding a substantial quantity of MnO to Mo powder, this enabling metallization of ceramic bodies at lower temperatures, circa 1250°C. The composition of the metallizing coating was given as a suspension of 80% Mo powder and 20% MnO in an amyl acetate/acetone/binder mixture. This was the forerunner of the now well established technique known as the Mo–Mn metallization process, used extensively for joining metal and alumina parts, and depicted schematically in Figure 76.

Many additional modifications have subsequently been made to this basic process. For example, the suspension may contain additional constituents to Mo and MnO, depending on the particular ceramic to be metallized. Such additions may include Fe, SiO_2, CaO, TiH_2, etc. (The use of titanium hydride alone to coat a ceramic prior to brazing was reported by Bondley in 1947 and Pearsall in 1949.) In the process developed at Telefunken for bonding to alumina ceramics a suspension of Mo and MnO together with various glass-forming additives is painted onto the alumina surface and fired in wet hydrogen. The glassy material densifies the metallic coating and bonds it to the alumina. The metallized surface is subsequently Ni plated, the purpose of the Ni layer being to improve the wetting characteristics of the metallized layer by brazing alloys. The suspension may be applied by brushing or by mass production techniques such as spraying or silk screen printing. Conventional brazing alloys may include Cu–Ag and Au–Ni. Advantages of the Mo–Mn process for metallization of alumina include the fact that it has a well established technological base, it is easily automated, and brazing can be carried out in a variety of atmospheres. Its major disadvantage is that it is only applicable to alumina and a small range of additional ceramics.

Alternatively, and more recently, reactive brazes have been employed which will wet and react with both the metal and ceramic components without the need for metallization of the ceramic. Such brazes are based on Cu or Ag and contain one or more of the reactive metals Ti, Zr, Hf or Al, although many alternative alloys have been studied including those based specifically on Ti and Zr (Akselsen, 1992) and those based on Au (Stephens and Hosking, 2003). The solubility of the reactive metal Ti is relatively low in Ag but much higher in Cu. Consequently, common reactive braze alloys are often based on the Cu–Ag–Ti system. When Ti is present in sufficient quantity, usually 10–30%, these alloys will readily wet and bond to such ceramics as alumina, silicon nitride and silicon carbide. Additional alloying elements including Sn, In, Al, Au, Co and Nb may be added to improve certain properties or brazing characteristics. Aluminium-based brazing alloys containing Cu additions have also been developed and employed to bond a variety of ceramics including alumina, zirconia, silicon nitride and silicon carbide to metals. One disadvantage in the use of reactive brazes is that brazing operations require joining to be carried out in vacuum or a protective atmosphere in order to prevent the active metals from reacting with oxygen, although an oxidation resistant ceramic-to-metal braze has recently been developed for use in YSZ-based electrochemical devices (Weil and Paxton, 2002). An attempt has also been made to join tungsten to a SiC/SiC composite with possible applications in fusion reactor technology, using copper and titanium as the bonding medium (Kurumada et al, 2003).

Brazing alloys of composition outside the normal range can now also be prepared as microcrystalline or ductile amorphous ribbons by rapid quenching techniques, ideal for use in braze applications. The amorphous braze alloys have the particular added advantage of high bend ductility, so that they can be deformed into the geometries required for brazing complex shapes.

Solid state diffusion bonding, in which the metal and ceramic components are heated together in contact and usually under pressure, can also be achieved for some systems, as reviewed by Derby (1990). Specific studies aimed at investigating the bonding characteristics of ceramics directly to metals are less well documented than is the case for glass-to-metal bonding, but are nevertheless very extensive. As in the case of glass-to-metal bonding various interactions are possible in ceramic-to-metal systems. These range from reactions at the atomic level which give rise to chemical bonding, to larger scale reactions involving the

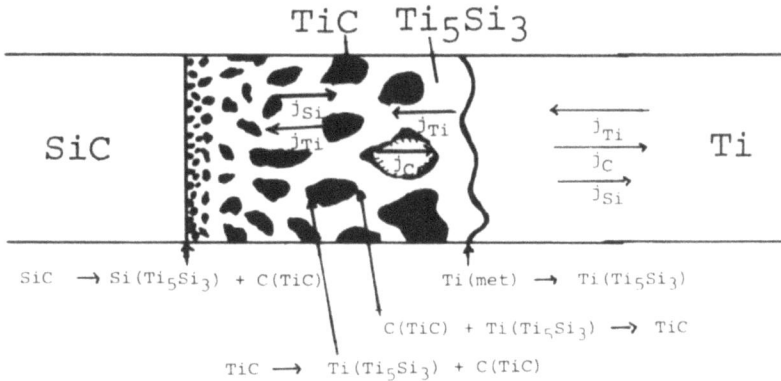

Figure 77: Schematic representation of ceramic-to-metal solid-state bonding (after Backhaus-Ricoult, 1990)

formation of reaction products and intermediate phases. Various types of interface are possible in ceramic-metal systems including banded structures, layered or aggregate morphologies, and interfaces containing precipitates. In one study Backhaus-Ricoult(1990) examined the solid state bonding of SiC to Ti. Reference to the Ti–Si–C phase diagram at 1200°C suggested that new phases, Ti_5Si_3 and TiC_{1-x}, should be formed in the reaction zone. It was also noted that the diffusion of C in Ti is around ten times faster than the diffusion of Si, and that the solubility of C in SiC is ≈1% compared with ≈3·7% for Si. This results in a predominance of Ti_5Si_3 close to the Ti interface and TiC_{1-x} closer to the SiC interface, as observed experimentally, and illustrated schematically in Figure 77.

The solid state reaction between SiC and Ni- and Fe-based alloys has been studied by Schiepers *et al*(1990). In the case of the SiC/Ni system it was noted that banded structures were formed consisting of regular layers of $NiSi_2$, Ni_5Si_2 and Ni_3Si. This contrasted with the Fe system for which a mixture of Fe_3Si and carbon precipitates were formed. Solid state bonding between Ni and alumina has been noted to result in the formation of a nickel aluminate spinel phase, $NiAl_2O_4$, as an interlayer (Trumble and Ruhle, 1990). Silicon carbide ceramics have also been bonded to Nilo-K metal using a solid state process (Komori *et al*, 2002).

The microstructure and fracture strength of joints between silicon nitride and vanadium produced by solid state diffusion bonding has been investigated by Maeda *et al*(2003). Microstructural analysis indicated that up to five different stages of bonding could occur, depending on the bonding time at 1200°C, in which different interfacial layers were formed and subsequently dissolved. It was noted that the fracture strength of joints increased with bonding temperature, but after reaching a certain maximum value, which depended on the precise temperature, strength then decreased with time. Bonding of W to SiC has been achieved by hot pressing (Son *et al*, 2004) where it was observed that reaction to give a mixture of WSi, W_2Si, WC and W_2C occurred. Despite significant interfacial reaction, joints with flexural strengths as high as 90 MPa were formed, although if excessive grain growth within the reaction layer was allowed to occur strength was seriously reduced.

Alumina has been successfully bonded to silver using the solid state technique despite the large difference in thermal expansion (Serier *et al*, 2004). This was achieved by use of silver samples which were saturated with oxygen (620 ppm). Similarly, alumina has been bonded to nickel by solid state diffusion bonding (Wesynczuk and Ruhle, 1986). Alumina has also been bonded to niobium using a solid state diffusion process at 1400°C (McKeown *et al*, 2005). In this study it was noted that the bonding strength could be improved significantly by depositing a 3 μm thick Cu film onto the Nb surface prior to bonding, this process being described as "liquid-film-assisted joining". It has been reported that ceramics may also be joined to metals at ambient temperature by the "surface activation" method in which two clean surfaces are brought together (Suga *et al*, 1990). This method involves first scrupulously cleaning the surfaces to be joined and then bringing them together in a clean atmosphere, typically ultra high vacuum.

Various ceramics have been successfully bonded to copper using a variety of techniques. In the case of bonding alumina to Cu, for example, it has been noted that either solid state or liquid phase bonding can be achieved (Juve *et al*, 1990). At a bonding temperature of 1000°C in the presence of oxygen $CuAlO_2$ is initially formed at the interface but subsequently decomposes to form alumina and Cu_2O precipitates in the metal near the interface. This results in a strong bond between the alumina and Cu. If a higher bonding temperature is employed, e.g. 1065°C, in addition to the formation of $CuAlO_2$ at the interface, a $Cu–Cu_2O$ eutectic phase is formed which wets both the Cu and the $CuAlO_2$ phases. This forms a strong bond on cooling. This reaction involving copper and Cu_2O has been employed as a practical means to wet and bond to alumina in the manufacture of copper-bonded alumina substrates for power semiconductor applications. Laser cladding of copper to alumina has also been reported (Shepeleva *et al*, 2000) in which copper powder was fed into the alumina substrate melt zone. It was noted that reaction between the copper and alumina is essential in order to promote the formation of a strong chemical bond. If the bonding was carried out in air, only weak bonding occurred due to the formation of an oxide layer on the copper particles, this layer hindering the diffusion of Al into the Cu. On the other hand, laser bonding under argon resulted in strong bonding by direct reaction between the alumina and copper to form $CuAlO_2$ at the interface. Copper has also been joined to alumina by eutectic bonding after first oxidizing the copper to give a layer of CuO around 3 μm thick on the surface to be bonded to or, alternatively, forming a layer of $CuAlO_2$ on the surface of the alumina by reacting the alumina surface with CuO (Seager, *et al*, 2002). It was noted that in order to provide good bonding the $CuAlO_2$ layer formed must be < 2 μm thick. Successful bonding of copper to AlN has been achieved by pre-oxidizing both the copper and the AlN prior to sealing to give thin layers of Cu_2O and Al_2O_3, respectively (Jarrige *et al*,

2004). Bonding subsequently occurs by wetting of the ceramic by a Cu–O eutectic phase at 1065°C to give strong adhesion, with failure of the system always occurring through the AlN phase and not at the interface.

Ceramic-to-metal seals have been made using an intermediate solder glass as the sealing medium, although the use of a solder limits the operating temperature capability of such a system. The solder glass may be melted using conventional fusion techniques or, alternatively, microwave heating may be employed (Meek and Blake, 1985). In the microwave method a slurry of glass sealing powder, together with a coupling agent such as an organic oil, is applied to the ceramic and metal parts which are then heated in a microwave oven to melt the glass and seal the parts. It was noted that leak tight seals could routinely be produced using this technique. The benefits of using such a method are questionable, particularly the use of an organic material as coupling agent, but it was claimed that the process is easily automated, efficient and fast. Ultrasonic bonding has also been employed to join alumina to aluminium (Wolterdorf et al, 1995). The bonding is achieved via a very thin (< 20 nm) interlayer of amorphous oxide.

As in the case of glass-to-metal seals, the effect of thermal expansion mismatch between ceramic and metal parts is a very important consideration. In the case of brazed joints some plastic deformation of the braze layer may be able to accommodate some of the stress, so that these residual stresses may be less of a problem than is the case with glass-to-glass, glass-to-ceramic or ceramic-to-ceramic systems. Similarly, as also is the case for glass-to-metal seals, chemical bonding is an important issue in the formation of strong ceramic-to-meal seals, as highlighted by Nicholas(1986). Direct chemical reaction between ceramics and metals is therefore important in order to form strong chemical bonds at the interface. Klomp(1986) has highlighted systems and processing conditions that are thermodynamically favourable in terms of suitable chemical reactions. As in the case of glass-to-metal seals, however, it is essential that any reactions that do occur are not so serious or extensive as to cause the properties of the interface to differ significantly from those of the bulk materials to be joined.

9.2   Glass-to-glass seals

As noted earlier, glass-to-glass joining would normally be carried out using a glass solder seal. This requires a suitably low melting temperature solder glass which is compatible with the glasses to be joined. As also noted earlier, low temperature solder and sealing glasses generally exhibit poor chemical durability, although many new compositions have now been developed with superior durability to the earlier systems.

Glass-to-glass joints have recently been prepared using ultrasonic welding (Wagner et al, 2001, 2003). In this method the glass components to be joined are subjected to simultaneous pressure and a high frequency ultrasonic field. One advantage of this welding technique is that only low temperatures are normally generated, below the glass transition region of the glasses, and this results in distortion-free joins. The preparation of glass-to-glass joints by ultrasonic welding has also been reported by Wagner et al(2001; 2003). A 3 kW ultrasonic metal welder operating at 20 kHz and with a variable oscillating amplitude of between 11 and 20 μm was employed in these trials. It was noted that care has to be taken in the choice of surface roughness, welding energy and amplitude, and contact force, otherwise either no bonding results or, alternatively, the glass disintegrates during the process. As the procedure does not involve melting of the bulk components no distortion of parts should occur. Typically, a glass surface roughness of < 20 nm is required. In the work reported by Wagner, borosilicate glass with a surface roughness of 14·9 nm and soda lime silica glass with a surface roughness of 6·4 nm were bonded to themselves using samples in the form of glass plates 20×20×3·3 or 2·4 mm thick. Unlike ultrasonic bonding of glass to metal, however, the glass does not bond over the entire welding area but, rather, at discrete points within this area. The reason for this behaviour is related to the fact that glass, unlike metals, cannot normally flow plastically below the annealing point (note that this is a debateable point as glass does, for example, undergo permanent deformation during microhardness testing at ambient temperature). Bonding between the glass/glass systems therefore only takes place at the areas of highest contact pressure, and this will be related to surface roughness and the consistency over large areas. At the present time it is not therefore possible to form hermetic joints by this method. A joint strength in shear of only 0·10 to 0·35 MPa was achieved in this study, but it was suggested that much stronger strengths should be achievable, with up to 45 MPa predicted.

9.3   Glass-to-ceramic seals

New joining techniques are being explored for the manufacture of miniature components and microsystems. For example, glass-to-silicon bonding is required in a wide array of opto-electronic components including laser sources, sensors, switches and multiplexes (used for telecommunications, biomedical devices and MEMS encapsulation). Typical adhesives which have been traditionally employed in these devices suffer from low bond strength and poor long term stability. Anodic bonding, another traditional method, is based on the transportation of alkali ions at high temperatures from one wafer to another, facilitated by an externally applied electrostatic field. This requires temperatures of the order of 500°C which cannot be applied locally and therefore have to be applied to the whole component, which may lead to some degradation in the properties. Bonding between pyrex glass wafers and silicon has nevertheless recently been achieved using an anodic bonding method at relatively low temperatures, i.e. 200–300°C (Wei et al, 2003). The resulting seals were monitored using a variety of techniques including tensile testing, SEM, SIMS, and scanning acoustic microscopy, SAM. It was noted that seal failure mainly occurred by cracking through the glass, rather than at the interface.

Other studies include the use of mixed silicate/phosphate glasses containing SiO_2, BPO_4, $AlPO_4$, Na_2O and CaO for bonding alumina laminates (Sroda and Stoch, 2001). In this system Na^+ and P^{5+} diffuse into the alumina, this helping to provide a good bond between the alumina and glass. Chalcogenide glasses based on As, Se and S have also been proposed for bonding together ZnS and ZnSe components to form multi-spectral infra-red transmitting windows (Hopkins *et al*, 1978). These glasses are ideal bonding media for this application due to their excellent infra-red transparency, low melting temperatures and good durability. A number of different glasses have recently been examined to determine their suitability for sealing applications involving high temperature SiC packages (Guinel and Norton, 2004). Selections were ultimately made based on thermodynamic calculations to assess the likelihood of reaction between constituents in the glass and SiC to form CO gas.

Novel bonding techniques include laser joining; for example the joining of glass to silicon wafers employed in opto-electronic components (Witte *et al*, 2002). Laser bonding is a promising alternative to the traditional methods, leading to higher bond strengths and good reproducibility. As laser bonding relies on localised heat input from the laser to initiate the bond it can, in principle, be applied to very small structures. A clean environment and close fitting of parts is, however, essential in operations of this nature in order to ensure high bond strength and hermetic sealing.

9.4   Ceramic-to-ceramic seals

Various ceramics have been joined either to themselves or to other ceramics by brazing. As in the case of ceramic-to-metal sealing, the ceramics may be metallized first and then joined using a conventional metallic braze filler alloy. Alternatively, active metal brazes may be used without the need to metallize the ceramics to be joined first. Microwave heating has also been used to join ceramic-ceramic and ceramic-glass parts (Meek and Blake, 1986). As also is the case for glass-to-metal seals, chemical bonding is an important issue in the formation of strong ceramic-to-ceramic bonding, as highlighted by Nicholas(1986), and as illustrated in Figure 12(b) which shows the presence of a $MgAl_2O_4$ spinel phase formed at the interface between MgO and alumina. Alumina-to-alumina bonding has been achieved recently using a thin B_2O_3 interlayer (Chang and Huang, 2004). The strongest bonded system, which exhibited a bending strength of 71 MPa, was noted by sealing for 15 hours at 800°C to give a fibrous aluminoborate interfacial region.

9.5   Coatings

The application of vitreous coatings (glazes) to ceramic ware is a very well established process and has been used for many years for artistic, domestic and industrial purposes; for example, pottery, dinnerware, sanitary ware, roof, wall and floor tiles, etc. The artistic glazing of pottery has, of course, been employed since ancient times (Rincon, 2002). More recently, there have been many studies aimed at the coating of a large variety of ceramic materials for diverse applications. Coating of biomedical materials in order to improve their compatibility with the body or to improve their mechanical or chemical properties has, for example, received considerable attention, e.g. glass coatings applied to titanium alloys (Chen et al, 2002). Alternatively, Li et al(2002) have coated titanium alloy substrates with hydroxyapatite and hydroxyapatite/titania mixtures using a high velocity oxy-fuel spraying technique, with the aim of improving the bonding properties to bone. In their study, they stressed the importance of the coating-substrate interface in determining the reliability of coated implants. Other studies include application of bioactive hydroxyapatite coatings onto Ti–6Al–4V alloy by a sol-gel dipping process (Hijon et al, 2004), and by a hydrothermal hot pressing route (Onoki et al, 2005). In addition, functionally graded hydroxyapatite coatings have also been applied to Ti (Chu et al, 2003).

Many studies have also been aimed at assessing and attempting to improve the properties of thermal barrier coatings, TBC, particularly their resistance to thermally activated time dependent deformation and failure. Rangaraj et al(2003) noted that addition of mullite to a zirconia based TBC improved the resistance to interface crack growth in both single layer and functionally graded coatings.

A range of abrasion and thermal fatigue resistant ceramic coatings have been applied to steel substrates using plasma spraying (Das et al, 2003). These have included coatings based on alumina, zircon, and yttria. It was noted that the wear resistance of the coatings improved significantly on annealing, thus minimizing stresses introduced into the coatings during deposition. Abrasion-resistant zirconia coatings have also been applied to stainless steel by plasma spraying (Chen et al, 2002). Coatings around 600 μm thick were applied to clean stainless steel substrates that had been grit blasted in order to increase the surface roughness and improve adhesion. Both micrometre and nanometre size ceramic starting powders were used and it was noted that the nanosize material gave better wear resistance coatings than the more conventional micrometer size material. Ceramic coatings have also been applied to diesel engine components in order to improve overall performance including fuel efficiency (Buyakkaya et al, 2004). Coatings of $CaZrO_3$ 0·35 mm thick were applied over a NiCrAl bond coat to the cylinder head and valves of a standard six cylinder diesel engine by plasma spraying. It was noted that fuel consumption improved by around 6% and exhaust emissions were lower than for the standard engine.

In an attempt to improve the shortcomings associated with plasma spray coatings, which can include the weakening presence of microcracks in the ceramic coating, coupled with poor substrate-coating adhesion, alternative techniques for applying coatings have been sought. One alternative is laser cladding. Coatings of Al_2O_3–ZrO_2 eutectic have been applied to low expansion ferritic steels by Bourbon *et al*(1999) using the laser cladding method. In order to achieve satisfactory coatings it was stressed that the thermal expansion mismatch between coating and substrate should be as small as possible. In the case of the Al_2O_3–ZrO_2 eutectic coatings on steel there was a mismatch of $2{\cdot}5{\times}10^{-6}$ K^{-1} (eutectic $\alpha \approx 9{\cdot}0{\times}10^{-6}$ K^{-1}, steel $\approx 11{\cdot}5{\times}10^{-6}$ K^{-1}). The coating was applied by first creating a molten pool on the substrate surface with the laser, followed by feeding ceramic powder into this molten layer via a specially designed nozzle system. It was noted that the quality of the resultant coatings depended strongly on the process parameters including substrate preheat, laser beam power and coating thickness. The best coatings were around 600 μm thick and exhibited higher toughness and improved wear resistance relative to similar plasma spray applied coatings. The laser cladding method would therefore appear to offer considerable scope for alternative ceramic–metal systems.

Sol-gel techniques have also been developed for applying ceramic coatings. For example, alumina coatings have been applied to stainless steel substrates using sol-gel techniques in order to improve the thermal properties and resistance to corrosion (Parola *et al*, 2003).

For many applications involving protective coatings it has been noted that multiple coatings offer far greater protection than a single coating with a specific material. For example, it has been shown that the resistance to wear and abrasion of CrN coatings on stainless steel can be significantly enhanced through the use of nanometric CrN/Cr multilayers applied using magnetron sputtering (Martinez *et al*, 2003). As an alternative to multiple coatings, the use of functionally graded coatings is also receiving significant attention. Ceramic coatings on superalloys, for example, require mismatch strain compatibility in order to minimise interfacial stresses and this can be achieved through the use of graded coatings. Oruganti *et al*(2003) have suggested using a superalloy/ceramic graded composite as a functionally graded coating on Rene 95 superalloy. In this system the coating consists of a graded distribution of the ceramic phase within the superalloy matrix. In addition to the graded coatings, composite coatings are also being employed. For example, composite coatings based on Al_2O_3–SiC have been applied to stainless steel (Jiansirisomboon *et al*, 2003). These coatings were applied by plasma spraying onto a CoNiCrAlY ground coat. It was observed that the coatings consisted of nanometre-size SiC particles dispersed in an alumina matrix consisting of both metastable γ- and δ-phases. Hard, dense and weather resistant coatings based on Ni and Cr carbides have also been applied to stainless steel substrates by high velocity oxy-fuel thermal spraying (Ak *et al*, 2004), and coatings of Ti(C,N) have been applied to both ceramic and metal substrates, including silicon nitride, using a novel pulsed high energy density plasma process (Miao *et al*, 2004). This process was found suitable for producing dense, fine grain size (circa 100 nm) coatings which exhibited high strength.

Many different coating techniques have been employed specifically in order to improve the mechanical properties of glass. These coating methods can be broadly classified into surface or bulk methods. Surface methods including chemical ion-exchange and coating with a protective layer are aimed at minimizing the influence of surface defects, whilst bulk methods such as fibre reinforcement are aimed at increasing the resistance of the glass to crack propagation (Donald, 1989; 1995). Since fracture of a brittle solid is almost invariably initiated from the surface due to the presence of stress intensifying defects it is feasible to increase the strength by placing the surface into a state of compression. In this instance, strength increases are achieved because the compressive stress at the surface must first be overcome before defects are subjected to tensile forces. As a very rough guide, the strength of a glass article with a surface compressive layer is equal to the magnitude of the compressive stress plus the normal fracture strength of the untreated material. In order to impart a useful increase in strength the depth of a compressive layer must generally be greater than the size of typical flaws, i.e. > 50 μm. Surface compression can be achieved by thermal treatments, chemical ion-exchange, surface crystallization, or by coating the glass surface with another glass with a lower thermal expansion coefficient (Donald,1989). Alternatively, the glass may be coated with a suitable material which seals existing flaws and helps to protect the surface from further damage. For example, coating with polymeric materials has been used for many years in the glass fibre industry for improving the resistance of glass fibres to surface damage during handling and storage (Loewenstein, 1983). Coating a glass surface with organically modified silicates (ORMOSIL) has also been used to improve the strength of bulk glass articles, and modest increases in strength have, for example, been reported for glass microscope slides (Verganelakis *et al*, 2000).

Coatings have also been employed to improve the resistance of glass to the effects of moisture. In the case of glass-to-metal seal terminals for compressors used in refrigerators Wen *et al*(1999) have used hydrogen containing methyl silicone oil to improve the moisture resistance. Coatings were applied by dipping followed by heating at 160°C.

Additional information on ceramic coatings is given elsewhere (e.g. Wachtman and Haber, 1993).

10.   SPECIFIC APPLICATIONS AND R&D SUPPORT

It is important when considering any application that may require a glass-, glass-ceramic-, or ceramic-to-metal seal to examine thoroughly what is already available from commercial suppliers, before embarking on what may prove to be a costly new seal development programme. As has been explained by Buckley(1979), it is often possible to use an existing seal component that is available off-the-shelf, or to make minor modifications to a system design in order to accommodate an off-the-shelf seal. On the other hand, if a new seal design is required for a given application, it may be possible to modify an existing seal at moderate cost, rather than develop an entirely new component, or alternatively to use a simpler design of seal that will still serve the intended purpose. Buckley gives an example of a 14 to 1 cost saving by a gyroscope manufacture from using a simpler seal design than that which was originally envisaged for a brand new component.

In order to emphasise the wide range and diversity of glass- and glass-ceramic-to-metal seal and coating applications, a number of generic and specific areas are reviewed below. Some applications for alternative ceramic systems are also briefly covered. Where there is on-going research and development in particular areas some examples of the type and scope of work in progress are given. Examples of the range of components covered are shown in Figure 78. A summary of some of the typical applications which employ glass-to-metal seals is given in Table 14 and a small selection of commercial suppliers of glass-to-metal seals and/or components which utilise seals is provided in Table 15.

10.1   Electrical feed-through seals

Glass-to-metal seals are widely employed in the electrical and electronics in-dustries as electrically insulated feed-through connectors and related devices. Examples include lamp envelopes, vacuum tubes, radar and microwave com-ponents, television tubes, reed and relay switches, magnetic sensor switches, movement and tilt sensors, together with the very important area of electronic packaging. Glass-to-metal seals are also employed in such devices as switching and high frequency diodes, photomultiplier tubes and scintillation detectors. Specific applications are considered in detail in the following sections with elec-tronic packaging, for example, covered in Section 10.5. Some typical examples of glass-to-metal seal feed-through components are shown in Figure 79.

Electrical feed-throughs with potential applications as microprobe arrays for high density data storage, sensors and electronics packaging have been reported by Li *et al*(2001). A novel method for producing these was proposed involving

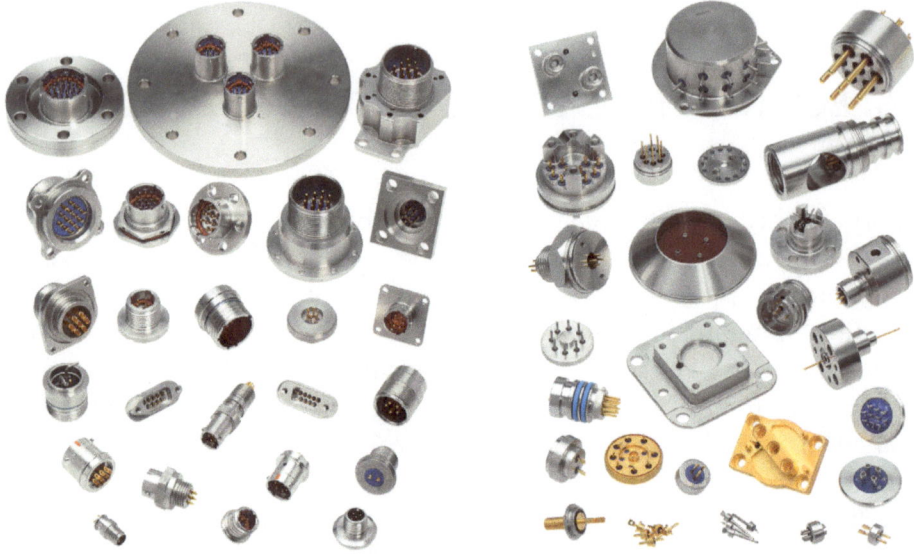

(a) Glass-to-metal seal components (courtesy of Martec)

(b) Glass-to-metal seal components (Courtesy of Latronics Corporation)
Figure 78: Photographs of a selection of components employing glass- and glass-ceramic-to-metal seals illustrating the wide variety of component types

deep reactive ion etching of pyrex plates 150–200 μm thick to produce an array of accurately spaced holes 40–60 μm in diameter which are then filled by electroplating on the inside with nickel to yield an array of electrical feed-through connections. It was noted that the only other method capable of producing a fine array of holes is by laser drilling, but this is not suitable for batch fabrication of components, whereas the proposed deep reactive ion etching is. The etching is carried out in a SF_6 plasma after suitably masking the pyrex plate with a thin layer of electroplated nickel.

Glass-ceramic-to-metal seals are a more modern invention used for more arduous applications. These include vacuum and vacuum interrupter envelopes,

Figure 78: (continued)

(c) Glass-ceramic-to-metal seal components (Courtesy of Ceramaseal)

high pressure explosive actuator components, together with related components requiring higher operating temperatures than is possible with glass-to-metal seals, or when precise matching of thermal expansions is required to metallic alloys which are difficult to bond to conventional glasses.

10.2   Hermetic connectors

Hermetic connectors are employed in numerous applications ranging from sensors, valves, transducers, microwave components, to aircraft, missile and spacecraft instrumentation, gyroscopes, engine control systems, and high pressure fuel tank connectors. Some typical examples are shown in Figure 80.

In recent years there has been an increasing drive toward the miniaturization of components. This has resulted, for example, in the successful development of smaller and smaller electrical feed-through components, including hermetic co-axial connectors for high frequency and ultra-high frequency (up to 110 GHz) microwave applications. Co-axial connectors with glass-to-metal seals as small as 1 mm diameter are now available for use in automotive radar systems, high bit rate optical-to-electrical converters, and ultrabroadband test systems (Powers, 2000). These connectors are available in both screw-in and flange mounted versions, as illustrated in Figure 81.

Glass-ceramic-to-metal seals are also employed in the manufacture of lightning arrestor connectors, an example being shown in Figure 82. The purpose of

(a)

(b)

(a) Pressed seal to Cr-Fe alloy pins for a vacuum tube (after Partridge, 1949)
(b) Tungsten seals for an electric light bulb (after Partridge, 1949)

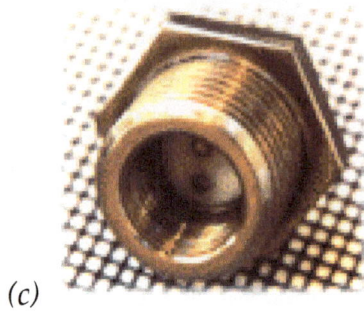

(c)

(d)

(c) High pressure seal (courtesy of Concept Group, USA)
(d) Schematic illustration of feed-through seals for a halogen lamp (after Mizuhara, 2000)

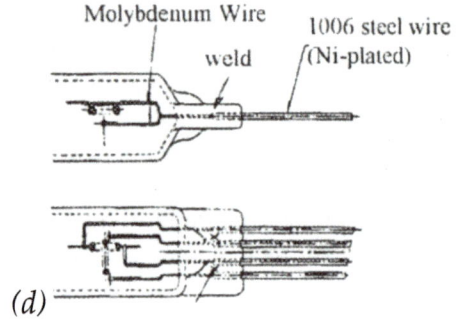

(e)

(f)

(e) Transmitting valve employing a glass feed-through seal to Cu-coated Fe–Ni alloy (after Mizuhara, 2000)
(f) High amperage feed-through seals (after Hermetic Seal Technology, Inc.)
Figure 79: Photographs of some electrical feed-through components employing glass-to-metal seals

such an electrical feed-through connector is to prevent a lightning strike from causing accidental operation of the device that is being protected by the connector. It does this by preventing any current induced by the lightning strike from being transmitted through the connecting pins directly to the device by transferring the

Figure 79: (continued)

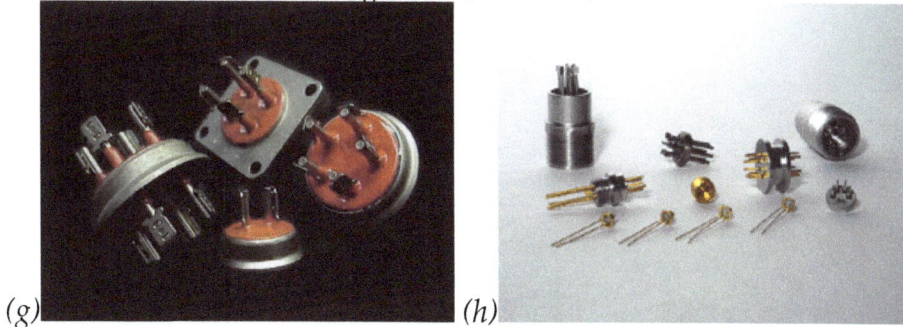

(g)   (h)

(h) Selection of feed-through seals (courtesy of Accra Tronics Seal Corporation)
(g) Feed-through seals (courtesy of Vitrus)

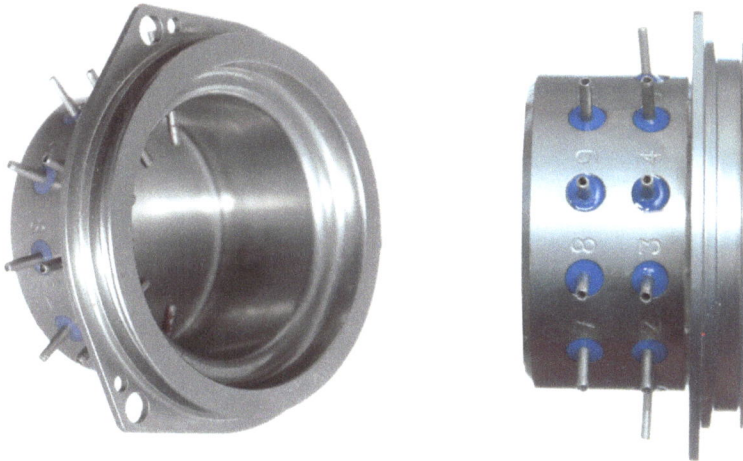

(i) Aerospace feed-through seal employed in Concorde aircraft
(courtesy of Martec)

induced current away from the pins and harmlessly to earth via the metal housing of the connector. Figure 82 also shows a component before and after a simulated 200 kA lightning strike. It may be noted that the component has survived this ordeal by remaining effectively intact. The front face of the device is seen to have undergone extensive damage but the strike has been successfully transferred to the metal housing, and away from the connecting pins at the back face.

10.3   Lamps, lasers and related devices

Glass-to-metal seals are employed in lamps, lasers, and related devices both as electrical feed-through seals and for joining metal and glass component parts together. They are, for example, used in the ordinary incandescent light and torch bulbs which come in all shapes, sizes and operating voltages and wattages, halo-

(a) Selection of connectors (Courtesy of Hermetic Seal Technology Inc.)

(b) Fuel tank connectors
(Courtesy of Deutsch)

(c) Multi-pin connector
(Courtesy of Deutsch)

Figure 80: Photographs of some connectors employing glass-
and glass-ceramic-to-metal seals

gen lights, mercury and xenon arc lights, and fluorescent lights. These systems often employ borosilicate or lead oxide based sealing glasses. Automobile halogen and similar lights, on the other hand, employ more refractory aluminosilicate glass envelopes with glass-to-metal feed-through seals, whilst high wattage security lighting may employ fused silica envelopes. Other applications include linear flashlamps and lasers. Linear flashlamps employ capacitors and are designed to provide high intensity light pulses. Their operating temperatures can be very high, with up to 600°C excursions, and they normally use fused silica envelopes. The seal design for such a device is critical, requiring reliable joining of the fused silica envelope to the metal electrode supports by means of a graded thermal expansion seal. Typical seal constructions for a xenon flashlamp are shown in Figure 83. Lasers may range in power from a few mW or less to several kW and they also employ glass-to-metal seals in their construction. The design of a typical low power He–Ne laser is shown in Figure 84. The borosilicate glass plasma

Figure 80: (continued)

(d) Multi-pin connector
(Courtesy of Deutsch)

(e) Circular and D-Type Series connectors
(Courtesy of Ceramaseal)

(f) Circular connector and header
(Courtesy of Ceramaseal)

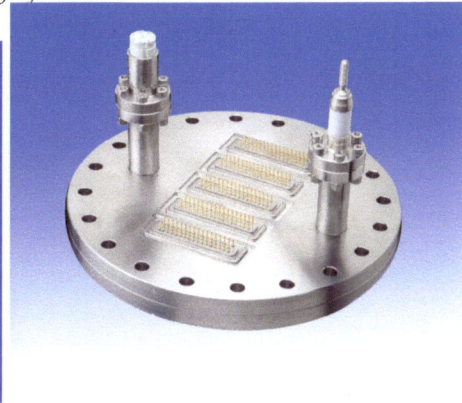

(g) Flange component incorporating
five multi-pin connectors (Courtesy of
Ceramaseal)

tube assembly contains glass-to-metal feed-through seals in addition to mirrors mounted directly to the tube with a solder glass.

10.4   Sensors, switches, relays and transducers

Glass-to-metal seals are employed in many sensor, switch, transducer and relay applications. For example, reed switches, proximity switches, proximity sensors, thermostatic switches and sensors, pressure transducers and transmitters, and in vacuum systems including Pirani gauges and diaphragm monometers. Relays, including vacuum and gas filled relays and heavy duty mercury displacement relays, also employ glass-to-metal seals and feed-throughs.

Precision pressure transmitters are available employing a miniature diaphragm electrostatically bonded to a glass substrate which is bonded into a glass-to-metal seal assembly (Druck Incorporated). The transmitter unit, incorporating sensor

Figure 80: (continued)

(h) *High pressure aerospace connector (Martec)*

(i) *Corrosion monitoring connector for an off-shore application (Martec)*

(j) *Connector assembly for an MRI scanner (Martec)*

components, has been designed to operate under a wide variety of conditions, typically with operating pressures from around 35 MPa to 70 MPa using suitable pressure transmitting fluids compatible with the transmitter assembly which may be constructed from stainless steel and/or Ni-based superalloys. Pressure transducers have been manufactured using high stability pressure measurement elements micromachined from single crystal silicon mounted within a glass-to-metal seal which is isolated from the pressure media by a Ni-based diaphragm electron beam welded to the front of the glass-to-metal seal unit. This construc-

(a) Screw-in and flange-mounted co-axial 1 mm diameter microwave connectors

(b) Miniature 1 mm diameter   *(c) As above highlighting the*
glass-to-metal seal for the above   *miniature size of the seal*
Figure 81: Precision miniature connectors (after Powers, 2000)

tion provides minimum sensor size with high degrees of stability and reliability. A commercially available system is shown in Figure 85. Applications for such transducers include industrial, automotive and aerospace test cells. Combined pressure and temperature transducers employing glass-ceramic-to-metal seals are under development for subsea wellhead pressure and temperature control applications (Phaze Technologies AS). Miniature pressure switches employing glass-to-metal seals have also been designed to support a wide range of applications requiring alarm, shutdown and control functions, ranging from off-shore oil platforms and marine facilities, chemical plant, paper mills, outdoor plants and off-road vehicles, and medical facilities, which may employ such equipment as pumps, compressors, heat exchangers, hydraulic systems, food processing, sterilizers, etc. Switches which employ glass-to-metal seals offer such attributes as small size, corrosion resistance, explosion resistance, and sanitary advantages when used in the food process or medical industries. Reliable performance under diverse operating conditions is also more readily assured with glass-to-metal seal systems, not always possible with alternative systems which may rely on potting or o-ring seals which may fail under pressure.

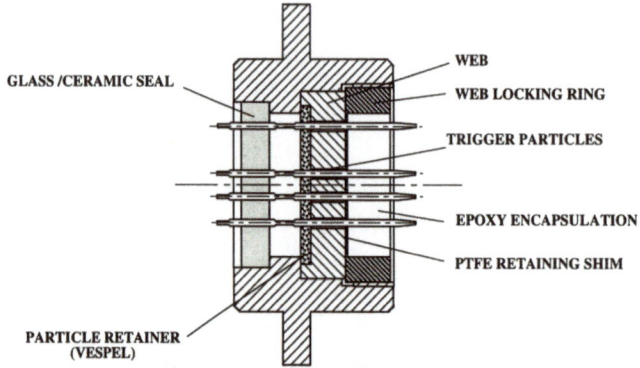

(a) Schematic illustration of a component

(b) Photograph of an actual component which employs a lithium aluminosilicate glass-ceramic sealed to a stainless steel body and Hastelloy C276 alloy pins. Also shown is the metal fixture employed in assembly of the component prior to sealing

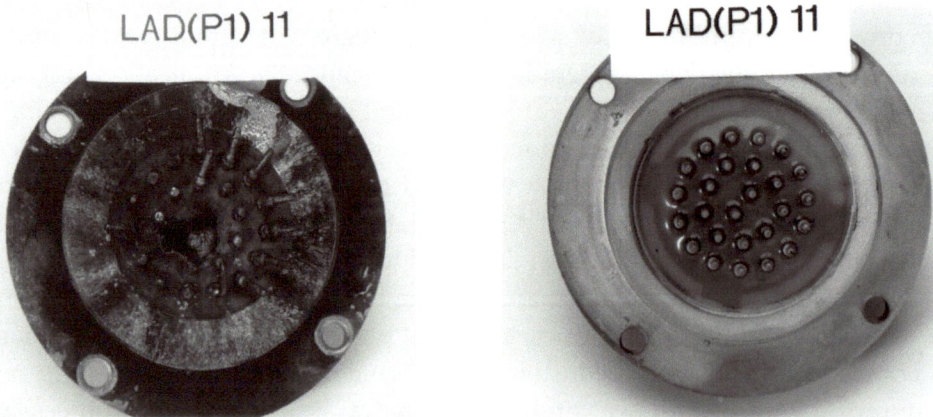

Front face (directly hit by lightning strike)

Back face (note that this has survived the strike intact)

(c) Photographs of prototype lightning arrestor component after a 200 kA simulated lightning strike

Figure 82: Prototype lightning arrestor electrical feed-through component (AWE)

(a) End cap seal

(b) High power graded seal

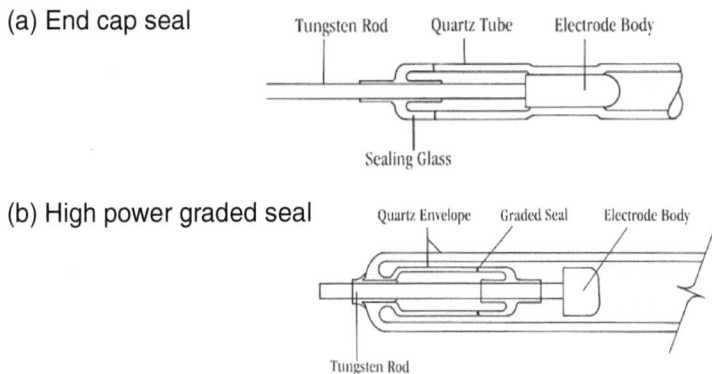

Figure 83: Schematic diagrams showing typical xenon flashlamp seal designs (www.polytech-pi.fr/EGG)

Simple reed switches consist of ferromagnetic contacts hermetically sealed into a glass envelope which is filled with an inert gas. The switch is operated by an externally applied magnetic field either from an electromagnetic coil or a permanent magnet, as illustrated in Figure 86(a). They have been employed widely and for many years for all types of relay circuits and have become popular again more recently as a more reliable substitute for the less reliable snap-acting mechanical switches for computer input circuits in many industries. Advantages include long life, low operating power, and electrical contacts isolated from the environment. They are normally designed for specific voltage and current applications. Reed switches with tungsten contacts, for example, typically operate at 1–100 Watts and < 250 volts, whilst Ru and Rh contacts are designed to operate well in low energy, low voltage applications, i.e. 120 volts AC and 24 volts DC. Applications

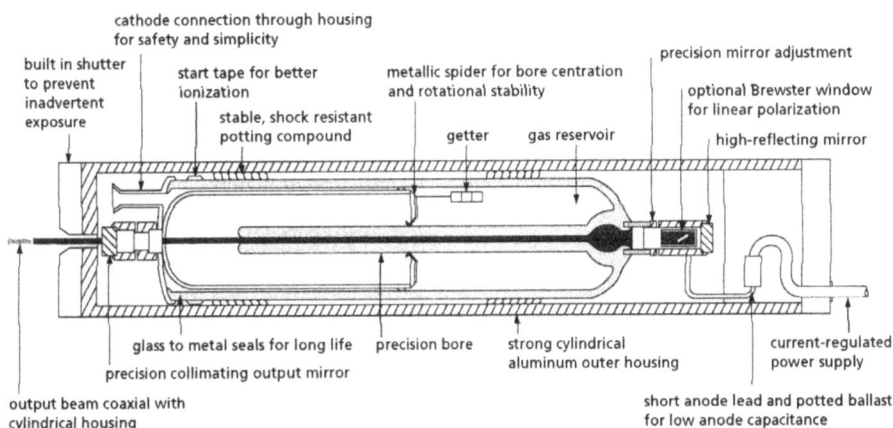

Figure 84: Schematic diagram of a He–Ne laser construction showing glass-to-metal seals (www.mellesgroit.com)

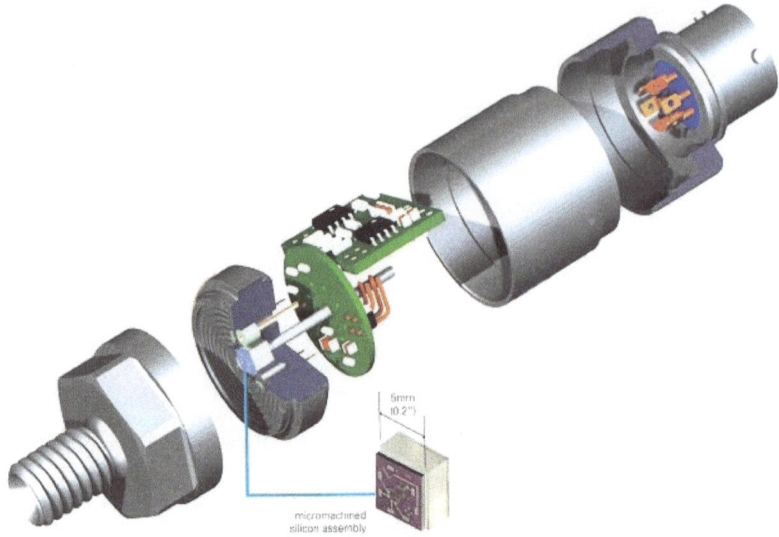

(a) Pressure transducer

(b) Glass-to-metal seal unit
Figure 85: Amplified output pressure transducers and transmitters
(Courtesy of Druck)

are far ranging, and include the computer applications noted above; automotive components, e.g. level and light detectors, seat belt and air bag sensors and cruise control systems; security systems, e.g. door, window and location sensors; audio equipment; proximity switches, e.g. for use with conveyors, escalators and lifts and robotics; together with flow and float sensors, thermostats and domestic equipment. Some typical switches are shown in Figure 86(b).

There are also many types of additional sensors and detectors employing glass-to-metal seals, ranging from temperature sensors, vibration sensors, movement and tilt sensors, accelerometers, magnetic sensors, to pyroelectric detectors, and optoelectronic components which include optocouplers, emitters and detectors. Some examples are shown in Figure 87. The optoelectronic detector, for example, consists of an NPN Si phototransistor in a glass lensed metal package, whilst the pyroelectric detector consists of a lithium tantalate pyroelectric

Figure 85: (continued)

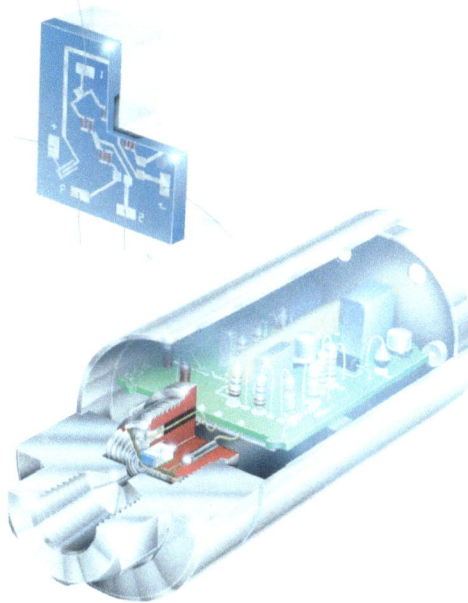

(c) Pressure transmitter

(d) Glass-to-metal seal unit

component housed in a metal package with a glass-to-metal feed-through seal. Another type of sensor, an automobile tyre monitoring sensor system, is shown in Figure 88. This consists of a sensor package which transmits signals via an antenna to a receiver module located within the vehicle which, in turn, provides signals to the instrument display panel. Individual sensor units are mounted, as shown, on each wheel. The package itself consists of a sealed unit housing a thin metal diaphragm which in conjunction with a metal dome acts as a pressure sensitive switch. If the pressure within the tyre falls below a pre-set value the diaphragm bows outwards due to the internal pressure within the package and opens the switch formed between the diaphragm and the dome. A signal is then sent to the receiver via the receiver antenna wire. The sensor package itself incorporates a glass-to-metal seal header unit to bring the small transmitter antenna out of the package.

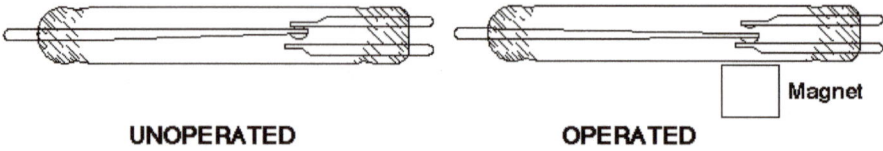

(a) Schematic diagram illustrating operation (after Kunz, 2005)

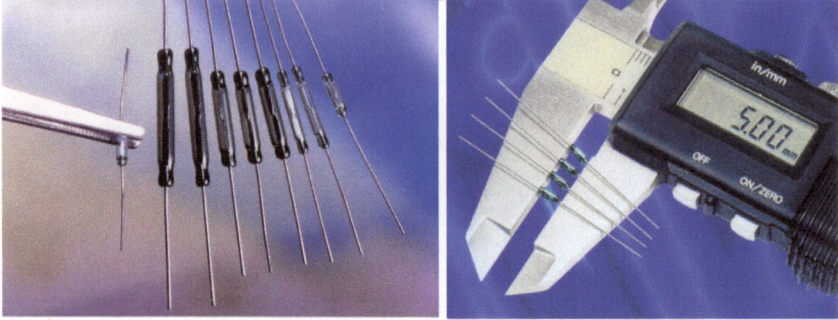

(b) Actual reed switches (Courtesy of Rhopoint)
Figure 86: Reed switch employing glass-to-metal seals

10.5   Electronic substrates and packaging

In its simplest form, an electronic package is a container for electronic components with terminals included to provide electrical connections from the outside to the inside of the container. An electronic substrate is the material onto which the circuits and electronic components are laid down or attached to. There has been a steady advancement in semiconductor device technology and in the miniaturization of integrated circuits over a number of years which has made electronic and microelectronic packaging an extremely important and topical issue. These packages are now employed in a multitude of diverse and everyday applications in the aerospace, medical, military, radio and television, computer, telecommunications, automobile and transport, and marine industries.

One of the main functions of the package is to protect the circuits from the environment, which may be corrosive and/or humid in nature. The housings require electrical interconnections in order that the transfer of electrical signals into and out of the housing can be made, and these interconnections must also be hermetic. These interconnections usually employ glass-to-metal seals where the lead wires are hermetically sealed and electrically isolated from the package body. As noted, the package body itself serves to protect the microcircuit device or devices inside. Photographs of a selection of microelectronic packaging housings and components employing glass-to-metal seals are shown in Figure 89. There are various types of packaging geometries including, for example, "flat pack" configurations where the circuits are packaged in thin rectangular or square shaped containers with the connecting pins projecting from the edges of the package; "flip chip" semiconductor devices in which the connecting pins are situated on one face which permits flip (face down) mounting of the package to the required

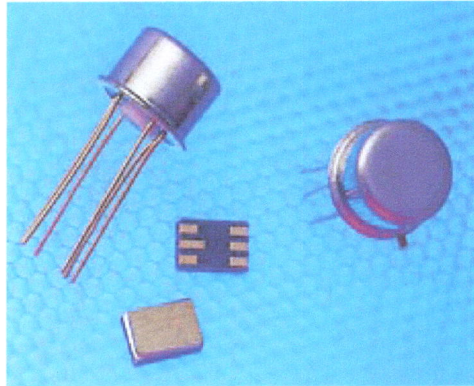

(a) Optoelectronic sensor (Courtesy of Minco Technology Laboratories)

(b) Sensor to regulate cabin pressure in commercial aircraft (Courtesy of Concept Group)

(c) Sensor header used in aerospace applications (Courtesy of Concept Group)

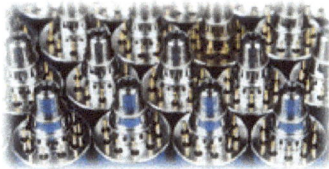

(d) Pressure sensor packages to regulate air and fuel pressure in jet engines (Courtesy of Concept Group)

(e) Pressure sensor (Martec)

Figure 87: Pictures of some sensors employing glass-to-metal seals

external circuit interconnects; "header" packages which are cylindrical containers with a metal lid with connecting leads emanating from the package base; and "single-in-line" packages which are plug in packages with a single row of pins.

The manufacture of a package comprises a number of steps including preparation of the package housing incorporating the necessary holes and slots for the glass-to-metal seals, assembly of the feed-throughs, often with the assistance of graphite jigging, and oxidation of the metal parts either before or during the sealing operation. Sealing often takes place using a belt furnace with neutral or mildly oxidizing atmospheres. Glass-to-metal seals employed in packaging have often used Nilo-K and borosilicate glass, these being well established sealing

Figure 87: (continued)

(f) Glass-to-metal package used in an
infra-red optical sensor
(Courtesy of Concept Group)

(g) Feed-through seals in an anodized
aluminium plate used in pressure
sensors (Courtesy of Concept Group)

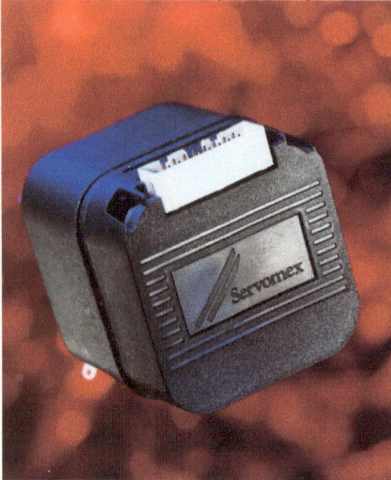

(h) Gas analysis sensor package employing
glass-to-metal seals to ensure that
contamination of the gas does not occur
(Courtesy of Deutsch)

(i) Aerospace Sensor
(Martec)

(j) Aerospace sensor
(Martec)

(k) Humidity sensor
(Martec)

materials. In this case, preparation of the seals normally involves three separate
steps: decarburization of the Nilo-K, followed by oxidation and finally sealing,
although it has been shown that high-quality seals can be made successfully
in a single operation involving *in-situ* oxidation of the Nilo-K during sealing

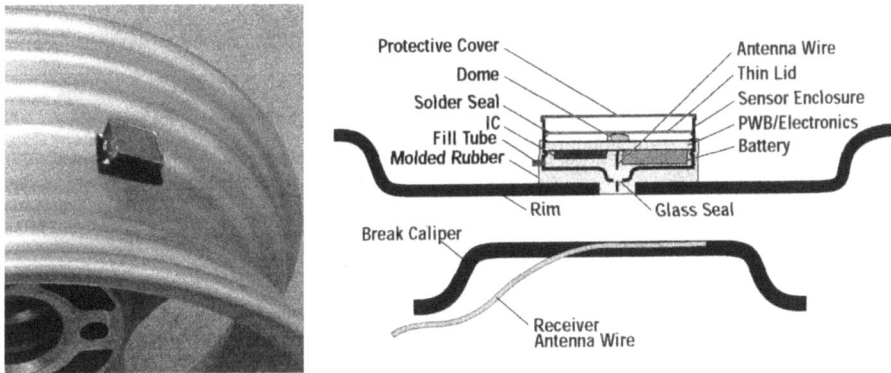

(a) Showing sensor position within the wheel rim
(b) Schematic of the sensor unit showing the position of the glass-to-metal seal
Figure 88: Automobile tyre pressure sensor employing glass-to-metal seals
(Courtesy of Dyco)

(Bandyopadhyay *et al*, 1989). For service, the Nilo-K would normally be gold-plated over a nickel undercoat.

In addition to Nilo-K, commonly used metals include cold-rolled steel, Cu and Mo. The choice of metal is very much dependent on the specific application and in particular the importance of thermal conductivity, mechanical integrity and chemical durability of the device. Thermal dissipation, for example, can be critical in power packages which therefore favour the use of such metals as Cu, Cu–W, Cu-clad Mo and Cu-clad Invar, whilst for less arduous applications Nilo-K with its lower conductivity is perfectly adequate.

Radio frequency and similar electronic circuit components are currently packaged in hermetically sealed steel housings. The electrical feed-through seals are typically manufactured from stainless steel and silicate-based insulating glasses. Such electronic packages are used in a variety of aerospace and related applications including communications satellites, microwave communications, and radar systems. Packages employing stainless steel tend to be heavy thus increasing the overall weight of the final system. This has led to a drive to identify alternative metals, and aluminium is an obvious candidate; however, aluminium suffers from the disadvantages of low melting temperature and high chemical reactivity. This makes sealing to Al difficult, and it cannot normally be successfully bonded to conventional silicate-based insulating glasses. In order to overcome this difficulty, some microwave packages employ aluminium bodies with transition joints comprised of seals between steel rings and pins. These seals are bonded to an aluminium ring which is subsequently welded into an aluminium housing thus forming the package. Although such packages are lighter in weight than the common steel packages, the complex design involved tends to be less reliable due to the large mismatch in thermal expansion between the aluminium housing and the electrical feed-through seals and connectors. Relatively low melting

(a) Photographs of electronic package
housing which employ glass-to-metal feed-
through seals (Courtesy of Schott)

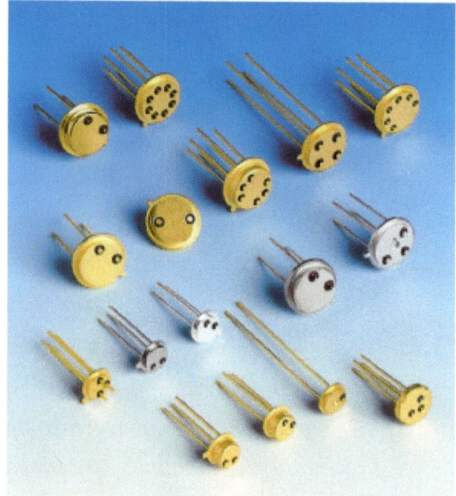

(b) Photographs of electronic packages
(Courtesy of Schott)

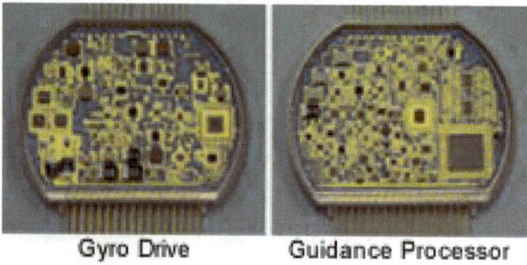

Gyro Drive   Guidance Processor

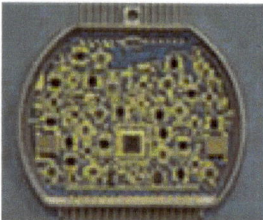

Front End Processor

(c) Signal hybrid electronic packages with a
double row of glass-to-metal seals
(Courtesy of NATEL Engineering Co. Ltd.)

(d) Glass-to-metal package used
in the GEOS weather satellite
(Courtesy of Concept Group)

Figure 89: Electronic packaging

temperature phosphate-based glasses with high thermal expansions have been
developed that can be successfully bonded to aluminium and which exhibit satisfactory chemical durability, mechanical and electrical properties. For example,
Wilder(1980) has reported alkali barium phosphate glasses which can be sealed
to aluminium, whilst Brow *et al*(1998) and Kilgo *et al*(1999, 2000) have provided

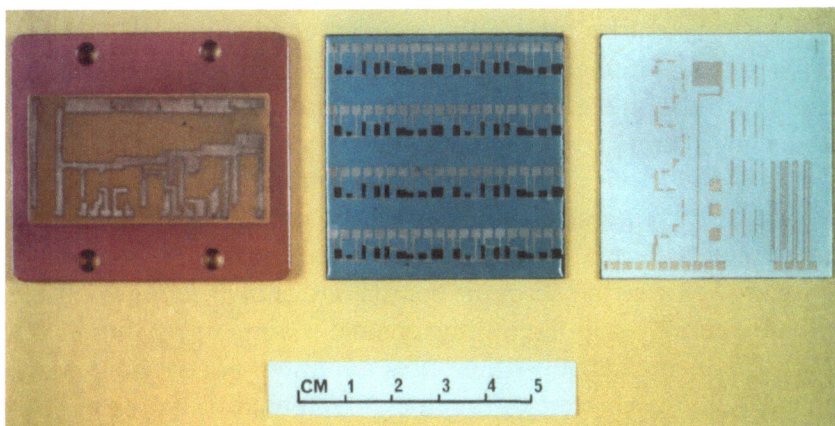

*Figure 90: Enamelled substrates for printed circuit boards
(Courtesy of former GEC Alsthom)*

details of alkali aluminophosphate based glass compositions. Light weight binary Al–Si alloys have also been proposed for use in electronic packaging applications (Jacobson *et al*, 2005). These binary phase alloys are prepared by a spray-forming process to yield fine uniform microstructures with Si contents from 27 to 70% and thermal expansion coefficients in the range 7·4 to $16·0 \times 10^{-6}$ K^{-1} (25–500°C) making them suitable for use with a variety of substrate materials including alumina and gallium arsenide. They can be nickel or gold plated and they are also noted to be readily wet by molten glass, providing glass-to-metal seals of sound quality (presumably using conventional silicate based sealing glasses).

As far as electronic substrates are concerned, electronic packaging devices normally include a substrate material which is usually an organic polymer or a ceramic such as alumina, aluminium nitride, beryllia, silicon nitride, silicon carbide or boron nitride, although recently interest has been shown in enamelled and glass-ceramic substrates. Together with the substrate are the semiconductor devices, capacitors, sensors, interconnections and protective coatings, usually of glass or polymer. A package may consist of many components and interconnections, all of which must work together successfully and reliably. Substrate materials need to possess good mechanical, thermal and electrical properties in order to satisfy the criteria for strength and robustness, heat dissipation and fast signal propagation, required of a substrate. Higher packing density of components places a greater demand on heat dissipation capability which cannot be met with epoxy based circuit boards. On final assembly electronics packages are usually laser welded closed.

It has been noted that glass-ceramics are ideally suited to a range of electronic substrate applications (Partridge *et al*, 1989; Kumta and Sriram, 1993) through a combination of useful properties including relatively high strength and a range of dielectric constants. Both thin and thick film technologies can be employed for laying down conductor tracks. A major disadvantage of glass-ceramics for applications of this nature is their low thermal conductivity. This makes them less suitable for substrate applications where the heat generating chip is in direct contact with the substrate, although their advantages can be exploited for alternative chip geometries. Glass-ceramic coated metal substrates have also been advanced for applications where mechanical robustness and good thermal dissipation are required. Enamelled steel or copper substrates used in the electronics industry for microelectronics applications and multi-layer electronic packaging, including printed circuit boards, are shown in Figure 90. In these applications, circuitry is applied to the enamel coating using thin or thick film technology. An enamelled substrate offers a number of advantages over more conventional alumina ceramic or organic polymer substrate materials, including lower dielectric constants (which allow higher operating frequencies), and higher thermal dissipation factors which is important in close-packed electronic systems (Garland, 1986). Glass-ceramics employed to coat metal substrates not only provide a mechanically stronger system, but also allow closer matching of thermal expansion characteristics, and allow higher temperatures to be employed in the application of thick film circuitry; and this higher temperature capability allows the use of conventional screen printing inks to be employed (Partridge *et al*; 1989a, 1989b). It has been observed that certain glass-ceramic compositions doped with CuO or NiO when heat-treated under suitable conditions may produce a layer of metallic copper or nickel on the surface (McMillan and Hodgson, 1963; Donald, 1993), as seen in Figure 91. It has been suggested that such materials may be useful for producing printed circuit boards.

It has been noted by Shulz-Harder(2003) that direct bonded copper, DBC, substrates are becoming important electronic circuit boards for multichip power semiconductor modules and are replacing complicated assemblies based on leadframes and refractory metallized substrates. The advantages of DBC substrates include ease of assembly and the low thermal expansion coefficient of these substrates which matches silicon in spite of thick copper metallization. This technology allows bonding of copper to alumina or aluminium nitride, and also allows for the production of efficient water cooling devices with internal micro channel structures for cooling power laser diodes and similar high power density electronics. Applications include automotive, avionics and space applications where low weight, mechanical stability and temperature cycling reliability are required.

Hermetic packaging of micro-electronics mechanical systems, MEMS, and micro-opto-electromechanical systems, MOEMS, is a less mature technology. The current methods for manufacturing windows for MOEMS applications involve attaching a metal frame to glass either by heating the glass to a temperature $>T_g$ to bond it directly to the frame or by metallizing the glass and attaching it to the frame using a metallic solder or, alternatively, by use of a glass solder to join the metal and glass parts. Diffusion bonding has also been proposed as an alternative to soldering (Stark, 2003), with benefits including lower temperature processing

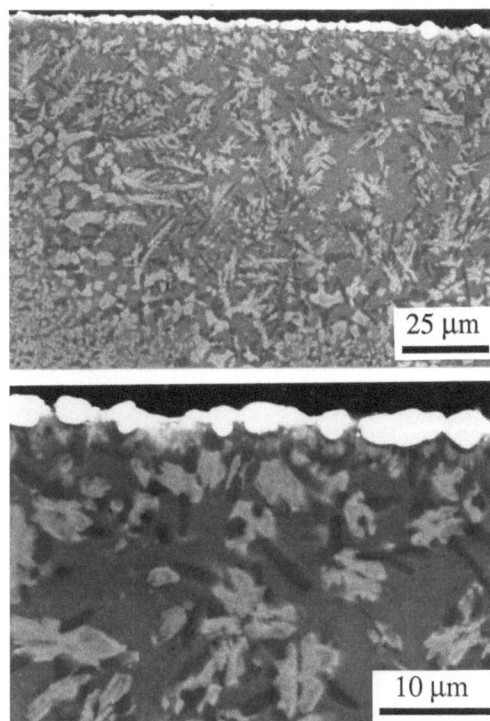

Figure 91: SEM micrograph of a lithium zinc silicate glass-ceramic doped with 2 mol% CuO. A thin layer of copper is visible on the surface due to reduction of CuO to metallic Cu at the surface (after Donald, 1993)

and potentially lower cost processing. Wafer-to-wafer bonding in MEMS has been achieved employing a glass frit (Ser *et al*, 2003).

In addition to electronic packages which operate close to ambient temperature, high temperature packages have been developed which may operate at temperatures anywhere within the range 150 to 1000°C and with applications in aerospace, nuclear power, oil, gas and geothermal instrumentation (NMAB, 1995a). The most common failure mode for such packages is failure of the glass- or ceramic-to-metal seals. For many high temperature applications ceramic packages are employed (in preference to metal packages) based on alumina or aluminium nitride, and these avoid the use of glass-to-metal seals. Components may be bonded to the package substrate material using a variety of techniques including attachment with suitable high temperature glass solders or, for improved resistance to thermal stresses, a metal-loaded glass solder, e.g. silver/glass, may be employed.

Assessment and testing of hermetic packages have led to the development of many standards. For example, MIL-M-38510 is a general specification for microcircuits and fully assembled devices, whilst MIL-STD-883 includes checking the hermeticity of the glass-to-metal seals employed in the package. A device would normally be rejected if the measured He leak rate was found to be $>10^{-8}$ cm^{3}s^{-1}. A method for measuring residual stresses in glass-ceramic substrates with multi-layer copper circuits has been reported using a microhardness technique (Doi and Akio, 2003). Lifetime assessment of packages and package materials may also be carried out under various ageing conditions. In one such study, for example, Shapiro et al(2001) have reported ageing of a Ag/glass-ceramic electrode system for high frequency electronic integrated multilayer packaging applications. A calcium borosilicate glass-ceramic crystallized to varying degrees of crystallinity was employed with silver electrodes embedded at regular intervals. The aim of the investigation was to assess the lifetime behaviour of this combination, as migration of silver is known to limit package life in such components. It was noted that the best lifetime behaviour was found for the most highly crystallized glass-ceramics, probably due to lower diffusion rates in the crystalline material. Testing of high temperature packages includes both short term and long term methods including thermal cycling tests (NMAB, 1995b).

Additional information on electronic packaging materials is given elsewhere (e.g. Gupta, 2005).

10.6   Battery applications

There are a number of different battery designs that employ ceramic or alternative electrolytes and for which use of reliable glass-to-metal seals is common practice. These include Na–S (NAS), Na–SO$_2$, Li–SO$_2$, Li–SOCl$_2$, and Li–MnO$_2$ batteries.

The rechargeable Na–S battery with an output voltage of 2·1 volts, for example, employs a solid β-alumina electrolyte with a Na negative electrode (cathode) and a S positive electrode (anode) and operates at a temperature in the range 300–350°C in order to achieve significant diffusion of Na$^+$ in the electrolyte. The cell operates through the following reactions, which are reversible:- $2Na \rightarrow 2Na^+ + 2e^-$ and $xS + 2e^- \rightarrow Na_2S_x^{2-}$ (overall $2Na + xS \rightarrow Na_2S_x$). Individual cells can be used in series in order to furnish a system capable of providing higher voltages. These types of battery offer high power density (three times higher than that of lead-acid batteries) over wide temperature ranges, high charge–discharge efficiency and low self discharge characteristics, coupled with long term durability and the capability of undergoing up to 2500 charge/discharge cycles. They require hermetic and corrosion resistant seals to provide containment for the sulphur or SO$_2$ and liquid sodium electrodes, together with electrical isolation between the battery case and the cathode. Silicate glasses are not suitable for such applications (O'Hara, 1988) due to reaction between the glass and Na or Li to form electrically conducting silicide phases which leads to short battery lifetimes. Alkaline earth aluminoborate glasses which are chemically more resistant have therefore been developed for these battery sealing applications (Brow and Tallant, 1997). NAS batteries have been under development by Tokyo Electric Power Company and NGK Insulators of Japan since 1984 with work culminating in their use in the 2 MW and 6 MW

NAS battery installations at the Shinigawa and Ohito substations, respectively. They are also employed at the Chichibu, Matsuo and Kamiyama substations for load-levelling purposes, and at the Fujitsu Akiruno Technology Center and Tokyo Dome City LaQuc amusement facility for load levelling and emergency power. The type of battery stack employed in these applications is completely sealed with operating temperatures achieved with internal electric heaters and natural air convection, thus yielding a virtually maintenance free system. A NAS battery facility has also been installed at the American Electric Power Office in Columbus, Ohio, and at the Dolan Technology Center in Groveport, Ohio. Other applications have included their use as emergency DC power sources and uninterruptible power supplies, ideal for backing up communications and computer systems.

Non-rechargeable lithium batteries offer considerable advantages over more conventional battery designs. These include small size, higher cell voltages, i.e. 2·7 to 3·6 V compared to 2 V for batteries that employ aqueous electrolytes, higher energy density, low self-discharge rates, i.e. a long shelf life, the ability to operate under extreme conditions, e.g. over the temperature range $-60°C$ to $+85°C$, and the ability to use a variety of cathode materials that can be used with a Li anode. In this respect, there are a range of Li battery options available. These include $Li–SO_2$, $Li–MnO_2$, Li–I, $Li–Cr_2O_3$ and Li thionyl chloride ($Li–SOCl_2$). In the case of the $Li–SOCl_2$ battery operation proceeds through the sequence of reactions $4Li \rightarrow 4Li^+ + 4e^-$ and $2SOCl_2 + 4e^- \rightarrow S + 2SOCl_2 + 4Cl^-$, the overall reaction being $4Li + 2SOCl_2 \rightarrow 4LiCl + S + SO_2$. These advantages make Li batteries ideal for applications in the medical field such as heart pacemakers, defibrillators, and portable diagnostic equipment, and in aerospace and military applications. In addition, they are extensively used in commercial and domestic equipment such as personal computers and mobile phones. $Li–SO_2$ batteries are also employed in beacon and emergency location transmitters, sonobuoys and portable radio communications equipment, and military and aerospace applications, whilst $Li–SOCl_2$ batteries are used in tracking and identification systems, wireless alarm systems, portable instrumentation, transponders for electronic toll collection, various pressure monitoring systems, automobile air bags, tyre-pressure sensors and computers, and various alternative safety-related systems. Battery driven solid state processors are now routinely employed in the oilfield industry for well logging operations which require measurements to be taken *in-situ* down bore holes and wells and these systems utilise lithium batteries to power them (Hensley, 1998). $Li–MnO_2$ batteries are also highly reliable with a long shelf life and capable of operating continuously or in pulse mode at temperatures down to $-30°C$. This makes them ideal for applications such as emergency location and survival gear in cold climates. Glass-to-metal compression seals are employed in many of these Li battery systems, including Li–I batteries which are used in heart pacemakers and which need to be extremely reliable over their life span.

An early study of glass seal corrosion in $Li–SOCl_2$ batteries was reported by O'Hara (1988). In this work, the compatibility of a number of different com-

(a) Commercial Li–SOCl₂, Li–SO₂ and Li–MnO₂ batteries

(b) Photograph of the exposed top of a SAFT battery showing the glass-to-metal seal

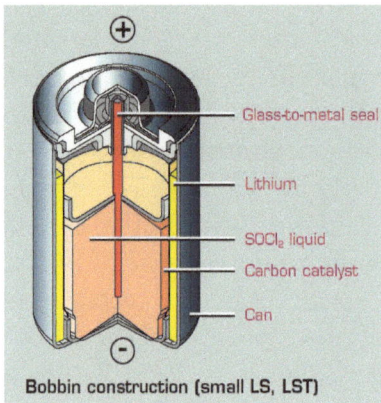

(c) Schematic Li–SOCl₂ battery design showing positioning of glass-to-metal seal

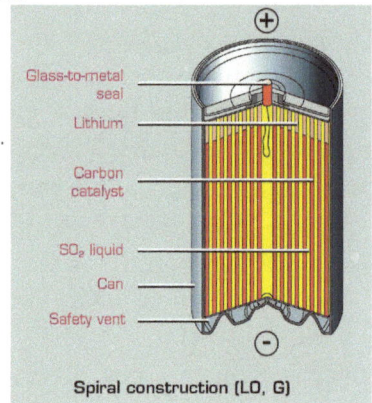

(d) Schematic Li–SO₂ battery design showing positioning of glass-to-metal seal

Figure 92: Lithium batteries (courtesy of SAFT)

mercial glass compositions with stainless steel, Mo and Fe–Ni alloys, typically used in battery construction, was evaluated. It was noted that high silica content glasses were particularly susceptible to glass corrosion in these systems and that SNLA's TA-23 glass and Fusite's 402/425 composition which have relatively low silica contents provided the most resistant seals.

A Li battery design is shown in Figure 92 with the positioning of the glass-to-metal seals highlighted. Bobbin cells with limited anode–cathode surface area are used for low power and low-to-moderate current applications, whilst spiral designs with a high anode–cathode surface area are better suited to high power and high current use.

Thermal batteries also employ glass-to-metal seals. A thermal battery is designed to provide electrical power at high energy density under a variety of environmental conditions (EaglePitcher Technologies). A battery typically consists of a thermal source, which may be a mixture of iron powder and potas-

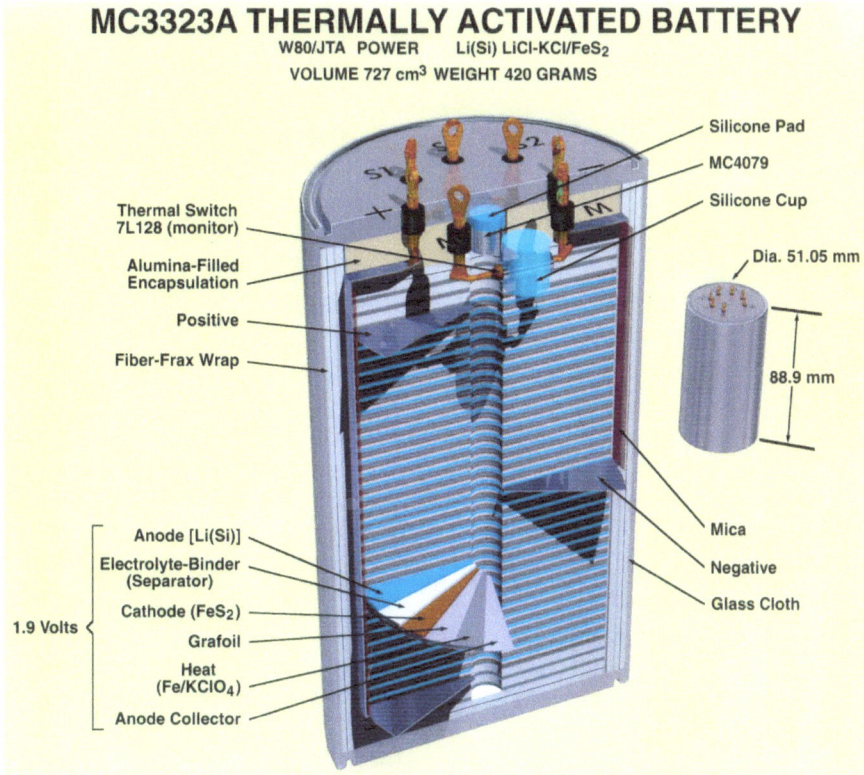

Figure 93: Thermal battery design (Courtesy of SNLA)

sium perchlorate, which provides heat to an electrolyte comprised of an eutectic mixture of inorganic salts contained in a cylindrical metal case with a lithium alloy anode, stainless steel electrodes and an iron disulphide cathode. The battery is completely inert until activated and has the advantage that it can be stored indefinitely until use without loss of performance. For use, the thermal source is ignited by an electrical initiator or alternatively by mechanical means. The electrolyte subsequently melts and becomes conductive, thus allowing the cathode to interact with the anode. The battery will continue to function until an active component is exhausted or the temperature falls below the melting point of the eutectic mixture. A typical thermal battery design is shown in Figure 93. A detailed description of Li battery designs, materials, operating fundamentals and applications is given in the book edited by Nazri and Pistoia(2004). It is, however, very disappointing that in a book of this nature that "provides the most recent knowledge and concepts on advanced lithium batteries", no mention is given to the use of glass-to-metal seals in these batteries.

A selection of individual battery seals used in Na, Li and thermal battery designs is shown in Figure 94.

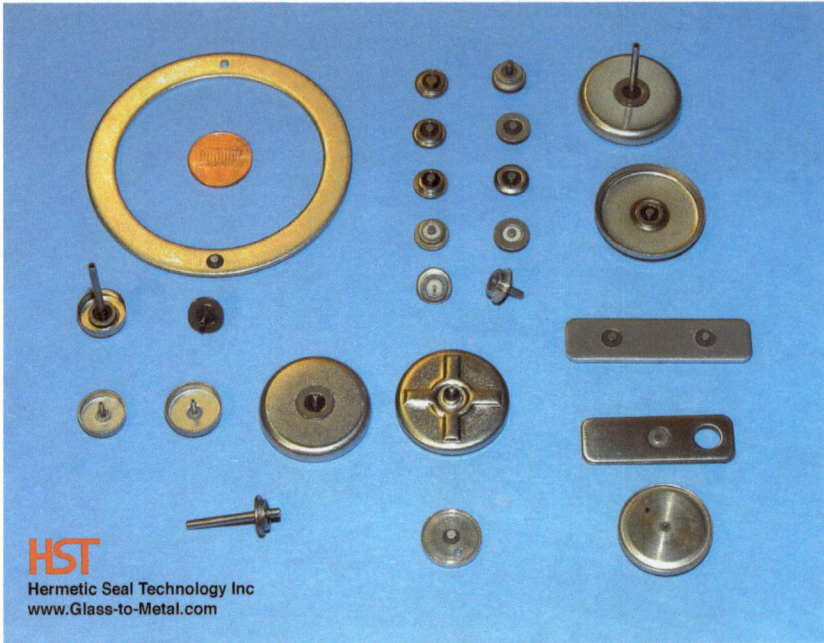

(a) Li battery headers (Courtesy of Hermetic Seal Technology Inc.)

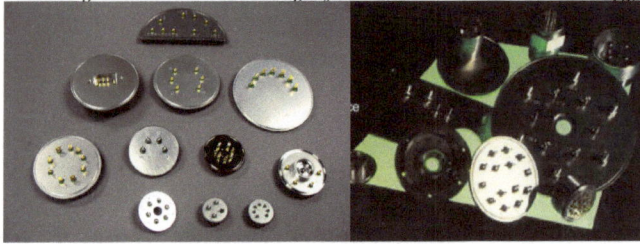

(b) Thermal battery headers (Courtesy of SNLA)
Figure 94: Battery glass-to-metal seals

10.7   Fuel cells

Fuel cells are designed to convert chemical energy to electrical energy, not unlike a battery. They work on the basis of reverse electrolysis. Unlike a conventional battery, however, the fuel to drive the battery (fuel cell) is continuously replenished. The ideal fuel is hydrogen, although in practice fossil or other fuels may be employed including coal gas, methane, methanol and naptha. These are normally broken down to provide hydrogen to the cell as part of the process. The fuel cell converts chemical energy directly into electrical energy without the need for combustion by combining gaseous fuel and an oxidizing gas via an ion-exchange medium. This gives rise to very high conversion efficiencies, which can be > 50%, coupled with very low emissions (mainly water, although use of hydrocarbons may give rise to some CO_2, SO_2 and nitrous oxides.). In its simplest form, a fuel cell consists of a porous anode and cathode separated by an electrolyte. The electrolyte may be phosphoric acid, a proton exchange membrane, a molten carbonate or a solid oxide. An oxidant (usually air) is provided to the cathode to supply oxygen whilst fuel is provided to the anode to supply hydrogen. A reverse electrolysis reaction subsequently takes place:- $2H_2 + O_2 \rightarrow 2H_2O$.

Because individual fuel cells usually generate only a small voltage, typically 0·6–0·8 V, individual cells must be joined together in series using an appropriate bonding medium in order to furnish a stack capable of providing higher voltages. A suitable sealant must meet a number of quite severe requirements, depending on the specific type and operating temperature. For example, the thermal expansion characteristics of the sealant must be compatible with the other fuel cell components, it must possess good chemical stability at fuel cell operating conditions in both reducing and oxidizing atmospheres, and it must remain stable in the presence of the other materials composing the fuel cell. The commercial development of fuel cells has consequently led to the exploitation of a wide range of sealing media, including glasses and glass-ceramics, for the individual fuel cell elements. Work is underway at various research organizations and laboratories around the world including, for example, PNNL and SNLA in the United States. This is confirmed by the many reports in this area, with some of the more important being highlighted in the following sections. At the present time there is no ideal sealing medium meeting all the requirements. Materials research is therefore an area attracting considerable interest with high potential rewards for the identification and successful development of glass and glass-ceramic materials offering better performance.

10.7.1   Molten carbonate fuel cells

Molten carbonate fuel cells, MCFC, utilise a molten carbonate as the electrolyte, usually $LiCO_3$ or KCO_3 or a mixture of both, as illustrated schematically in Figure 95. Practical fuel cells consist of a number of individual fuel cell elements stacked together within an alumina frame/stainless steel housing and sealed together with a suitable sealant. As already noted, a suitable sealing medium must meet a number of specific criteria including compatibility with the alumina, stainless steel and molten electrolyte, good gas tightness, high electrical resistance and thermal stability (fuel cell operating temperatures are in the region of 650°C), compatible thermal expansion characteristics, and a useful lifetime. Early designs utilized zirconia felt as the sealing medium with lifetimes severely limited due to migration of electrolyte into the porous structure of the felt. An ideal sealing medium is glass, but one of the major areas of concern is the stability of glass in the presence of molten carbonate (most glass compositions are made by melting oxide and carbonate starting materials). Glasses that have been considered for this application include alkali and alkaline earth oxide borosilicate compositions (Pascual *et al*, 2001, 2002, 2003).

Pascual *et al*(2001, 2003) noted that the reaction between glasses and molten carbonates is not well documented, although it was known that attack may lead to phase separation and crystallization in addition to more general chemical

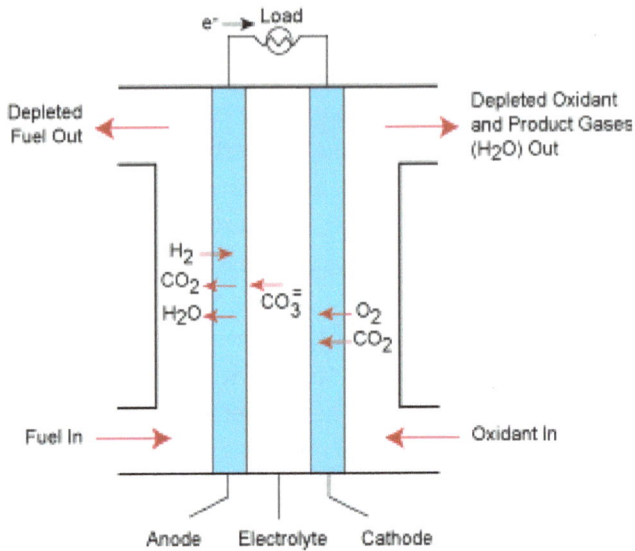

Figure 95: Schematic diagram of a molten carbonate fuel cell
(after www.dodfuelcell.com/molten)

corrosion. They examined the corrosion behaviour of a series of alkali oxide borosilicate glasses in molten Li_2CO_3–K_2CO_3 eutectic at temperatures in the range 500 to 700°C and observed that the glasses were indeed severely corroded, with corrosion taking place in three stages. First, there is interdiffusion giving rise to a gel layer; this is followed by congruent dissolution of the glass network; and this leads finally to the formation of additional reaction layers which may include Li_2SiO_3 and which minimised further attack. It was noted that the durability of the glass can be significantly improved by the addition of zirconia fibres to give a composite structure. Kedding *et al*(2001) have also studied the corrosion behaviour of borosilicate sealing glasses using impedance spectroscopy, a method which can be utilized to screen candidate glasses relatively easily.

10.7.2 Solid oxide fuel cells

Unlike earlier fuel cell designs, solid oxide fuel cells, SOFC, are made entirely from solid materials and employ a solid ceramic electrolyte coated on both sides with electrically conducting materials to act as anode and cathode. During operation at high temperature, typically in the range 800 to 1000°C although fuel cells operating at < 700°C are feasible, oxygen ions are conducted from the cathode to the anode where they react chemically with the fuel, which may be methane or hydrogen. The passage of oxygen ions induces an electric charge which provides the electrical energy. A schematic representation illustrating the principle of operation of a solid oxide fuel cell is shown in Figure 96. The oxygen ion conducting electrolyte is typically made from yttria-stabilized zirconia, YSZ, whilst the cathode may be lanthanum manganate, $LaMnO_3$, and the anode a nickel–YSZ

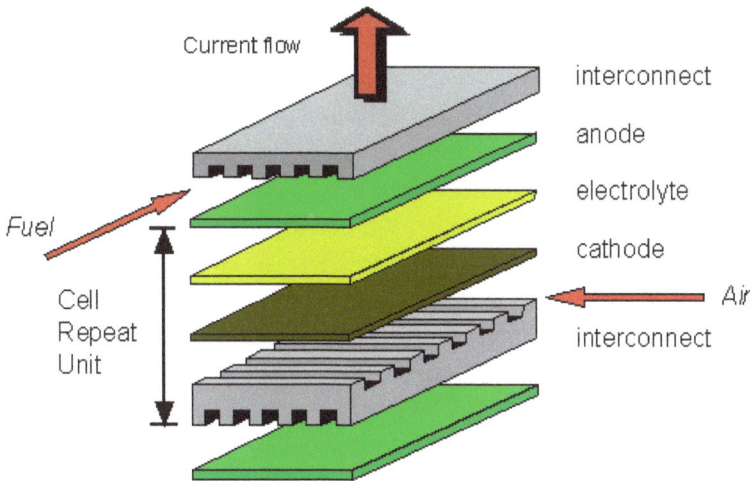

(a) Schematic exploded view (after www.spice.or.fisher/sofc)

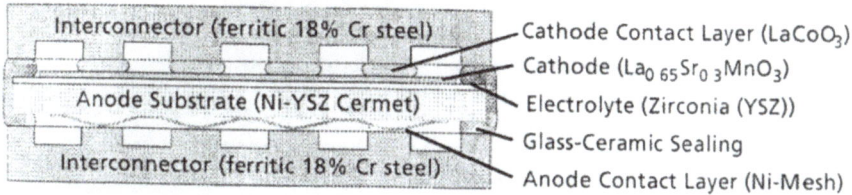

(b) Schematic showing the area requiring sealing

Figure 96: Schematic diagrams of a solid oxide fuel cell employing glass- or glass ceramic-to-metal seals

cermet. In the manufacture of a practical device, many of these cells are stacked together, as also illustrated in Figure 96, which shows the cell repeat unit. For high temperature operation each cell is connected via a doped lanthanum chromite interconnect and bonded to the next unit via a glass or glass-ceramic seal, whilst for lower operating temperature cells ferritic stainless steels may be used as the interconnect. The interconnect serves the purpose of carrying current between the individual cells and also acts as a separator between the oxidant (usually air) and the fuel. In many SOFC stack designs the interconnect is hermetically sealed to the ceramic positive electrode–electrolyte–negative electrode by a sealing glass. The whole unit is contained in a heat resistant alloy housing, which may be stainless steel or a nickel-based superalloy. The glass seal employed to attach the individual cells together must be heat resisting, must bond to but not react adversely with the fuel cell elements, and must be resistant to corrosion by the fuel cell gases, which are both reducing and oxidizing in nature. The glass seal therefore must operate reliably in an extremely hostile environment, and this imposes severe restrictions on the composition of the glass that can be used.

Increasing attention is being given to the mechanical properties of SOFC components and materials in a drive to scale up the size and power density of SOFC cells and stacks, and in an attempt to improve the lifetime behaviour of these systems. As the materials are rigidly bonded in the multilayer system of planar SOFC's which consist of three major layers, i.e. anode, electrolyte and cathode, differences in materials properties can result in the generation of residual stresses. These stresses may arise due to a number of different reasons including differences in the thermal expansion behaviour between the individual components, thermal gradients generated during operation, and chemical gradients of the diffusing species (Adamson and Travil, 1997; Hendriksen *et al*, 1995). Additional stresses may also be introduced during the assembly of the final fuel cell due to the different properties of glass or glass-ceramic materials employed in the bonding arrangements. A number of investigations have been carried out to determine the properties of SOFC component materials (Atkinson and Selcuk, 1997, 2000; Lowrie and Rawlings, 2000; Soerensen and Primdahl, 1998; Selcuk *et al*, 2001; Fergus, 2005). The number of studies on the final cells and stacks are, however, rather limited, as also are studies of the mechanical properties of the interfaces in these systems.

A range of glasses and glass-ceramics have been investigated as potential sealing media for use in SOFC designs (Horita *et al*, 1993; Yamamoto *et al*, 1996; Ley *et al*, 1997; Eichler *et al*, 1999; Schwickert *et al*, 2000; Sohn *et al*, 2002; Yang *et al*, 2003; Malzbender *et al*, 2003; Pevzner and Klyuev, 2003; Sohn and Choi, 2004; Anon, 2004; Budd, 2005). Compositions investigated include, for example, refractory glasses and glass-ceramics based on $SrO–La_2O_3–Al_2O_3–B_2O_3–SiO_2$ (Ley *et al*, 1997), $BaO–Al_2O_3–B_2O_3–SiO_2$ glass-ceramics containing smaller additions of La_2O_3, ZrO_2 or NiO (Sohn and Choi, 2004), $BaO–La_2O_3–SiO_2$ glass-ceramics (Budd, 2005), and glasses and glass-ceramics based on $MgO–CaO–BaO–SiO_2$ containing additions of Al_2O_3, La_2O_3, B_2O_3 and MnO (Schwickert *et al*, 2000). Sohn *et al*(2002) concluded that the most suitable glass from the barium aluminium borosilicate system for sealing to the yttria stabilized zirconia electrolyte was the composition $35BaO–10Al_2O_3–5La_2O_3–16·7B_2O_3–33·3SiO_2$ (mol%). This glass exhibited good wetting and bonding characteristics, a low thermal expansion mismatch, and showed no sign of undesirable interfacial reactions at the SOFC operating temperature of 800–850°C. Certain aluminosilicate glass compositions have, however, been shown to exhibit relatively poor thermal stability in SOFC environments, with Lahl *et al* (2000) and Eichler *et al* (2000) noting the occurrence of devitrification within very short periods of time.

Barium calcium aluminosilicate based sealing glasses have also been proposed for SOFC applications (Yang *et al*, 2003a, 2003b). In these studies, chemical interactions between the glass and a number of different austenitic and ferritic Cr_2O_3 forming Fe- and Ni-based alloys and an Al_2O_3 forming alloy employed in SOFC construction were investigated. It was observed that all these alloys reacted with the glass to varying degrees. It was observed in the case of the Cr_2O_3 forming alloys, for example, that at the edges of the interconnect joint, accessible to oxygen or air, interaction can occur to form $BaCrO_4$, as illustrated in Figure 97, whereas in the interior of the joints Cr or Cr_2O_3 dissolved into the glass to

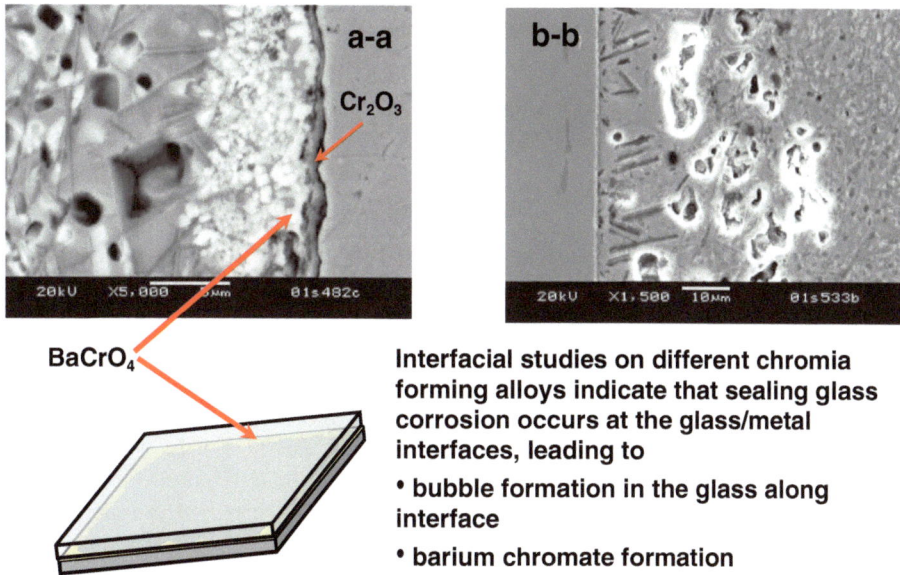

Figure 97: Seal between a barium calcium aluminosilicate glass and stainless steel showing the formation of a barium chromate phase near the seal edges due to the reaction:- $2Cr_2O_3 + 4BaO + 3O_2 \rightarrow 4BaCrO_4$ (Courtesy of Yang et al, 2003)

form a thin layer of a Cr-rich solid solution together with voids aligned along the interface. The presence of voids was attributed to reaction of Cr with water dissolved in the glass. It was noted that if the alloys were pre-oxidized prior to sealing to provide a well-defined oxide layer which prevented excessive diffusion of Cr into the glass, porosity was absent from the interfacial region. On the other hand, the alloy which formed an Al_2O_3 layer did not react to form $BaCrO_4$, but did result in many voids attributed to the reaction of such alloying elements as Al, Y and Cr in this Ni-based superalloy with water. These are further examples of the deleterious effect that water dissolved in glass can have on the resulting interfacial chemistry of practical systems.

The properties of an aluminosilicate glass-ceramic sealant employed to bond individual cells into a SOFC stack have been investigated by Malzbender *et al*(2003). The glass-ceramic was bonded to a steel interconnect and individual glass-ceramic-to-steel seal samples cut from the assembly were tested in four-point bending. It was noted that the strength of the samples increased with increasing sealing time from 1 day to 8 days at 850°C. In these tests, metal strips were tested in combination with aluminosilicate glass-ceramics. The glass-ceramic/metal sandwiches held under pressure in a ceramic jig were heated in air at 800–850°C for various periods of time in order to achieve bonding. Notched samples were subsequently tested in four-point bending. A maximum fracture energy of around 23 J/m^2 and fracture toughness of 1·6 MPa m$^{-1/2}$ was achieved.

Applications for SOFC's range from large scale power generation to small scale units for domestic use. A large 250 kW SOFC has been installed at the University of Toronto in Canada, a joint venture between Siemens Westinghouse Power Corporation and Kinectrics. This plant operates using natural gas as the fuel and is designed to produce both electricity and heat.

10.8   Electrical power and Telecommunications

Glass-to-metal or ceramic-to-metal seals are employed in many components used in the electrical power and telecommunications industries including capacitors and thyristors used in motor speed controls, inductive heating, power switching and rectifier applications. Tantalum hybrid capacitors have been reported as amongst the most powerful capacitors available (Evans, 2001). This unique design of capacitor combines the high cell voltage capacity and low resistance of an electrolytic capacitor with the increased energy density of an electrochemical capacitor. In this design, the case of the capacitor is part of the cathode with isolation for the positive feed-through made using a glass-to-metal seal.

Examples of some electrical power and telecommunications components which employ the newer glass-ceramic-to-metal seals are shown in Figure 98. These include a glass-ceramic/metal vacuum interrupter envelope used in the electrical power supply industry, a glass-ceramic-to-copper vacuum envelope, and glass-ceramic insulators with mild steel fixing studs. Some alternative electrical components are also shown in Figure 98. Glass-ceramic ferrules for telecommunications applications have also been reported by Mitachi *et al*(1998). Optical connectors have been developed both for indoor and outdoor use in severe environments. Currently, SC-type optical connectors use zirconia ferrules to connect optical fibres very precisely; for example, zirconia ferrules are manufactured with an outside diameter of 2·499 mm ±0·5 μm and with a central hole eccentricity of < 0·7 μm. It has been noted that zirconia can degrade in hot and humid environments and this may lead to deteriorating optical performance. Glass-ceramic ferrules are therefore under development offering improved outdoor durability. In their study Mitachi *et al*(1998) employed a lithium aluminosilicate glass-ceramic composition containing smaller quantities of MgO, TiO_2, ZrO_2, K_2O and B_2O_3. Metal/glass-ceramic ferrules were subsequently prepared and subjected to environmental testing in water at 45°C and in a hot humid environment (85°C and 85% RH) for up to 2000 h. They were shown to exhibit superior durability to the equivalent zirconia based ferrules. Direct sealing of optical fibres to provide hermetic feed-throughs has also been accomplished using glass preforms with low temperature sealing properties (Dietz, 2004)

10.9   Detonators and explosive actuators

Detonators and explosive (pyrotechnic) actuators are used in many applications. These range from initiating explosive devices, to operating mechanical systems or flow control devices. In the case of actuators, when the pyrotechnic charge is

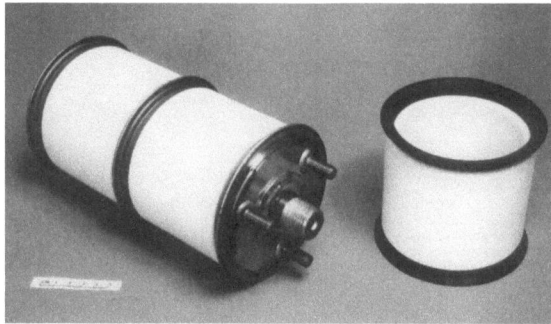

(a) Glass-ceramic/metal vacuum interrupter envelope. This component consists of a lithium aluminosilicate glass-ceramic bonded to a 17% Cr/Fe alloy body (Courtesy of former GEC Alsthom)

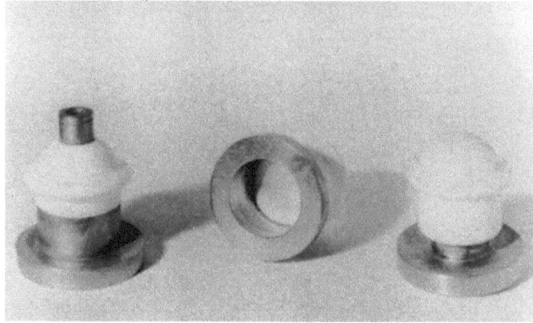

(b) Glass-ceramic-to-copper vacuum envelope; the diameter of the metal ring is ~3 cm (Courtesy of former GEC Alsthom)

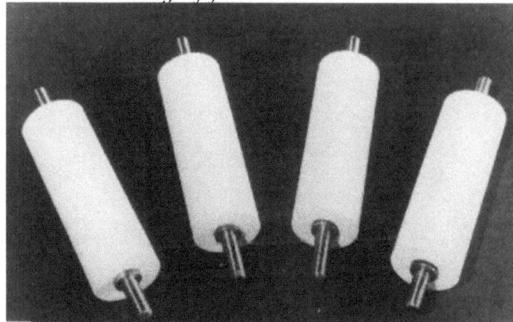

(c) Glass-ceramic insulators with mild steel studs; the glass-ceramic cylinder is ~10 cm long (Courtesy of former GEC Alsthom)
Figure 98: Photographs of a selection of electrical components

ignited it produces gas pressure that can be used to perform mechanical work; for example, rapid response opening or closing control valves, operating remote latches on spacecraft and missile systems, and even operating automobile air bags. These actuators often require complex electrical feed-through seals which can withstand high pressures. As the successful operation of an actuator is dependent on the reliable delivery of gas pressure it is essential that the seal is robust and leak tight. Many actuators employ glass-to-metal seals, although actuators for more arduous applications, where short duration pressures pulses

Figure 98: (continued)

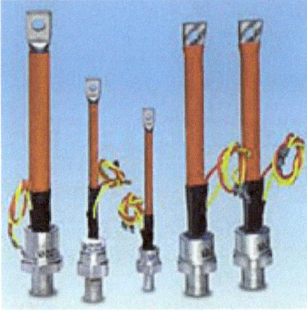

(d) Stud type thyristors employing glass-to-metal seals (Courtesy of Aegis Semicondutores)

(e) Disc type thyristors employing ceramic-to-metal seals (Courtesy of Aegis Semicondutores)

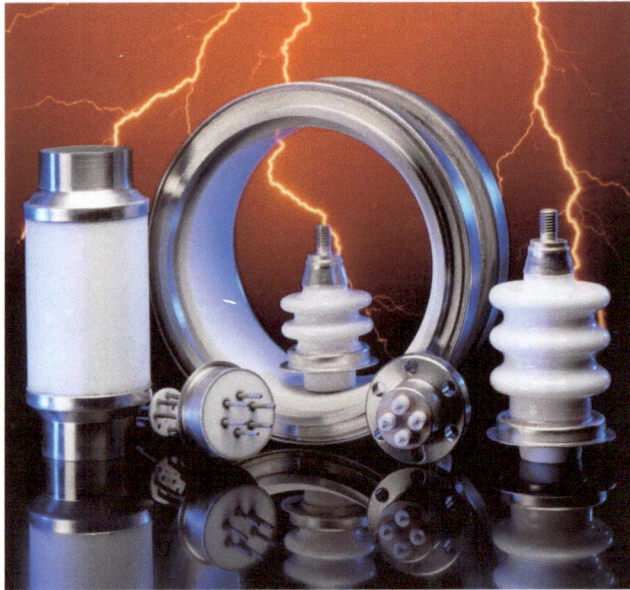

(f) High voltage components for use at voltages up to 200 kV (Courtesy of Ceramaseal)

in excess of 700 MPa may be experienced, employ brazed alumina-to-metal seals or glass-ceramic-to-metal seals. Examples of some of the glass-ceramic-to-metal components developed at AWE and at SNLA are shown in Figure 99. These consist of stainless steel or Ni-based superalloy shells and Ni-based pins bonded using lithium zinc silicate or lithium aluminosilicate glass-ceramics. A selection of alternative components including automotive airbag and seatbelt initiators, explosive bolts, squib headers, cartridges and detonators is shown in Figure 100.

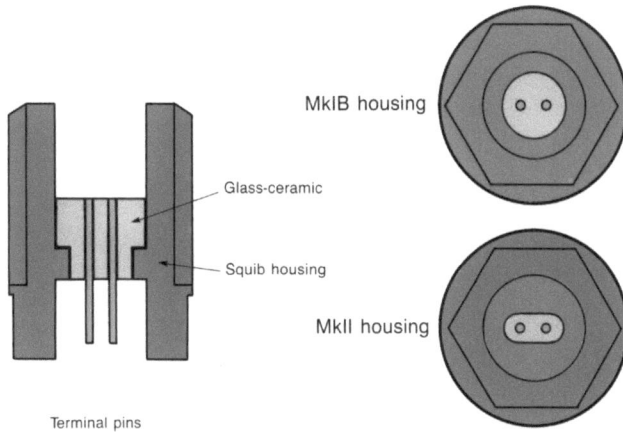

(a) Schematic representation of a two-pin electrical bridge-wire explosive actuator
squib developed at AWE

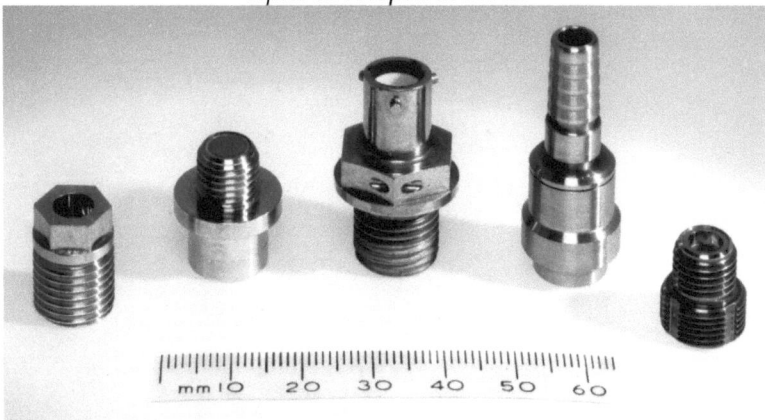

(i)   (ii)   (iii)   (iv)   (v)

(b) A selection of explosive actuator squibs:-
(i), (ii) & (iv) glass-ceramic-to-metal seal components developed at AWE;
(iii) brazed alumina seal component; (v) glass-ceramic-to-metal seal actuator
developed at SNLA

Figure 99: Some examples of explosive actuator and detonator components developed
at AWE and SNLA

10.10 Applications involving protective coatings and coatings to improve properties

Much of the early work in the area of glass-to-metal coatings was concerned with
the preparation of enamelled metals. As noted previously, the enamelling of metals such as gold, silver, copper and bronze was thought to date back to ancient
Egyptian times. It is now believed, however, that the earliest objects known to be
enamelled using fusion techniques were made in Cyprus around the thirteenth
century BC (The Institute of Porcelain Enamellers). In the later Byzantine era the
art of enamelling spread to Europe and elsewhere where it was broadened to

Figure 99: (continued)

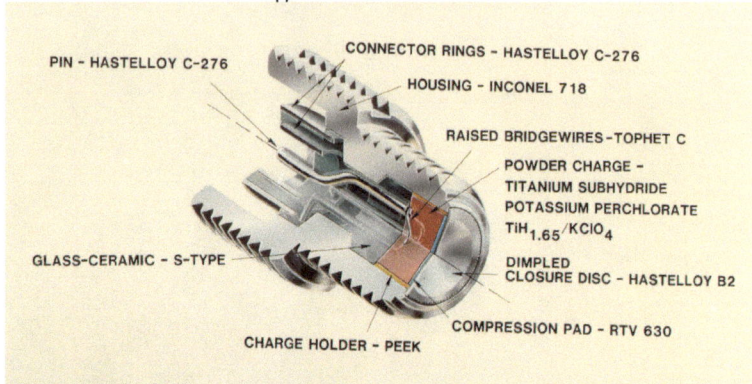

(c) The MC3753 Explosive Actuator Squib developed at SNLA (an actual component is shown in (b)-(v)

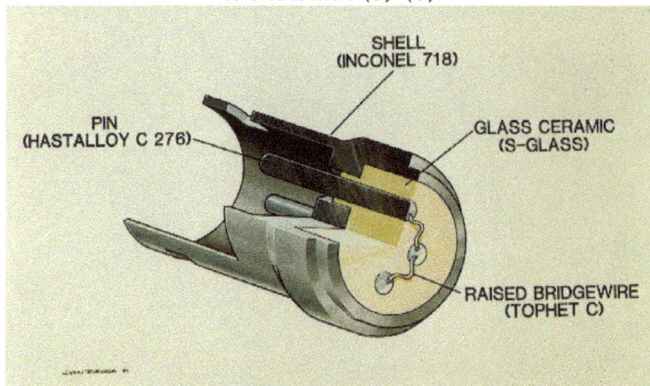

(d) The MC3478-3479 Explosive Actuator Squib developed at SNLA

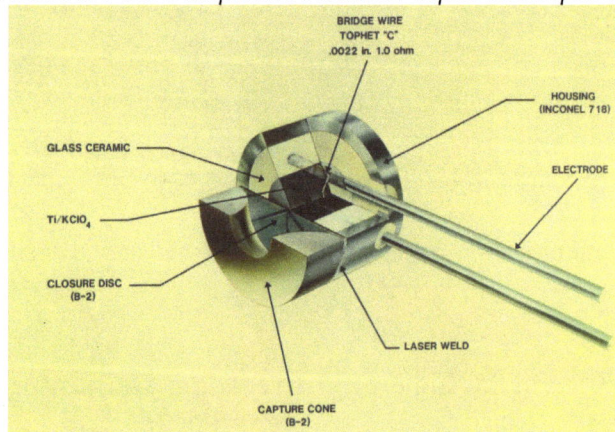

(e) The MC4170 Igniter developed at SNLA

encompass the decorative enamelling of larger items, although it was still largely limited to the preparation of objects of art, rather than applied specifically to protect metal surfaces. By the middle of the nineteenth century, the large scale enamelling of metallic objects was introduced, originally in Germany. This included the

Figure 99: (continued)

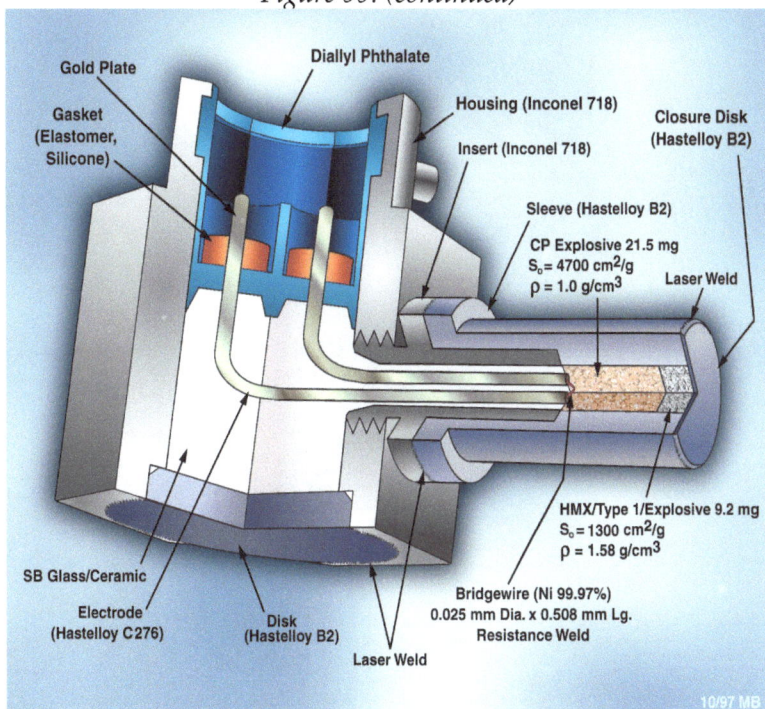

(f) The MC4217 Detonator developed at SNLA

industrial production of enamelled cast and wrought iron ware, including cooking utensils, baths and underground piping, the enamel being applied in order to provide a corrosion and abrasion resistant protective surface on the metal. More recently, vitreous enamelled metals have been employed in a multitude of large volume/low technology applications, both in the home and industry. Examples of the use of enamelled metals in the home include their application in cookers, sink units, washers, cooking utensils, water and storage heaters, and gas fires and stoves. Industrial applications include their use in architecture, e.g. building panels and tunnel walls. Other industrial applications include enamelled vessels, pipes, valves, stirrers, baffles, nozzles, etc, used in the chemical industry; for the protection of metals used in agricultural applications including storage silos; and for use in heat-exchangers and related devices (Andrews, 1961; Maskall and White, 1986; Garland, 1986; Anon, 1991; Hairston *et al*, 2002).

More recently, glass-ceramic coatings exhibiting superior mechanical and other important properties have been applied for the protection of, for example, metal pipes and vessels in the chemical industry. A glass-ceramic coating enamel exhibiting relatively high electrical conductivity has been reported based on the system $Na_2O–B_2O_3–TiO_2–SiO_2$. This enamel is suitable for coating stainless steel where its higher electrical conductivity relative to conventional enamels, i.e. 10^5 Ωcm compared to around 10^{12} Ωcm, makes it ideal for the protection of chemi-

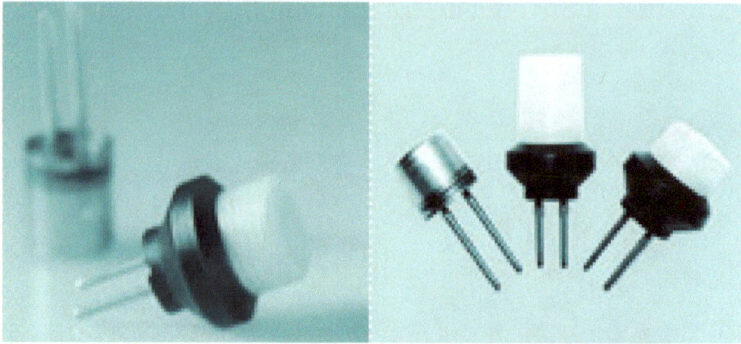

(a) Automotive airbag and seatbelt initiators (Courtesy of Hirtenberger)

(b) Explosive bolts (Courtesy of Accra Tronics Seal Corporation)

(c) Squib headers (Courtesy of Accra Tronics Seal Corporation)
Figure 100: Some examples of squib, header and initiator components employing glass-to-metal seals

cal and related equipment which may be susceptible to the build up of static electricity (Tavgen, 2000). Enamelling of ferrous metals with a porous (foamed) glass-ceramic, utilizing SiC as the foaming agent, has also been reported to be of interest for construction applications, with these systems exhibiting good thermal and acoustic insulating properties (Gomez de Salazar *et al*, 2001). Enamel and

Figure 100: (continued)

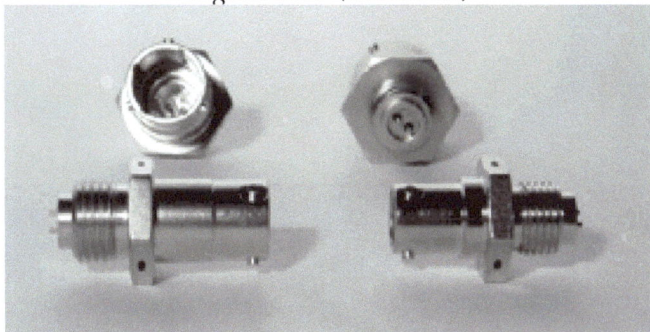

(d) Cartridges (Courtesy of Accra Tronics Seal Corporation)

(e) Detonator (Courtesy of Accra Tronics Seal Corpor

glass-ceramic coatings have also been employed in the aerospace industry for the protection of metals against oxidation; for example, coating of after-burners and low-pressure turbine blades in jet engines, as shown in Figure 101 (Garland, 1986). In addition, they have been examined as protective coatings for medical and dental prostheses (Krajewski *et al*, 1985).

An area of increasing interest is the application of refractory ceramic coatings to metals operating under severe service environments. This is particularly true in the case of gas turbine and jet engine components (Nicholls, 2003; Eskner and Sandstrom, 2003). This need has arisen due to the continuing demand to improve operating performance whilst reducing emissions, and this has led to normal turbine blade operating temperatures of around 1050°C with peak temperature excursions in excess of 1150°C. The improvements in performance have been achieved through the use of cooled coated-blade technology. Coatings are required in order to improve the oxidation and corrosion resistance of the blade material in the harsher environments now experienced. Early coatings were achieved by forming a layer of β-NiAl on the surface of the component through, for example, chemical vapour deposition and heat treatment techniques. This intermetallic coating provides a source for the formation of a protective alumina ceramic layer at the surface of the component, examples are shown in Figure 102. The latest developments include thermal barrier coatings which are designed,

(a) Clamshell air-to-air heat exchanger for domestic heating applications fabricated from steel coated with acid and heat resisting vitreous enamel

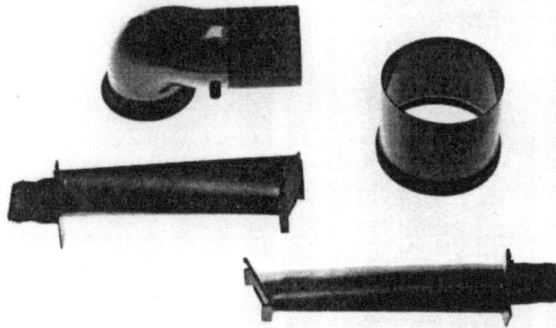

(b) Enamelled turbine components (exhaust system components from a Bristol Siddeley Proteus turbo-prop engine and low pressure turbine blades from a Rolls Royce RB211 engine)
Figure 101: Photographs of components employing enamelled substrates (after Garland, 1986)

in conjunction with cooling, to lower the metal surface temperature by forming a low thermal conductivity surface in contact with the hot gases. Such coatings include zirconia with the tetragonal phase partially stabilized by addition of 8 wt% yttria, applied initially by plasma spraying but more recently by electron beam physical vapour deposition. The microstructure of a coating applied by electron beam physical vapour deposition consists of columnar grains, and this exhibits a good surface finish and offers the necessary erosion resistance (Nicholls, 2003). This is in contrast to coatings applied by plasma spraying which do not

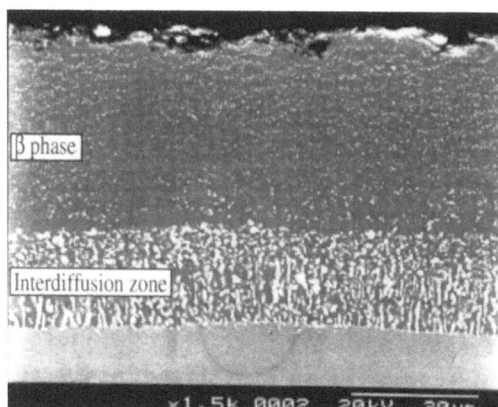

(a) Nickel aluminide coating on Inconel 738 alloy (the β-phase is an intermetallic compound based on NiAl formed by diffusing Al into the surface of the alloy)

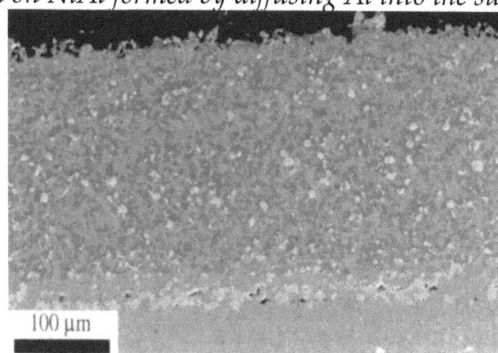

(b) Electroplated NiCoCrAlY coating on MarM002 Ni-based superalloy

Figure 102: Micrographs of coatings developed for the protection of gas turbine blades (after Nicholls, 2003)

meet the stringent requirements for turbine blades, but which has been used to coat the internal surfaces of combustion chambers which are subject to less harsh conditions. Additional studies include application of thin coatings (2·5–5·0 μm thick) of TiN and TiAlN by arc-ion plating onto turbine rotor steels to improve the fatigue properties (Suh et al, 2002), and aluminide and CoCrAlY coatings applied to TiAl intermetallic alloys to improve oxidation resistance (Tang et al, 1998).

Coatings are also required to protect infra-red windows and domes employed in heat-seeking missile technology from the harsh environments found in service, e.g. frictional heating and rain erosion. Zinc sulphide is a conventional window material but suffers from poor rain erosion resistance. Various coatings have therefore been applied to this material in an effort to improve its resistance. Diamond and diamond-like coatings have been successfully applied using CVD techniques (Fujii et al, 2001). Diamond itself is under consideration as a window material for operation in the 8 to 14 μm waveband, but requires suitable anti-reflection coatings. Most far infra-red coatings are not suitable due to poor durability so alternatives based on transition metal oxide coatings have been considered; for example, Yb_2O_3, Y_2O_3, ZrO_2 and HfO_2, applied by magnetron sputtering. Recently, gallium arsenide has also emerged as a candidate infra-red transmitting window and dome material. This has been successfully coated with boron phosphide using a plasma assisted CVD process in order to improve its durability (Clark and Haddow, 2001).

Micromesh copper or gold coatings have been applied to the surfaces of infra-red seeking missile sensor windows as electro-magnetic interference shielding. More recently, it has been found that continuous infra-red transparent conductive ceramic films can be applied which offer better off-axis performance (Johnson and Moran, 2001). The composition of these films, of general formula ABO_2, where A is a monovalent metal, e.g. Cu, Ag, Au, etc, and B is a trivalent metal, e.g. Al, Fe, etc, can be tailored by adjusting the oxygen content to vary the electrical conductivity over a wide range. Films of formula $Cu_xAl_yO_z$ have been applied by magnetron sputtering from copper and aluminium sources.

Many types of coating have been applied to glass substrates in order to improve the optical, mechanical or other properties. Both anti-reflection and reflective coatings have been applied for many years for optical and architectural applications. Sol-gel coatings have been applied to glass to enhance the mechanical properties by protecting the surface and sealing existing flaws (Nikolic and Radonjic, 1998; Hand et al, 1998). More recently, titania coatings have been applied to glass in order provide self-cleaning architectural glass products, e.g. window glass (Sanderson et al, 2003), aimed at reducing the amount of maintenance required. Such coatings reduce the contact angle of the glass surface, resulting in increased wetting and a layer of water being formed on the glass surface rather than discrete droplets. This improves the visibility during periods of rain. The coating, being a semi-conductor, also absorbs incident light at wavelengths of < 320 nm and promotes oxidation and reduction reactions (photocatalysis) of organic materials which may form on glass surfaces and impair transparency.

The increased levels of pollutants, and in particular SO_2, in the atmosphere have led to the deterioration of many glasses of historical interest due to surface degradation. There are many studies aimed at the protection of these ancient glass surfaces through the development and application of compatible coatings (e.g. Carmona, 2004).

Thin glasses are very important for display screen applications where high strength is required, ideally coupled with low cost. It has been noted that silica coatings a few micrometres in thickness can be applied to 0·5 mm thick display glass using sol-gel technology (Mennig et al, 2001), with bending strength increased from around 450 to 1100 MPa.

Other coatings that have been reported include the use of an inexpensive glass-ceramic prepared from metallurgical wastes which contain a mixture of the oxides of Si, Al, Fe, Ca, Mg, Ti, Mn, K and Na, as high temperature protective coatings for nichrome steel (Zubekhin et al, 2001). Nitride and carbide coatings including TiN and TiC have been applied using physical vapour deposition or a filtered arc ion plating technique to alloy steels to provide wear, heat and corrosion resistance (Kim *et al*, 2003). Barium oxide containing mixed silicate/phosphate glasses with molar ratio of $Al_2O_3:SiO_2:P_2O_5 \approx 1$ are also promising materials for heat and oxygen resistant protective coatings for Nb–Zr alloys with operating temperatures > 700°C (Sedmale *et al*, 2001).

Pictures or other markings including writing on glass are usually applied by melting in coloured enamels. The glass enamel, consisting of glass powder and pigments, may be applied to the substrate by silk-screen printing, for example. The system is then heated in order to melt the enamel and allow it to wet the substrate. An alternative method has been proposed involving applying the glass enamel directly on a cold substrate using laser technology (Seufert and Lenhart, 2001). The enamel can be applied to the substrate by spraying and then immediately fixed by laser. Certain conditions must, however, be met in order for this method to be successful; for example, a glass enamel must be used that will absorb the laser light sufficiently, and the laser must be scanned rapidly in order to avoid excessive heating of the substrate.

Conventional ceramic glazes, composed mainly of a vitreous material, generally give rise to a glazed surface with relatively poor mechanical properties. It has been noted that use of glass-ceramic glazes based on spinel can provide a more robust system, particularly in the case of glazing for porcelain stoneware tiles (Nebot-Diaz *et al*, 2001).

The application of vitreous or ceramic glazes to ceramics in order to improve mechanical and physical properties and to enhance aesthetic appeal has, of course, been employed for many years. Examples range from floor and wall tiles, to storage containers, household ceramic ware and objects of art.

10.11   Biomedical materials and devices

Glass-to-metal seals are employed extensively in a range of biomedical implant and monitoring components including heart pacemakers and pacemaker batteries, cochlea implants, exploratory probes, cardiac and respiratory monitors, etc. Glass seals to titanium and titanium alloys offering good corrosion resistance have also been proposed for use in biomedical implants including batteries, pacemakers, defibrillators and pumps (Brow *et al*, 1999, 2000). Some examples of seal components are shown in Figure 103.

Bulk metal-ceramic parts in which the ceramic and metal parts are joined by brazing have also been used in experimental medical devices (Breme *et al*, 1998). These include an instrumented hip prosthesis consisting of an alumina head and hollow Ti-alloy stem containing electronic components, and encapsulated implants composed of brazed alumina/stainless steel parts. Metal–ceramic dental restorations have been used for many years (Kelly *et al*, 1996). These normally comprise a glass, porcelain or ceramic outer layer bonded to a metal core. Porcelain fused to a platinum post was first used as long ago as 1885 by Logan (known as

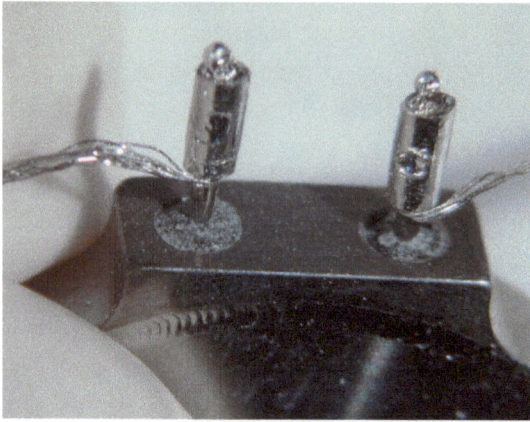

(a) Cochlea implants (Courtesy of IJ Research)

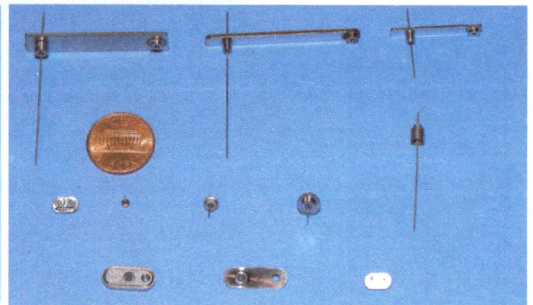

(b) Various implants (Courtesy of Hermetic Seal Technology Corporation)

Figure 103: Glass-to-metal seals employed in biomedical implant components

a Richmond crown). Porcelain dental materials often contain the mineral phase leucite in order to increase the thermal expansion of the porcelain to match more closely that of the casting alloys used in these restorations. Fairhurst *et al*(1985) have reported the bonding characteristics of porcelain to a number of dental alloys including those based on Ni–Cr, Co–Cr and Pd–Ag. Coornaert *et al*(1984) have described a long term technical study of porcelain-fused-to-gold dental restorations. These restorations combine the aesthetic qualities of porcelain with the strength and related properties of gold. An example of a ceramic–metal dental

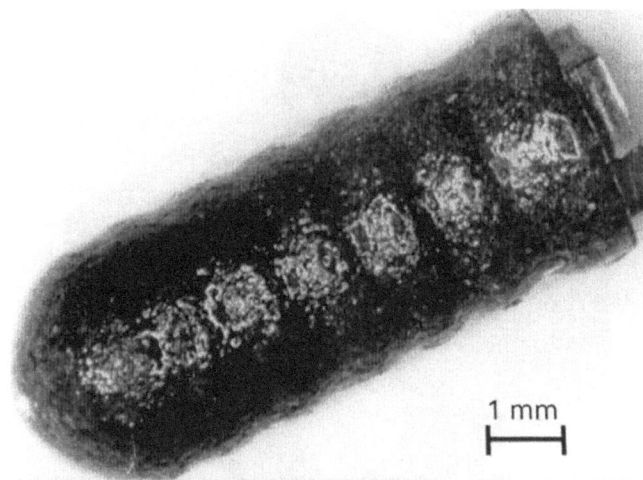

*Figure 104: Metal/glass dental prostheses employing a bioactive glass coating on a
Ti6Al4V oral implant (after Schrooten and Helsen, 2000)*

product is shown in Figure 104.

Bonding of dental porcelain to Ni–Cr dental alloys has also been studied by
Boyraz *et al*(2004). Bonding was achieved at temperatures in the range 905–955°C
under vacuum by diffusion of Cr and Ni into the porcelain. The mechanical
properties of porcelain/Pd alloy bonded systems employed in restorative den-
tistry as dental crowns and bridges have been determined by Zhao *et al*(2000).
Contact damage was monitored using an indentation technique. Cracking of the
porcelain layers was observed under different conditions and depended strongly
on the coating thickness. For example, for thin coatings < 300 μm thick damage
first occurred by yield in the Pd, whilst for thicker coatings cone cracks were
formed in the coating.

Recently, a variety of glass and glass-ceramic coatings have been applied to
medical implant materials in order to increase their resistance to chemical attack
or wear, or to improve their frictional resistance, or to improve their bonding characteristics to natural bone. Bonding of bone is becoming a particularly important
requirement as metals commonly used as biomedical implants including stainless
steels, Co–Cr alloys and Ti-based alloys are bioinert. In an untreated form, when
implanted into a body environment, these bioinert materials become surrounded
by fibrous scar tissue. This tissue subsequently impedes stress distribution at
the bone–implant interface which may result in loosening of the implant and
ultimate failure as fracture of the adjacent bone. It is therefore highly desirable
to coat implant materials with a suitable material.

With this consideration in mind, Bloyer *et al*(1999) successfully applied bioactive glass coating onto Ti–6Al–4V implant alloy. Bioactive glasses are materials
that undergo a series of chemical reactions in the presence of bodily fluids to

form hydroxycarbonate apatite on their surfaces. This compound is sufficiently similar to natural biological apatite that it is recognized as bone-like and new bone is deposited onto it. It may therefore be highly desirable to coat biological implants if the growth of new bone is desired; for example, when strong bonding between the implant and bone is required, as in the case of reconstructive surgery. In the study of Bloyer *et al*, coating of bioactive glass powder of composition $11Na_2O–3K_2O–15CaO–8·5MgO–6P_2O_5–56·5SiO_2$ (wt%) was first applied to the titanium alloy in the form of a slurry in isopropyl alcohol, followed by heating at 700–820°C. Excellent bonding was achieved between the coating and the substrate and this was believed to result from the formation of a thin reaction layer 50–100 nm in thickness of Ti_5Si_3. The adhesion of bioactive glass to Ti–6Al–4V alloy used in oral implants has also been examined by Schrooten and Helsen(2000). It was stressed that the coating is only satisfactory if it adheres strongly to the metal surface and is strong enough to transfer all loads anticipated during the lifetime of the implant. Samples were prepared by plasma spraying glass onto the metal alloy. A combination of mechanical testing, finite element analysis, acoustic emission, and microstructural analysis was employed to monitor the effectiveness of the system. Bioactive glass-ceramic coatings based on fluorapatite have been applied to Ti–6Al–4V alloy substrates by electrophoretic deposition and by magnetron sputtering (Bibby *et al*, 2004). The influence of glass composition on the bonding behaviour of bioactive glasses to AISI 316L stainless steel used in typical surgical implants has been studied by Krajewski *et al*(1985). It was noted that quite small changes in the glass composition markedly affected the bonding characteristics. It has been observed that the adhesion of hydroxyapatite coatings to a CoNiCrMo implant alloy can be improved by oxidation of the substrate prior to applying the coating (Tao, 2004).

Graded coatings 25–150 μm thick of glass and of glass–hydroxyapatite mixtures have also been applied to implant alloy surfaces including Ti-based alloys and Co–Cr alloys using a relatively simple enamelling technique (Saiz *et al*, 2000). Glasses employed were based on the $Na_2O–CaO–MgO–K_2O–P_2O_5–SiO_2$ system with softening temperatures in the range 503–644°C and thermal expansion coefficients in the range $8·8–15·6×10^{-6}$ K^{-1} (200–400°C). It is claimed that use of graded coatings offers the advantage of programmed dissolution of the coating with optimization of compatibility and long-term stability. Use of graded coatings using a corrosion resistant glass composition adjacent to the metallic implant surface and a surface low in silica for biocompatibility would therefore appear to offer excellent prospects for the development of new implant coatings with superior properties. In addition to graded coatings, dual-layer glass-ceramic coatings have been applied to Ti–6Al–4V alloy employed in dental implants (Verne *et al*, 2004). The first, ground coat, was applied to the metal by dipping and firing, this coat providing strong bonding to the metal. The second, bioactive glass-ceramic coating containing alkali metal fluorides, was subsequently applied to this ground coat. As an alternative to plasma-sprayed hydroxyapatite coatings which are now

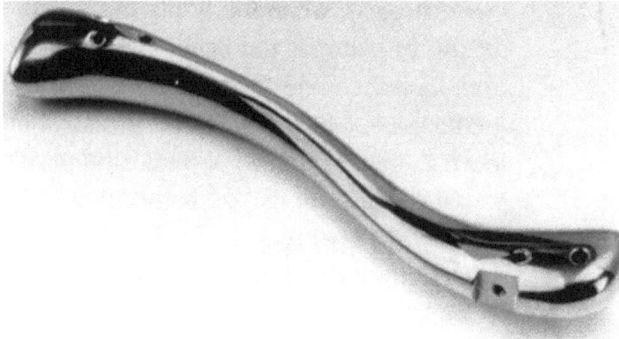

(a) TiN coating on a titanium collar bone implant

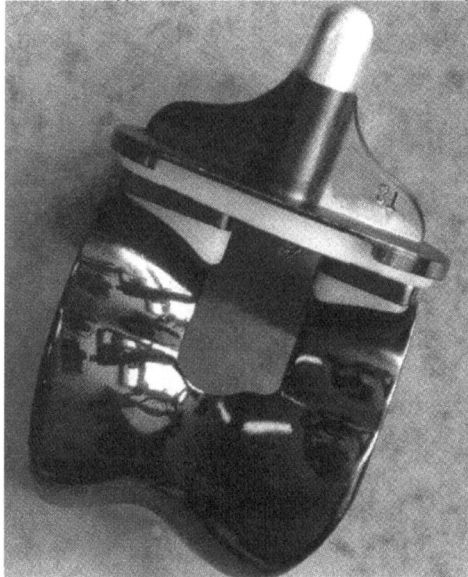

(b) Diamond-like carbon coating on a knee joint prosthesis
Figure 105: Ceramic coatings to improve the properties of biomedical surgical implants (after Burslem, 2004)

often used to coat Ti–6Al–4V hip prostheses Stanton *et al*(2004) have applied a glass-ceramic composed of apatite and mullite crystal phases using a sedimentation and heat-treatment process. It was stressed that this process is significantly less expensive than plasma spraying and offers additional advantages including good adhesion and close match of thermal expansion characteristics. The starting glass used, based on the $CaO–Al_2O_3–SiO_2–P_2O_5–CaF_2$ system, on crystallization yields fluorapatite, which is biocompatible, together with mullite. Examination by SEM, TEM and selected area diffraction indicated that a complex non-planar interface was produced containing titanium silicide nano-size precipitates.

Ceramic coatings including, for example, TiN, have also been applied to titanium implants including a collar bone replacement, and diamond-like carbon coatings have been applied to hip and knees prostheses (Burslem, 2004). Examples are shown in Figure 105.

10.12   Microwire

There has been a renewal of interest in glass-coated microwire in recent years, and this is particularly true in the case of magnetic alloys (e.g. Wiggins *et al*, 2000; Zhukov, 2002; Deprot *et al*, 2002; Kim *et al*, 2004). Glass-coated Fe- and Co-based microwires exhibit excellent magnetic properties which include soft ferromagnetic behaviour. They may also exhibit the phenomenon of giant magnetoimpedance, GMI. The high sensitivity of the GMI ratio to small mechanical loads makes microwire ideal for applications involving the detection of small stresses. This includes their use in sensors and security devices.

For example, a pen for identification of signatures has been proposed based on the magnetic response to induced stresses during use (Zhukov *et al*, 1998). Signatures are characterised by a typical series of stresses. The sequence and magnitude of these stresses are a characteristic feature of any personal signature, which can therefore be employed for identification of signatures. The pen consists of a ferromagnetic microwire with positive magnetostriction together with a miniature secondary coil and a mechanical system inside the pen containing a spring which transfers the applied stresses to the wire. The corresponding signal produced during writing with this pen is reproducible for any given signature, with the major characteristics being time of signature and the sign and sequence of the detected stress peaks. A schematic illustration of the device and the corresponding traces obtained during the writing of a signature are shown in Figure 106.

Magnetic tags and other types of sensor also utilise microwire. The magnetic characteristics of microwire consist of a large change in the impedance of soft magnetic conductors, driven by a high frequency current, when placed in a static magnetic field. This characteristic can be put to advantage in a wide range of applications including magnetic sensors. Magnetic tags (anti-shoplifting devices), for example, may employ several microwires with well-defined coercivities characterized by rectangular hysteresis loops (Larin *et al*, 1996). Once the magnetic tag is submitted to a magnetic field each microwire remagnetizes at different magnetic fields giving rise to an electrical signal on a detecting system. The extended range of switching fields possible with Fe-based microwires makes it possible to employ a large number of combinations, also making them useful for applications involving magnetic codification.

Glass-coated microwire prepared using the Taylor-wire process is available commercially from a limited number of suppliers. One such manufacturer is Global Microwire Technologies Ltd. (Global Wire Group). They manufacture microwire with a range of metallic cores and with stated applications ranging from miniature electrical and electronic components, where good electrical and magnetic properties are important, to military applications. Novel devices also include EMI filters capable of providing high common-mode noise attenuation through a wide frequency range. For this type of application microwire can exhibit high dissipation losses to much higher frequencies than possible with

(a) Schematic representation of a magneto-
elastic pen for confirming signatures, and
(b) plots obtained from two signatures
(after Zhukov, 2002)

(c) Spools of glass-coated
microwire (after Global Micro
Wire Technologies Ltd)

(d) High-performance EMI filter panel
assembly (after Global Micro Wire
Technologies Ltd)

(e) EMI suppression filter
(after Global Micro Wire
Technologies Ltd)

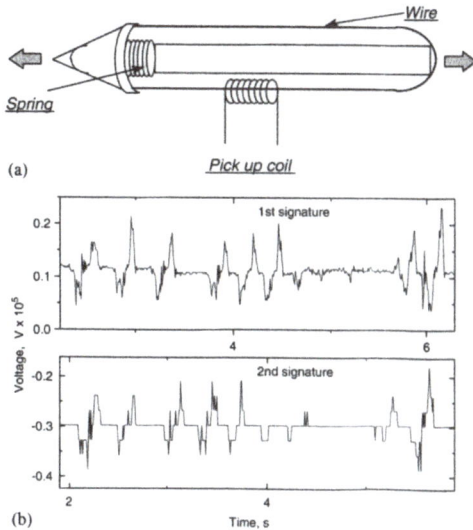

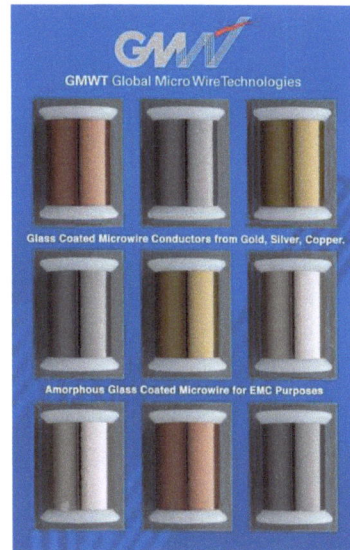

Figure 106: Components employing glass-coated microwire

known ferrite materials. Examples are depicted in Figure 106. EMI protection
can be provided for a wide range of low and high speed data communication
interfaces in military, medical, automotive and airborne equipment. Advanced
Coding Systems, ACS, manufacture magnetic microwire for applications in the
anti-shoplifting, product authentication and brand protection markets, where the
unique magnetic signature of the microwire enables unambiguous identification.
One advantage offered by microwire for this type of application is its very small
diameter (7–10 μm) which makes it hard to detect. It also offers non line-of-sight
detection through layers of various packaging materials including non-ferrous
metals up to several centimetres in thickness.

(a) Sight glasses (Courtesy of Hermetic Seal Corporation)

(b) Rotary motion mechanical feed-through component (Courtesy of Ceramaseal)
Figure 107: Some miscellaneous components which employ glass- and
ceramic-to-metal seals

Alternative applications for microwires include their use in a diverse range of sensor systems including eddy current, proximity, electrochemical, biomedical, implant, flow, pressure and electromagnetic sensors, in addition to use as precision resistors, infra-red detectors, thermistors, miniature heaters, coils and transformers, multi-chip connection, and even as reinforcement for composite materials. Microwire has also been employed in various specialised military applications.

10.13   Miscellaneous applications

Glass- and ceramic-to metal systems are used in many other applications. Only a small selection is presented here. For example, glass-to-metal seals have been employed in the construction of permeation devices used as standards for a range of applications, and have also been used in commercial and domestic flat plate solar panel systems for providing hot water. Additional applications include sight glass components, mechanical feed-throughs, tube seals, cryogenic components, and filament supports and related devices. Glass-to-metal seals have also been employed

Figure 107: (continued)

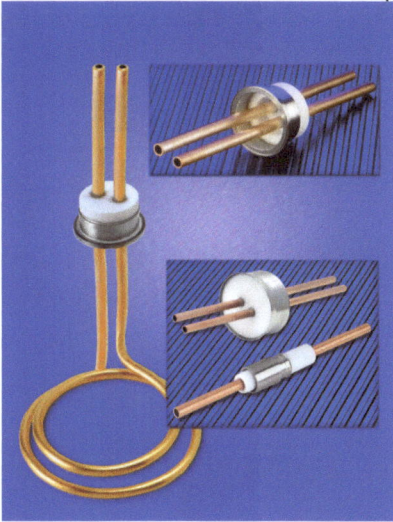

(c) Tube seals (Courtesy of Ceramaseal)

(d) Cryogenic components for operation down to −269°C (Courtesy of Ceramaseal)

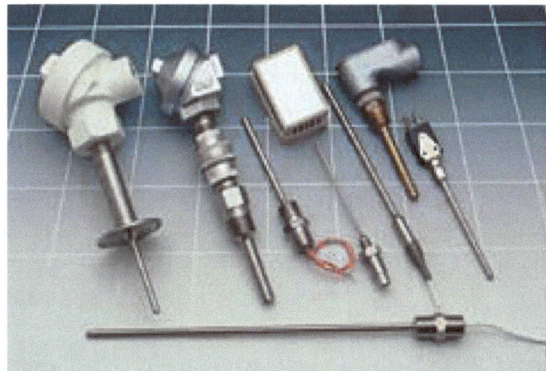

(e) A selection of ceramic-to-metal resistance temperature detectors (Courtesy of Kavlico Corporation)

in a novel corrosion monitoring system for radioactive high level waste storage tanks (Terry *et al*, 2002) and in apparatus for creating Bose–Einstein condensates employing a magneto-optical trap which utilizes a cylindrical glass cell attached to glass-to-metal feed-through seals (Lewandowski, 2003). Ceramic-to-metal seals have been employed in such applications as high temperature cartridge heaters, and in ceramic RF windows which are utilized in such areas as high power microwave applications. These windows, which may be made from BeO, high purity alumina or more recently AlN, are attached to metal frames usually by metallization and brazing procedures. In the case of AlN windows, Bartkowski *et al*(2004) have successfully employed refractory Mo–Mn metallization of the window followed by brazing to copper and Nilo-K structures using standard brazing alloys to form hermetic seals. Some of these additional applications for glass-, glass-ceramic- and ceramic-to-metal seals are shown in Figure 107.

TABLE 14
SOME TYPICAL APPLICATIONS AND COMPONENTS THAT EMPLOY GLASS-TO-METAL SEALS AND COATINGS

Application/Component	Specific Examples/Comments
Aerospace industry	• Radar and communications components • Instrumentation • Actuators • Gyroscopes • Safe and arm devices • High pressure connectors
Battery applications	• Seals in Li–S, Li–SO$_2$, Li–SOCl$_2$, and Li–MnO$_2$ batteries • Seals in Na–S, and Na–SO$_2$ batteries • Thermal battery seals • Used in numerous applications from heart pacemakers to location transmitters, emergency beacons and monitoring systems
Biomedical components	• Heart pacemakers and pacemaker batteries • Cochlea implants • Cardiac and respiratory monitors • Exploratory probes • Coatings for implants
Coatings	• Enamelled metals • Domestic ware including enamelled cooking utensils, cookers, sink units, water and storage heaters, and sanitary ware • Jewellery and objects of art • Protective coatings for underground pipes • Protective coatings for vessels and pipes used in the chemical industry • Agricultural applications, e.g. storage silos • Heat exchangers • Aerospace components, e.g. low pressure turbine blades and after-burners; also ceramic coatings for high temperature turbine blades • Protection of aircraft and missile radomes and infra-red windows • Coatings for biomedical implants, e.g. dental prostheses and artificial hip joints • Building and architectural items including building panels and tunnel walls

(Table 14 continued)

Detonators and explosive actuators	• Actuators for rapid response opening or closing control valves • Actuators for operation of remote latches on spacecraft and missile systems • Explosive bolts • Automobile systems, e.g. airbag actuators • Detonators to initiate explosive devices
Domestic applications	• Enamelled metals • Heating and lighting systems • Solar panels • Audio and television • Security systems
Electrical components and electrical power industry	• Relays, chokes and coils • Transformers • Capacitors • Thyristors • Vacuum interrupter envelopes • Lightning arrestors
Electronics and microelectronics packaging	• Transistor packages • Hybrid electronic housings and packages • Optical packages • MEMS packaging • MOEMS packaging • Sensor packages • Electronic black boxes • Crystal clock oscillators • Used in domestic, transport, aerospace, military, environmental and industrial applications • Substrates employed in electronic packaging, e.g. enamelled substrates, glass-ceramic/metal substrates
Fuel cell applications	• Seals in molten carbonate and SOFC fuel cells • Used in small and large power generation, emergency back-up facilities, aerospace, military and transport
Hermetic connectors	• Multi-pin connectors • Co-axial connectors and terminators • Sensors • Valves • Automobile, aircraft, missile and spacecraft instrumentation

(Table 14 continued)

	• Radar systems
	• Microwave systems
	• Lightning arrestor components
	• High pressure fuel tank connectors
	• Vacuum connectors and feed-throughs
	• Electrical feed-throughs
Lamps, lasers and related devices	• Lamp and bulb envelopes and feed-through seals
	• Halogen, mercury, sodium and xenon lights and arc lights
	• Fluorescent lighting
	• Linear flashlamp seals
	• Photomultiplier tubes
	• Laser tubes and feed-throughs
Microwave applications	• Microwave equipment
	• Hermetic co-axial connectors for high frequency microwave applications
Microwire	• Used in sensors and switches, e.g. proximity sensors, pressure sensors, electromagnetic sensors, flow sensors, magnetic tags and security devices, magnetic codification and advanced coding systems, EMI filters and protection; data communication interfaces, infra-red detectors
	• Miniature transformers, coils, resistors and heaters
	• Multi-chip connection
	• Reinforcement for composite materials
	• Military applications
Reed switches	• Proximity switches for robotics, conveyors and escalators
	• Thermostats
	• Test equipment
	• Smoke detectors
	• Audio and video equipment
	• Computer equipment, including keyboards
	• Telecommunications
	• Detectors and sensors
Scientific and technical research	• Scientific and technical apparatus
	• Glass ware
	• Monitoring systems

(Table 14 continued)

Security devices	• Reed switches, relays and sensors employed in security systems, e.g. door and window sensors, sensor switches, magnetic switches, etc.
Sensors, switches, relays, detectors and transducers	• Pyroelectric and optoelectronic detectors • Vibration, tipover and tilt sensors • Proximity switches • Thermostatic switches • Pressure switches, transmitters and transducers for alarm, shutdown, control and interlock applications • Hydraulic systems • Test cells • Scintillation tubes • Vacuum and gas-filled relays • Mercury displacement relays • Liquid level and optical sensors • Aircraft, spacecraft and missile instrumentation • Chemical and nuclear industry, including waste tank corrosion monitoring systems • Off shore oil platform, paper mill, and food processing equipment
Telecommunications	• Vacuum tubes • Television tubes • Fibre-optic components and packages • Optical switches, attenuators and wavelength lockers
Transport	• Reed switches, relays and sensors employed in automobile components including air bag and seat belt sensors and initiators, light controls, and cruise controls • Tyre pressure monitoring systems • Engine control systems • Load cells used in truck scales

TABLE 15
SOME COMMERCIAL SUPPLIERS AND MANUFACTURERS OF GLASS-
AND GLASS-CERAMIC-TO-METAL SEALS AND COMPONENTS

Supplier	Products
Accra Tronics Seals Corporation, USA	Electrical feed-throughs, squib headers, detonators, cartridges, explosive bolts
Acroname, USA	Pyroelectric detectors
Aegis Semicondutores LTDA, Brazil	Thyristors
Agilent Technologies, USA	Co-axial connectors, adapters and terminators including 1 mm diameter connector, diodes
ASB Group, France	Feed-through seals, electronics package housings, thermal batteries
Astro Seal Inc., USA	Connectors and feed-throughs, multi-pin headers, medical devices, thermal battery headers, glass preforms
Ceramaseal, USA	Glass-ceramic-to-metal feed-throughs and hermetic connectors; ceramic-to-metal components
Ceramic Seals Ltd, UK	Ceramic-to-metal feed-through seals and connectors for ultra high vacuum, high pressure, high temperature and cryogenic applications
Comus International, USA	Feed-through seals, electronic package housings, headers, diode packages, medical devices, fibre optic packages, sensors
Concept Group, USA	Sensors and sensor packages, feed-through seals
Century Seals, UK	Feed-through seals, electronic packaging, connectors
Deutsch Ltd, UK	Hermetically sealed fuel tank connectors, miniature connectors, gas analysis components
Druck Incorporated, USA	Pressure transmitters and transducers
Eagle Picher, USA	Li and thermal batteries
Electron Tubes Ltd, UK	Photomultiplier tubes; x-ray tubes, light detector assemblies, sensors, feed-through seals
Fujitsu, Japan	Electronic and optoelectronic packages
Fusite, USA	Battery seals
GE, USA/UK	Sensor packages
Glenair, UK	Hermetic connectors; fibre optic components
Global Micro Wire Technologies Ltd., Israel	A range of glass-coated microwires together with woven cloth, and miniature filters, coils and cables

(Table 15 continued)

Hermetic Seal Technology, Inc., Cincinnati, USA	Medical hermetic seals, battery seals, hermetic connectors, sight glasses, high amperage feed-throughs, custom seals and feed-throughs
Hilgenberg, Germany	Custom glass-to-metal and glass-to-glass seal components
Hirtenberger-Schaffler, Germany	Automobile airbag and seatbelt initiators
HMEL, UK	Electronic components and packaging
IJ Research Inc., USA	Biomedical components including heart pacemakers, artificial cochlea, electrical feed-throughs, terminals and heaters
Kyocera, USA	Fibre optic components and packages; connectors
Latronics Corporation, USA	Electrical and electronic components, feed-throughs and housings, ceramic-to-metal seals
Mansol (Preforms) Ltd,UK	Glass preforms
Martec, UK	Hermetic connectors, electrical feed-through seals and custom seal components
Maxwell Technologies, USA	Vacuum and optical feedthrough components
Micrometics, UK	Feed-through seals, electronic packaging
Minco Technology, USA	Optoelectronic detectors, emitters and couplers
Morgan Advanced USA/UK	Glass preforms; glass-to-metal seals for IC Ceramics, packages, fibre optic connectors and relays, ceramic-to-metal seals, medical devices
MPF Products Inc., USA	Ultra high vacuum feed-throughs, connectors, and custom seal components
Natel Engineering Co. Inc., USA	Data acquisition and signal hybrid electronic packages
Networks Electronic Corp., USA	Miniature squib actuated switches, thermal fuses and relays, and custom glass-to-metal seal components
Pure Energy Technology Ltd., UK	Flat plate solar water heating systems
Rhopoint Components Ltd., UK	Magnetic sensors and reed switches
SAFT, France	Lithium batteries
Schott, Germany	Glass-to-metal feed-throughs, electronic package housings; glass preforms
SCM International, Inc, USA	Vibration, movement and tilt sensors
SeaCon Phoenix, USA	Glass-to-metal feed-throughs and connectors, lithium battery seals

(Table 15 continued)

Shinko Electric Industries, Japan/UK	Electronic package housings, laser diodes, sensors
Sinclair, USA	Electronic package housings and custom packages; connectors
Sonnenschein Lithium GmbH, Germany	Lithium – thionyl chloride batteries
StratEdge, USA	Microwave electronic packages
Tekna Seal, USA	Feed-through seals; custom seals; hermetically sealed optical windows
TEPCO/NGK Insulators, Japan	Na-S batteries
Ultratech, France	Feed-through seals; multi-pin headers, connectors, hermetic microelectronic packages
Victory Lighting, UK	Lamps
Vitrus, USA	Hermetic terminal assemblies and feed-throughs
Willow Technologies, UK	Feed-through seals
Wesley Coe, UK	Connectors, battery seals, detonators and igniters, custom seals

11. CONCLUDING REMARKS, RECENT ADVANCES AND OUTLOOK FOR THE FUTURE

Since the era of Partridge(1949) a very wide range of glasses have been developed for use in the preparation of glass-to-metal seals and coatings. These new systems have made possible numerous and diverse applications covering many exciting areas of interest, ranging from microelectronics packaging to biomedical implants. The major advantages offered by glasses for these applications are their refractoriness and thermal stability, relative to organic sealants and coatings, coupled with the comparative ease of preparation of well-bonded glass-to-metal systems, when contrasted to the use of alternative ceramic materials. Many new glass compositions have been and continue to be developed and their unique properties exploited, not only from the more conventional silicate and borosilicate systems but also by utilizing phosphate and borate based materials, for many years regarded as academic curiosities.

The newer glass-ceramic-to-metal seals and coatings, not available in Partridge's day, offer all the advantages associated with their glassy counterparts, including ease of fabrication and excellent bonding characteristics, in addition to a further improvement in refractory behaviour. In this respect the performance of glass-ceramics parallels that of ceramic-to-metal seals, but without many of the difficulties and shortcomings associated with the preparation of these systems. Glass-ceramics are also mechanically stronger than their glassy counterparts, are more resistant to static fatigue, and are more resistant to chemical attack. In addition, one of the major advantages offered by glass-ceramics is undoubtedly the far greater range of practical thermal expansion characteristics possible, compared to what can be obtained with glasses or ceramics alone. It is now clear, however, that when selecting a glass-ceramic for bonding to a particular metal or alloy it is essential that a full simulated sealing/heat-treatment schedule, including the high temperature sealing stage, is employed for determining the thermal expansion characteristics of the material prior to its use in a glass-ceramic-to-metal seal application. Failure to carry out this procedure can result in misleading thermal expansion data and lead ultimately to seal failure. This is due to the fact that nucleation and growth of crystals often occurs during either the heating up or cooling down stages, rather than being confined solely to the specific nucleation and crystallization cycles. This may subsequently give rise to different proportions and types of crystals in the final sealed product, relative to those obtained during the manufacture of bulk glass-ceramics where a high temperature sealing stage is not included. It is also equally important to identify processing windows accurately for the preparation of glass-ceramic-to-metal seals, in particular crystal-lization temperature ranges. Failure to do so can also lead to unreliable products. This is because in many instances the variation of thermal expansion coefficient of glass-ceramics with crystallization temperature (and time) is neither constant nor linear, except over relatively narrow temperature ranges. If the temperature range over which a constant thermal expansion coefficient is achieved is too narrow, it will be very difficult to achieve a reliable product unless temperature control can be very accurately maintained, not always possible in a large scale manufacturing environment. The best systems from a manufacturing viewpoint are therefore those systems that are the most process tolerant and which provide wide processing windows.

The nature of the bond between a metal and a glass has been the subject of considerable controversy over many years, but a greater knowledge of the factors responsible for achieving strong bonding has gradually emerged. Great advances in the understanding of bonding between dissimilar materials in general have been made through the application of modern analytical techniques including spectroscopic analysis, SEM/EDS, TEM, and more recently HRTEM. It is now recognised that strong chemical bonding between a metal and a glass (or glass-ceramic) can only be achieved if the conditions during sealing are such that the glass at the interface becomes and remains saturated with the appropriate substrate metal oxide. Traditionally, and in practice, this is usually achieved by pre-oxidizing the metal substrate prior to sealing or coating so that a ready supply of the appropriate substrate metal oxide is available. It is now appreciated, however, that suitable conditions can also be achieved in many instances either through appropriate choice of the sealing atmosphere or directly through redox reactions between the glass and the substrate, without the need to carry out a separate pre-oxidation stage. In addition to direct reaction between the substrate and glass or glass-ceramic, desirable interactions can also be promoted by the use of appropriate additives to the glass; for example, in the case of Fe-based alloys, use of the classical adherence promoting oxides, CoO and/or NiO which react preferentially with the metal substrate to promote the formation of a suitable substrate metal oxide favourable to bonding, e.g. $2Fe + 3CoO \rightarrow Fe_2O_3 + 3Co$. In the case of Cr-containing alloys, suitable glass additives include not only CoO and NiO, but also additional metal oxides that will react preferentially with Cr (which diffuses into the interfacial region from the substrate) to form Cr_2O_3, e.g. $2Cr + 6CuO \rightarrow Cr_2O_3 + 3Cu_2O$, this specific oxide being particularly conducive to the promotion of strong chemical bonding.

It is now recognized that there are many factors which influence the quality of a seal or coating. The formation of bubbles or voids at the metal/glass interface is a particularly troublesome occurrence which not only weakens the seal or coating but may also prevent it from being hermetic. Bubbles occur for a number of reasons including the formation of gaseous products due to the presence of volatile contaminants, e.g. traces of cutting fluids, carbon dust from jigging, and even finger print grease and salt deposits. Such effects can be substantially eliminated by careful seal preparation and good house-keeping practices. In addition, many Fe-based alloys, together with modern high strength non-ferrous alloys, contain carbide phases and precipitates which, if present in or close to the surface, can react during sealing to produce gaseous CO or CO_2, again leading to the formation of bubbles in the seal. The detrimental effect of such carbides is well known and steps can be taken to prevent or minimise such deleterious reactions; for example, by decarburization treatments prior to sealing.

Although the detrimental effects of dissolved water on the optical properties of, for example, infrared optical components is well known (e.g. Donald and McMillan, 1978), perhaps less well appreciated is the influence that water dissolved in the glass can have on the overall glass-to-metal sealing characteristics. Glasses produced using conventional techniques invariably contain not insignificant quantities of dissolved water, usually in the range 0·02–0·06%, although amounts can be considerably higher. This dissolved water may react with metallic species diffusing into the glass from the metal substrate to form hydrogen gas, and this may lead to the formation of bubbles in the seal. Thermodynamic data suggest that hydrogen gas formation may be responsible for bubble formation in many metal/glass and metal/glass-ceramic systems, and in particular for those alloys containing significant quantities of such metallic elements as Hf, Y, Zr, Al, Ti, Ta, Nb, Cr or Mn; for example, $2Cr + 3H_2O \rightarrow Cr_2O_3 + 3H_2$. Two different approaches have been adopted to minimise the influence of dissolved water. The first is to produce the glass under dry conditions, which surprisingly is not necessarily an easy task. The second is the use of suitable additives to the glass, additives that will react preferentially with metals diffusing into the glass from the substrate to form non-gaseous reaction products; for example reaction of CuO with Cr to form Cr_2O_3 and Cu_2O. This reaction is more favourable thermodynamically than the reaction between Cr and H_2O. In addition to possible reaction with water, it is also important in the case of glass-ceramics to avoid reaction between diffusing metal species and the nucleating agent present in the glass; for example, reaction of Cr with Li_3PO_4 nuclei to give a mixture of $Cr_{12}P_7$, Li_2O and CrO/Cr_2O_3. Reactions of this nature remove the nucleating agent from the interfacial region and may result in a glassy region or a region of coarse microstructure with thermal expansion characteristics that differ markedly from those of the bulk glass-ceramic.

As a consequence of these and similar reactions, and in order to have confidence in an adequate component lifetime, it is necessary not only to ensure compatible bulk thermal expansion behaviour between metal and glass or glass-ceramic, but also to minimise interfacial reactions and diffusion of bulk metallic species, whilst retaining strong chemical bonding at the interface. The formation of reaction products and precipitated phases within the interfacial region will invariably lead to induced residual stresses due to mismatch of thermal or elastic properties of these phases with the bulk glass or glass-ceramic, and these can indeed have serious implications for the lifetime behaviour of glass- and glass-ceramic-to-metal systems.

In the case of ceramic-to-metal sealing, coating or joining, the Mo–Mn ceramic metallization process is a well established technique that has been used for many years, mainly for use with alumina, although it can be applied to a limited number of alternative ceramics. The formation of a metallized layer on the ceramic allows conventional brazing of the parts to be performed in order to form the ceramic-to-metal bond. Reactive metal brazing, on the other hand, employing alloys containing such reactive metals as Ti, Zr, Hf or Al, is a more recent conception which is being used more and more widely, and which can be employed with almost all existing ceramics including, for example, silicon nitride and silicon carbide. It is therefore expected that the technology of reactive metal brazing will continue to advance, and may soon supersede the Mo–Mn process as the major ceramic-to-metal bonding method in use (Mizuhari, 2000).

Partridge(1949) reported the bonding of a small variety of glasses to a limited number of metals and alloys, which included Cu, Pt, Mo, W, together with relatively simple two or three component Fe-based alloys. These metals and alloys also had limited applications, mainly in the electrical industry; for example, in lamps, radio components and x-ray tubes. Relative to Partridge's era, many more glass compositions are now available and new and more exotic compositions continue to be researched and developed, enabling sealing to a wider and wider range of metals and alloys, in addition to other glasses and ceramics. The scope has also been expanded considerably by inclusion of the versatile glass-ceramic materials. The range of applications for seal components has also expanded astonishingly to include such diverse areas as radio and television, electrical power, battery and fuel cells, sensors, electronics, microelectronics packaging, telecommunications and biomedical materials, in addition to their use in automobile, transport, aerospace, architectural, security, and military systems, coupled with a multitude of different types and sizes of hermetic connectors which support these and related applications. Glass-to-metal seals are also finding use in such topical environmental areas as radioactive waste tank monitoring systems, and in the construction of novel experimental apparatus; for example, for collecting Bose–Einstein condensates.

There is also a continuing focus on the miniaturization of electrical and electronic components and related devices. Glass-to-metal seals only 1mm in diameter are now made routinely and it is likely that sizes will decrease even further as the demand for miniature components continues to increase. New applications in the biomedical, aerospace and military fields also continue to arise requiring smaller and more sophisticated components which include glass- and glass-ceramic-to-metal seals. A particular issue of topical interest is the development of sealing media for fuel cells. Fuel cells offer highly efficient systems, but their operating environment is extremely hostile to the sealant materials currently in use. Glass-ceramics, in particular, offer considerable scope for the development of more suitable sealants which will be resistant to the fuel cell environment. Fuel cell technology involves bonding glass-ceramics to metals such as stainless steels and Ni-based superalloys, areas which already have a strong background due to the earlier development of high performance components, including explosive actuators, detonators and related devices. There is considerable scope in this area to extend studies to include glass and glass-ceramic composite materials in order to achieve enhanced properties more suitable and adaptable to SOFC operating conditions.

Fusion techniques have traditionally been employed in the preparation of glass and glass-ceramic seals and coatings and these methods usually involve subjecting the entire component to high temperatures. Many new or novel processes are currently being developed, however, for sealing or bonding in cases where high temperatures may not be desirable, or where heating must be kept very localized in order to avoid damage to the final component or assembly. New sealing methods range from laser and electron beam techniques which can be very localized, to acoustic and ambient temperature solid state diffusion bonding, to plasma spraying and related techniques.

The use of vitreous enamels as protective coatings for metals used in domestic and industrial applications has been widened to include glass, glass-ceramic and ceramic coatings for the protection of high performance materials and components used, for example, in aerospace technologies, including turbines and jet engines. Undoubtedly, there is considerable scope for further improvement of such materials, with drives towards ever higher operating temperatures for improved fuel efficiency and towards more environmentally friendly systems.

There have been many advances over recent years in the characterization of seals and coatings and of the materials that make them up. One example is the use of HRTEM which has enabled the detailed atomic bonding characteristics of ceramic–metal and ceramic–ceramic systems to be studied. These studies have highlighted, for example, the importance of dislocation structures at interfaces and their mobility under sealing conditions. This technique undoubtedly offers promise for further investigations of alternative materials and systems, including glass-ceramic/metal components where interfacial dislocation structures also may be important. In addition, there is also still considerable scope for examining the detailed bonding characteristics of glass, ceramic and metal systems and the role of interfacial chemistry in determining the overall bonding behaviour, performance and lifetime behaviour of systems which employ glass-, glass-ceramic- and ceramic-to-metal components. In this respect, life assessment studies of bonded systems are still very much in their infancy and much remains to be done before a deeper understanding and more accurate predictive capability of the ageing characteristics of these systems emerges. The issue of life assessment and ageing characteristics is becoming an ever more important topic with the drive toward materials systems exhibiting greater reliability and extended lifetimes, and this is undoubtedly an area worthy of continuing and expanded research.

Glass-to-metal seals have now been in use for over one hundred years and their place in modern society and into the foreseeable future is still very much assured. Glass-to-metal coatings in the form of vitreous and porcelain enamels have been in use for much longer, with technological applications dating back to the eighteenth century and earlier, whilst the use of coatings for jewellery and objects of art date back further still. Ceramic-to-metal and glass-ceramic-to-metal seals and coatings are more recent additions with unique properties and diversity of applications, and which offer considerable scope for further developments and improvements.

In conclusion, glass-to-metal seals and coatings, together with the more recent glass-ceramic and ceramic systems, have made and continue to make important contributions to science and technology. Their continuing contribution in these areas is undoubtedly assured well into the foreseeable future.

ACKNOWLEDGEMENTS

The author is extremely grateful to many of his colleagues at AWE, and in particular to B. L. Metcalfe, for experimental data and for constructive discussions over many years. Thanks are also due to R. S. Greedharee, J. L. McGrath and D. J. Bradley for experimental data, to library staff at AWE for provision of literature search data, and to C. Ruckman and G. Lawrence, also lately of AWE, for scanning electron microscopy and microprobe analysis. The author is indebted to J. Marlin of Ceramaseal, D. Harnett of Martec Ltd., J. Wendeln of Hermetic Seal Technology Inc., I. Shepherd of GE Infrastructure Sensing, P. Matthiae of Rhopoint, D. Kramer of the University of Dayton, R. Yoon of IJ Research Inc, G. Yang of PNNL, D. Cherry of Mansol Preforms, R. Stone and R. Headley of SNLA, D. Holland of Warwick University, R. M. Charnah and M. I. Budd of the former GEC Engineering Research Centre, R. Poteaux of SAFT Batteries, J. Leib and H. Weimann of Schott Glaswerke, and R. Yurco of Latronics Corporation, for supplying photographs and details of components. The author also wishes to thank members of the Basic Science and Technology Committee of the Society of Glass Technology for assistance in this work and for constructive discussions, with particular thanks due to J. Henderson and A. C. Wright. Thanks are also due to D. Moore of the Society's editorial office for valuable advice and for preparing the manuscript in a publishable form.

The author is also particularly grateful to D. Holland and B. L. Metcalfe for critical reading of the manuscript and for constructive comments and advice.

Finally, he would like to thank B. Bowsher and G. Nicholson, Director of Research and Applied Science and Manger of the Materials Science Research Division, respectively, at AWE, for permission and facilities to carry out this undertaking.

GLOSSARY

ASTM	American Society for the Testing of Materials
AWE	Atomic Weapons Establishment
bcc	body centred cubic (crystal structure)
CBS	Calcium Borosilicate (glass)
CVD	Chemical Vapour Deposition
DBC	Direct Bonded Copper
DTA	Differential Thermal Analysis
DMTA	Dynamic Mechanical Thermal Analysis
DSC	Differential Scanning Calorimetry
E	Young's modulus
EDS	Energy Dispersive Spectrometer
EMI	Electromagnetic Interference
ESR	Electron Spin Resonance
GM	Giant Magnetoimpedance
Hard glass	A glass with a high softening point and a low thermal expansion coefficient
hcp	hexagonal close packed (crystal structure)
HRTEM	High Resolution Transmission Electron Microscopy
ICTA	International Confederation for Thermal Analysis
LAS	Lithium Aluminosilicate (glass)
LZS	Lithium Zinc Silicate (glass)
MCFC	Molten Carbonate Fuel Cell
MEMS	Micro Electro-Mechanical System
MOEMS	Micro Opto-Electro Mechanical System
NAS	Na–S (battery)
NMR	Nuclear Magnetic Resonance
ORMOSIL	Organically Modified Silicate
OSEE	Optically Stimulated Electron Emission
SAM	Scanning Acoustic Microscopy
SANS	Small Angle Neutron Scattering
SC	Subscription Channel (fibre optic connector)
SEM	Scanning Electron Microscopy
SIMS	Secondary Ion Mass Spectroscopy
SNL(A)	Sandia National Laboratories (Albuquerque)
SOFC	Solid Oxide Fuel Cell

Soft glass	Glass with a low softening point and a high thermal expansion coefficient
TBC	Thermal Barrier Coating
TEOS	Tetraethoxysilane
TEM	Transmission Electron Microscopy
TEPCO	Tokyo Electric Power Company
TMO	Transition Metal Oxide
UTS	Ultimate Tensile Strength
XPS	X-ray Photoelectron Spectroscopy
XRD	X-Ray Diffraction
YSZ	Yttria Stabilized Zirconia

REFERENCES

Abe, Y. & Hosono, H., *Inorganic Phosphate Materials*, edited by T. Kanazawa, Elsevier Science Pub. (Amsterdam), 1989, pp 247–281.

Adamson, A. W., *Physical Chemistry of Surfaces*, 4th ed., John Wiley & Sons (New York), 1982.

Adamson, M. T. & Travis, R. P, *Proc. Electrochem. Soc.*, **18**, (1997), 691.

Agrawal, D. C. & Raj, R., *Acta Metall.*, **37**, (1989), 1265.

Aitken, B. G., Bookbinder, D. C., Greene, M. E. & Morena, A., US Patent No. 5246890, 1993.

Ak, N. F., Celik, E., Cetinel, H., Tekmen, C. & Soykan, H. S., Euro Ceramics VIII, *Key Engineering Materials*, vols. 264–268, edited by H. Mandal & L. Ovecoglu, Trans. Tech. Publications (Switzerland), 2004, pp. 529-532.

Akselsen, O. M., *J. Mater. Sci.*, **27**, (1992), 569.

Akselsen, O. M., *J. Mater. Sci.*, **27**, (1992), 1989.

Andrews, A. I., *Porcelain Enamels: the preparation, application, and properties of enamels*, Garrard Press (Champaign, Illinois), 2nd. edition 1961.

Anon, *Fuel Cells Bull.*, (February 2004), 6.

Anon, *Bull. Am. Ceram. Soc.*, **36**, (1957), 279.

Anon, *Bull. Am. Ceram. Soc.*, **70**, (1991), 1454.

Aranha, N., Alves, O. L., Barbosa L. C. & Cesar, C. L., *Proc. XVII Int. Congr. Glass*, vol. 7, Int. Ac. Pub. (Beijing), 1995, pp. 282–286.

Arora, A, Surface contamination and measurement control by non-destructive techniques, *3rd. Annual Technical Meeting of the Institute of Environmental Studies*, 1985.

Arya, A., and Carter, E. A., *Surf. Sci.*, **560**, (2004), 103.

Asahara, Y. & Izumitani, T., US Patent No. 3885974, 1975.

Ashcroft, A. I. & Derby, B., *J. Mater. Sci.*, **28**, (1993), 2989.

ASTM F394-78, Biaxial flexure strength distribution of thin ceramic substrates with surface defects, 1978.

ASTM F14-80(2000), Standard practice for making and testing reference glass-metal bead-seal, 2000.

Atkinson, A. & Selcuk, A., *Proc. Electrochem. Soc.*, **18**, (1997), 671.

Atkinson, A. & Selcuk, A., *Proc. Electrochem. Soc.*, **134**, (2000), 59.

Bach, H. & Krause, D. (editors), *Analysis of the composition and structure of glass and glass ceramics*, Springer-Verlag (Berlin Heidelberg), 1999.

Backhaus-Ricoult, M., in *Metal-Ceramic Interfaces*, edited by M. Ruhle, A. G. Evans, M. F. Ashby & J. P. Hirth, Acta-Scripta Metallurgica Proceedings Series 4, Pergamon Press (Oxford), 1990, pp. 79–92.

Bandyopadhay, N., Tamhankar, S., & Kirschner, M, in *Microelectronic packaging technology, materials and processes*, ASM International (Materials Park) 1989, pp. 41–47.

Bao, Y., and Zhou, Y., Surface Engineering 2002, Synthesis, Characterization and Applications, *Mat. Res. Soc. Proc.*, **750**, (2002), pp. 47-52.

Barin, I., Knacke, O. & Kubaschewski, O., *Thermochemical properties of inorganic substances*, Springer-Verlag (Berlin), 1973.

Bashev, V. F., *Phys. Met. Metall.*, **55**, (1983), 114.

Bar-On, I., in *Engineered Materials Handbook, Vol. 4, Ceramics and Glasses*, Technical Chairman S. J Schneider, ASM International, 1991, pp. 645–651.

Bartkowski, R. J., Pekrul, E. & Kirshner, M. F., *Fifth IEEE Int. Vacuum Electronics Conference*, 2004, pp. 27–29.

Bates, C. H., Foley, M. R., Rossi, G. A., Sundberg, G. J., and Wu, F. J., *Ceram. Bull.*, **69**, (1990), 350.

Bauer, I., Russek, U. A., Herfurth, H. J., Witte, R., Heinemann, S., Newaz, G., Mian, A., Georgiev, D. and Auner, G. W., *Proc. SPIE*, Vol. 5339, 2004, pp. 454–464.

Baynton, P. L., Rawson, H. & Stanworth, J. E. *J. Electrochem. Soc.*, **104**, (1957), 237.

Beall, G. H., *Glass Science and Technology, vol. 1, Glass-Forming Systems*, edited by D. R. Uhlmann & N. J. Kreidl, Academic Press (New York), 1983, pp 403–445.

Beall, G. H., *J. Non-Cryst. Solids*, **73**, (1985), 413.

Beall, G. H., *Rev. Solid State Sci.*, **3**, (1989), 113.

Beauchamp, E. K. & Burchett, S. N., in *Engineered Materials Handbook, Vol. 4, Ceramics and Glasses*, Technical Chairman S. J. Schneider, ASM International, 1991, pp. 532–541.

Bender, H., *Ceram. Age*, **63**, (1954), 15.

Bengisu, M., Brow, R. K. and White, J. E., *J. Mater. Sci.*, **39**, (2004), 605.

Bibby, J. K., Mummery, P. M., Bubb, N. & Wood, D. J., *Key Engineering Materials*, **354–256**, (2004), pp. 335–338.

Bibby, J. K., Mummery, P. M., Bubb, N. & Wood, D. J., *Glass Technol.*, **45**, (2003), 80.

Birkbeck, J. C., Cassidy, R. T., Fagin, P. N., & Moddeman, W. E., *Technology of Glass, Ceramic, or Glass-Ceramic to Metal Sealing*, edited by W. E. Moddeman, C. W. Merten & D. P. Kramer, Am. Soc. Mech. Eng. (New York), 1987, pp 15–24.

Biswas, P. K., Kundu, D. & Ganguli, D., *J. Mater. Sci. Lett.*, **8**, (1989), 1436.

Blokhuis, E. M., *Surface and Interfacial Tension*, edited by S. Hartland, Marcel Dekker Inc. (New York), 2004, pp. 149–193.

Blom, M. T., Chmela, E., Gardeniers, J. G. E., Elwenspoek, M., Tijssen, R. & Van den Berg, A., *J. Micromech. Microeng.*, **11**, (2001), 382.

Bloyer, D. R., Gomez-Vega, J. M., Mcnaney, J. M., Cannon, R. M. & Tomsia, A. P., *Acta Mater.*, **47**, (1999), 4221.

Bondley, R. J., *Electronics*, July 1947, 97.

Borom, M. P. & Pask, J. A., *J. Am. Ceram. Soc.*, **49**, (1966), 1.

Bourbon, S., Jansen, F., Hofmann, H., & Kurtz, W., *Z. Metall.*, **90**, (1999), 608.

Boyraz, T., Kilic, A., Ertug, B., Tavsanoglu, T. & Addemir, O., *Euro Ceramics VIII, Key Engineering Materials*, vols. 264–268, edited by H. Mandal & L. Ovecoglu, Trans. Tech. Publications (Switzerland), 2004, pp. 683–686.

Bradley, E. F., *Superalloys, a technical guide*, ASM International (Metals Park, OH), 1988.

Breder, K & Wereszczak, A. A., in *Mechanical testing methodology for ceramic design and reliability*, edited by D. C. Cranmer & D. W. Richerson, Marcel Dekker Inc. (New York), 1998, pp. 223–293.

Brennan, J. J. & Pask, J. A., *J. Am. Ceram. Soc.*, **56**, (1973), 58.

Breme, H. J., Barbosa, M. A. & Rocha, L. A., in *Materials and Biomaterials*, edited by J. A. Helsen & H. J. Breme, John Wiley & Son (Chichester), 1998, pp. 217–264.

Britchi, M., Jitianu, G., Olteanu, M. & Crisan, D., *Surface Modification Technologies XV*, edited by T. S. Sudarshan, J. J. Stiglich & M. Jeandin, ASM Int. (Materials Park, Ohio) and IOM Communications Ltd (UK), 2002, pp. 299–306.

Brow, R. K., *J. Am. Ceram. Soc.*, **70**, (1987), C-129.

Brow, R. K., Kovacic, L. & Loehman, R. E., in *Ceramic Manufacturing Practices and Technologies*, edited by H. Basavaraj, American Ceramic Society, 1998, pp. 177–187.

Brow, R. K., McCollister, H. L., Phifer, C. C. & Day, D. E., US Patent 5,648,302, July 1997.

Brow, R. K., McCollister, H. L., Phifer, C. C. & Day, D. E., US Patent 5,693,580, Dec. 1997.

Brow, R. K. & Tallant, D. R., *J. Non-Cryst. Solids*, **222**, (1997), 396.

Brow, R. K., Tallant, D. R., Warren, W. L., McIntyre, A. & Day, D. E., *Phys. Chem. Glasses*, **38**, (1997), 300.

Brow, R. K. & Watkins, R. D., *Technology of Glass, Ceramic, or Glass-Ceramic to Metal Sealing*, edited by W. E. Moddeman, C. W. Merten & D. P. Kramer, Am. Soc. Mech. Eng. (New York), 1987, pp 25–30.

Brown, M. E., *Introduction to Thermal Analysis*, Chapman & Hall ((London), 1988.

Brown, M. E. & Gallagher, P. K. (editors), *Handbook of Thermal Analysis and Calorimetry, vol. 2 Applications to inorganic and miscellaneous materials*, Elsevier (Amsterdam), 2003.

Buckley, R. G., *Ceram. Ind.*, (1979), 20.

Budd, M. US Pat. Appl., 20050130823, June 16, 2005.

Bull, S. J. & Rickerby, D. S., *Advanced surface coatings*, edited by D. S. Rickerby & A. Matthews, Blackie & Son (London and Glasgow), 1991, pp 315–342.

Burslem. R., *Mater. World*, **12**, (2004), 31.

Butt, H.-J., Graf, K. & Kappl, M., *Physics and Chemistry of Interfaces*, Wiley-VCH (Weinheim), 2003.

Buyukkaya, E., Demirkiran, A. S. & Cerit, M., *Euro Ceramics VIII, Key Engineering Materials*, vols. 264-268, edited by H. Mandal & L. Ovecoglu, Trans. Tech. Publications (Switzerland), 2004, pp. 517–520.

Cai, H. & Bao, G., *Int. J. Solids Struct.*, **35**, (1998), 701–717.

Cannon, R. M., Dalgleish, B. J., Dauskardt, R. M., Oh, T. S., & Ritchie, R. O., *Acta Metall. Mater.*, **39**, (1991), 2145.

Cannon, R. M., Dalgleish, B. J., Dauskardt, R. H., Fisher, R. M., Oh, T. S., & Ritchie, R. O., in *Fatigue of Advanced Materials, Materials and Component Engineering*, 1999, pp. 549–482.

Carmona, N., Villegas, M. A., Fernandez Navarro, J. M., *Thin Solid Films*, **458**, (2004), 121.

Cassidy, R. T., & Fagin, P. N., *Technology of Glass, Ceramic, or Glass-Ceramic to Metal Sealing*, edited by W. E. Moddeman, C. W. Merten & D. P. Kramer, Am. Soc. Mech. Eng. (New York), 1987, pp 9–14.

Cassidy, R. T. & Moddeman, W. E., *Ceram. Eng. Sci. Proc.*, **10**, (1989), 1387.

Chai, H. & Lawn, B. R., *J. Mater. Res.*, **19**, (2004), 1752.

Chalker, P. R., *Advanced surface coatings*, edited by D. S. Rickerby & A. Matthews, Blackie & Son (London and Glasgow), 1991, pp 278–314.

Chambers, R. S., Gerstle, F. P. & Monroe, S. L., *J. Am. Ceram. Soc.*, **72**, (1989), 929.

Chang, L-S. & Huang, C-F., *Ceram. Int.*, **30**, (2004), 2121.

Chapman, B. A., DeFord, H. D., Wirtz, G. P.and Brown, S. D., *Technology of Glass, Ceramic, or Glass-Ceramic to Metal Sealing*, edited by W. E. Moddeman, C. W. Merten & D. P. Kramer, Amer. Soc. Mech. Eng. (New York), 1987, pp 77–87.

Charalambides, P. G., Lund, J., Evans, A. G. & McMeeking, R. M., *J. Appl. Mech.*, **56**, (1989), 77.

Charles, R. J. & Hillig, W. B., in *Symp. Mech. Strength and Ways of Improving it*, Union Scientifique Continentale du Verre, Charleroi, Belgium, 1962, pp. 511.

Chen, D., McNaney, J. M., Saiz, E., Tomsia, A. P. & Ritchie, R. O., *J. Mater. Sci. Technol.*, **18**, (2002), 387-391.

Chen, H., Zhang, Y. & Ding, C., *Wear*, **253**, (2002), 885–893.

Chen, S., Lin, M., Shie, B. & Wang., *J. Non-Cryst. Solids*, **220**, (1997), 243–248.

Chen, Z.-X. & McMillan, P. W., *J. Am. Ceram. Soc.*, **68**, (1985a), 220.

Chen, Z.-X. & McMillan, P. W., *J. Mater. Sci.*, **20**, (1985b), 3428.

Cho, S.-J., Jeon, S.-B., Kim, J.-J. & Moon, H., *J. Mater. Sci. Lett.*, **9**, (1990), 596.

Chu, C., Zhu, J., Yin, Z. & Lin, P., *Mater. Sci. Eng. A*, **A348**, (2003), 244–250.

Chui, C. C., *J. Am. Ceram. Soc.*, **73**, (1990), 1999.

Clark, C. C. & Haddow, D., in *Window and dome technologies and materials VII*, edited by R. W. Tustison, Proc. SPIE, vol. 4375, 2001, pp. 307–314.

Clarke, R. A. & DerMarderosian, A., *Proc. SPIE Int. Soc. Opt. Eng.*, **3582**, 1999, pp. 828–832.

Claypoole, S. A., US Patent No. 2889952, 1959.

Cook, R. F., *Mater. Sci. Eng.*, **A260**, (1999), 29.

Corning Glass Works, British Patent No. 829447, 1960.

Coornaerts, J., Adriaens, P. & De Boever, J., *J. Prosthetic Dent.*, **51**, (1984), 338.

Courbiere, M., *Interfaces in New Materials*, edited by P. Grange & B. Delmon, Elsevier Applied Science (London and New York), 1991, pp. 29–41.

Craven, S. M., Kramer, D. P. & Moddeman, W. E., *Chemistry of glass-ceramic to metal bonding for header applications*, Technical Report MLM-3403, December 3, 1986.

Dale, A. E. & Stanworth, J. E., *J. Soc. Glass Technol.*, **33**, (1949), 167.

Dalton, R. H., US Patent No. 2392314, 1946.

Daniels, T., *Thermal Analysis*, Kogan Page (London), 1973.

Das, S., Bandyopadhyay, P. P., Ghosh, S., Bandyopadhay, T. K. & Chattopadhyay, A. B., *Metall. Mater. Trans. A: Physical Metallurgy and Materials Science V*, vol. 34, (2003), 1919.

Davidge, R. W., *Mechanical Behaviour of Ceramics*, Cambridge University Press (Cambridge), 1979.

Davidge, R. W., McLaren, J. R. and Tappin, G., *J. Mater. Sci.*, **8**, (1973), 1699.

Davison, R. M. & Redmond, J. D., *Mater. Des.*, **12**, (1991), 187.

De Santis, O., Gomez, L., Pellegri, N., Paradi, C., Marajafsky, A. & Duran, A., *J. Non-Cryst. Solids*, **121**, (1990), 338.

De With, G. & Wagemans, H. M., *J. Am. Ceram. Soc.*, **72**, (1989), 1538.

Dehm, G., Scheu, C., Mobus, G, Drydson, R. & Ruhle, M., *Ultramicroscopy*, **67**, (1997), 207.

Del Val, J. J., Gonzalez, J. & Zhukov, A., *Physica B.*, **299**, (2001), 242.

Delahaye, F., Montagne, L., Palavit, G., Ballif, P. & Touray, J. C., *Glasstech. Ber. Glass Sci. Technol.*, **72**, (1999), 161.

Deprot, S., Adenot, A. L., Bertin, F. & Acher, O., *J. Magn. Magn. Mater.*, **242–245**, (2002), 247–250.

Derby, B., in *Metal-Ceramic Interfaces*, edited by M. Ruhle, A. G. Evans, M. F. Ashby & J. P. Hirth, Acta-Scripta Metallurgica Proceedings Series 4, Pergamon Press (Oxford), 1990, pp. 161–167.

Dietz, R. L., *Proc. SPIE*, vol 5454, (2004), 111.

Dietzel, A., *Ceram. Abs.*, **13**, (1934), 250.

Dietzel, A., *Ceram. Abs.*, **14**, (1935), 107.

Dislich, H., *Sol-gel technology for thin films, fibers, preforms, electronics and speciality shapes*, edited by L. C. Klein, Noyes Publications (Pakk Ridge, NJ), 1982, pp 50–79.

Doi, H. & Akio, Y., *Trans. Jpn. Soc. Mech. Eng. A*, **69**, (2003), 1158 (in Japanese).

Donald, I. W., *J. Am. Ceram. Soc.*, **60**, (1977), 89.

Donald, I. W., *Inorganic Glasses and Glass-Ceramics: a Review*, AWRE Report No. O 19/84, December 1984.

Donald, I. W., *J. Mater. Sci.*, **22**, (1987), 2661.

Donald, I. W., *J. Mater. Sci.*, **24**, (1989), 4177.

Donald, I. W., *Encyclopedia of Materials Science and Technology*, 3rd Supplement, edited by R. W. Cahn, Pergamon Press (Oxford), 1993, pp. 1689–1695.

Donald, I. W., *J. Mater. Sci.*, **28** (1993), 2841–2886.

Donald, I. W., in *Key Engineering Materials*, Vols. 108–110, edited by G. M. Newaz, H. Neber-Aeschbacher & F. H. Wohlbier, Trans Tech. Publications (Switzerland), 1995, pp. 123–144.

Donald, I. W., in *Proc. Int. Conf. Ageing Studies and Lifetime Extension of Materials*, St. Catherine's College, Oxford, 12–14 July 1999, edited by L. G. Mallinson, Kluwer Academic/Plenum Publishing (New York), 2001, pp. 653–666.

Donald, I. W. and McMillan, P. W. *J. Mater. Sci.*, **13** (1978), 2301.

Donald, I. W., & Metcalfe, B. L., *Proc. 7th CIMTEC World Ceramics Congress, Satellite Symposium 2*, Montecatini Terme, Italy, 2–5 June, 1990, Elsevier Science Pub. (Amsterdam), edited by P. Vincenzini, 1991, pp 479–488.

Donald, I. W., Metcalfe, B. L. & Bradley, D. J., *Proc. 2nd Int. Conf. New Materials and their Applications*, University of Warwick, UK, 10–12 April, 1990, Institute of Physics Conf. Series No. 111, edited by D. Holland, IOP (Bristol), 1990, pp 207–216.

Donald, I. W., Metcalfe, B. L., Bradley, D. J., Hill, M. J. C., McGrath, J. L. & Bye, A. D., *J. Mater. Sci.*, **29**, (1994), 6379.

Donald, I. W., Metcalfe, B. L. & Fong, S. K., *Proc. XX Int. Congr. Glass*, 2004.

Donald, I. W., Metcalfe, B. L. & Morris, A. E. P., *J. Mater. Sci.*, **27**, (1992), 2979–2999.

Donald, I. W., Metcalfe, B. L. & Taylor, R. N. J., *J Mater. Sci.*, **32**, (1997), 5851.

Donald, I. W., Metcalfe, B. L., Wood, D. J. & Copley, J. R., *J. Mater. Sci.*, **24**, (1989), 3892.

Donald, I. W., Metcalfe, B. L., Bradley, D. J. & Battersby, S. E., *Ceramics: Charting the Future, Proc. 8th. CIMTEC World Ceramics Congress*, Florence, Italy, July 1994, edited by P. Vincenzini, Techna Srl., 1995, pp. 2239–2248.

Donald, I. W. & Metcalfe, B. L., *J. Mater. Sci.*, 31, (1996), 1139.

Dupré, A., *Theorie Mechanique de la Chaleur*, Gauthier-Villars, Paris, 1869.

Eichler, K., Solow, W., Otschik, P. & Schaffrath, W., *J. Eur. Ceram. Soc.*, **19**, (1999), 1101.

Ellis, J. L., US Patent No. 3564587, 1971.

Ely, K., *Issues in hermetic sealing of medical products*, Medical Device and Diagnostics Industry, 2000.

English Electric Co., French Patent No. 1328620, 1963.

Eppler, R. A., *Glass Science and Technology*, vol. 1, edited by D. R. Uhlmann & N. J. Kreidl, Academic Press (New York), 1983, pp 301–338.

Erz, M. & Hennicke, H. W., in *Ceramics in Advanced Energy Technologies*, edited by H. Krockel, M. Mertz & O. van der Biest, D. Reidel Pub. Co. (Dordrecht), 1984, pp. 139–156.

Eskner, M. & Sandstrom, R., *Surface and Coatings Technology*, **165**, (2003), 71–80.

Eubank, W. R. & Beck, W. R., US Patent No. 2863782, 1958.

Evans, A. G., *J. Mater. Sci.*, **7**, (1972), 1137.

Evans, D. A., *11th. Int. Seminar on Double Layer Capacitors and Similar Energy Storage Devices*, 2001, Deerfield Beech, Florida.

Evans, A. G. & Charles, C. A., *J. Am. Ceram. Soc.*, **59**, (1976), 371.

Fabes, B. D., Doyle, W. F., Silverman, L. S., Zelinski, B. J. J. & Uhlmann, D. R., *Science of Ceramic Chemical Processing*, edited by L. L. Hench & D. R. Ulrich, John Wiley & Sons (New York), 1986, pp 217–223.

Fahmy, M. F. & Subramanian, *Phys. Chem. Glasses*, **28**, (1987), 1.

Fahmy, M. F. & Subramanian, *Phys. Chem. Glasses*, **28**, (1987), 49.

Fairhurst, C. W., Mackert, J. D., Twiggs, S. W., Ringle, R. D., Hashinger, D. T. & Parry, E. E., *Ceram. Eng. Sci. Proc.*, **6**, (1985), 66.

Fawcett, N., *Mater. Sci. Technol.*, **9**, (1998), 2023.

Feipeng, Z., Tomsia, A. P. & Pask, J. A., *Proc. 33rd Pacific Coast Regional Meeting of the American Ceramic Soc.*, San Francisco, 1980, pp 76–78.

Ferber, M. K., Wereszzczak, A. A. & Jenkins, M. G., in *Mechanical testing methodology for ceramic design and reliability*, edited by D. C. Cranmer & D. W. Richerson, Marcel Dekker Inc. (New York), 1998, pp. 91–170.

Fergus, J. W., *J. Power Sources*, **147**, (2005), 46.

Fink, C. G., US Patent 1,498,908, 1915.

Floch, H. G. & Priotton, J-J., *Bull. Am. Ceram. Soc.*, **69**, (1990), 1141.

Francis, G. L. & Morena, R., US Patent No. 5281560, 1994.

Freiman, S. W., *Mechanical testing methodology for ceramic design and reliability*, edited by D. C. Cranmer & D. W. Richerson, Marcel Dekker Inc. (New York), 1998, pp. 1–16.

Fraser, R. P. & Cianchi, A. L., US Patent No. 2660531, 1953.

Froumin, N., Frage, N. & Dariel, M. P., *Proc. 10th. Int. Ceramics Congr.*, Part C, 2003, pp. 733–740.

Fu, H.-G., Jiang, Z.-Q. and Zhang, X.-H., *Trans. Nonferrous Metals Soc. China*, **14**, (2004), 106.

Fuerschbach, P. W., Electronic Packaging: Package Sealing and Ceramic Processing in *Handbook of Laser Materials Processing*, Laser Institute of America, editors J. F. Ready & D. F. Farson, 2001, Chapter 26.

Garland, B. T., *Mater. Des.*, **7**, (1986), 44.

Geodakyan, J. A., Petrosyan, B. V., Hambardzumyan, A. G. & Geodakyan, K. T., *Seventh European Glass Science and Technology Conference*, Athens, Greece, 25–28 April 2006.

Geoffrion, R., US Tech. Electronics Pub., Sept. 2002; www.nealloys.com.

Gerstle, F. P. & Chambers, R. S., *Technology of Glass, Ceramic, or Glass-Ceramic to Metal Sealing*, edited by W. E. Moddeman, C. W. Merten & D. P. Kramer, Amer. Soc. Mech. Eng. (New York), 1987, pp 47–59.

Gilbert, P. K., MSc Thesis, Thames Polytechnic, (1989).

Goktas, A. A., Neilson, G. F. & Weinberg, M. C., *J. Mater. Sci.*, **27**, (1992), 24.

Gomez de Salazar, J. M., Barrena, M. I., Soria, A., Menendez, M. & Gonzalez, A., *Bol. Soc. Esp. Ceram. Vidrio*, **40**, (2001), 295 (in Spanish).

Gomez-Vega., Saiz, E. & Tomsia, A. P., *J. Biomed. Mater. Res.*, **46**, (1999), 549.

Goto, T., *Sen-I Gakkaishi*, **34**, (1978), T237 (in Japanese).

Goto, T., Nagano, M. & Wehara, N., *Trans. Jpn. Inst. Met.*, **18**, (1977), 209.

Goto, T., *Trans. Jpn. Inst. Met.*, **20**, (1980), 219.

Goto, T., *Mater. Sci. Eng.*, **59**, (1983), 251.

Griffith, A. A., *Phil. Trans. R. Soc.*, **A221**, (1920), 163.

Griffith, A. A., *First. Int. Congr. Appl. Mechanics*, Delft, (1924), 55.

Grossman, D. G., *Advances in Ceramics 4, Nucleation and Crystallization in Glasses*, edited by J. H. Simmons, D. R. Uhlmann & G. H. Beall, Am. Ceram. Soc. (Columbus), 1982, pp 249–260.

Guedes, A., Pinte, A. M. P., Vieira, M. & Viana, F., *Mater. Sci. Eng.*, **A301**, (2001), 118.

Guglielmi, M., *J. Sol-Gel Sci. Technol.*, **8**, (1997), 443.

Guinel, M. J. F. and Norton, M. G., *J. Non-Cryst. Solids*, **347**, (2004), 173.

Gy, R., *J. Non-Cryst. Solids*, **316**, (2003), 1–11.

Hairston, D., Butcher, C. & Ondrey, G., *Chem. Eng.*, **109**, (2002), 39–40.

Hammetter, W. F. & Loehman, R. E., *J. Am. Ceram. Soc.*, **70**, (1987), 577.

Hand, R. J., Wang, F. H., Ellis, B. & Seddon, A. B., in ECF12 - Fracture from Defects, *Proc. 12th. Bienniel Conf. Fracture*, Sheffield, 1998, pp. 557–562.

Harper, H. & McMillan, P. W., *Phys. Chem. Glasses*, **13**, (1972), 97.

Harrison, W. N., Richmond, J. C., Pitts, J. W. & Benner, S. G., *J. Am. Ceram. Soc.*, **35**, (1952), 113.

Haws, L. D., Kramer, D. P., Moddeman, W. E. & Wooten, G. W., *High strength glass-ceramic-to-metal seals*, Technical Report MLM-3288(OP), December 1985.

He, Y. & Day, D. E. *Glass Technol.*, **33**, (1992), 214.

Healy, J. H. & Andrews, A. I., *J. Am. Ceram. Soc.*, **34**, (1951), 207.

Headley, T. J. & Loehman, R. E., *J. Am. Ceram. Soc.*, **67**, (1984), 620.

Headley, T. J., Loehman, R. E., Watkins, R. D. & Madden, M. C., *Proc. 44th Annual Meeting of the Electron Microscopy Section of America*, San Francisco Press (San Francisco), 1986, pp 856–857.

Henderson, W. R., Kramer, D. P. & Sullenger, D. B., Determination of the optimum crystallization conditions of a high thermal expansion glass-ceramic, Monsanto Research Corporation Report MLM-3136, March 1984.

Hendriksen, P. V., Carter, D. J. & Magensen, M., *Proc. Electrochem. Soc.*, **951**, (1995), 934.

Hensley, D., Milewits, M., & Zhang, W., *Oilfield Rev.*, Autumn 1998, 42.

Hesse, D. & Senz, S., *Proc. 10th. Int. Ceramic Congress*, Advances in Science and Technology vol 31, Part A, Techna Srl, Faenza, 2003, pp 133–146, edited by P. Vincenzini.

Hesse, D., Senz, S., Scholz, R., Werner, P. & Heydenreich, J., *Interface Sci.*, **2**, (1994), 221.

Hey, A. W., *Joining of Ceramics*, edited by M. G. Nicholas, The Institute of Ceramics, Chapman & Hall (London), 1990, pp 56–72.

Hijon, N., Cabanas, M. V., Izquierdo-Barba, I. & Vallet-Regi, M., *Chem. Mater.*, **16**, (2004), 1451.

Hikino, T. & Mikoda, M., Japanese Patent No. 46-3467, 1971.

Hill, R. & Gilbert, P. K., *J. Am. Ceram. Soc.*, **76**, (1993), 417–425.

Hill, R. G., Goat, C. & Wood, D., *J. Am. Ceram. Soc.*, **75**, (1992), 778.

Hiruma, H. & Shimizu, S., Japanese Patent No. 48-17848, 1973.

Hitachi Ltd., British Patent No. 1563790, 1980.

Hlavac, J., *The Technology of Glass and Ceramics*, Elsevier (Amsterdam), 1983, pp 228–243.

Hohne, G. W. H., Hemminger, W. F. & Flammersheim, H-J., *Differential Scanning Calorimetry*, 2nd. edition, Springer (Berlin), 2003.

Holand, W. & Beall, G, *Glass-Ceramic Technology*, The American Ceramic Society (Westerville), 2002.

Holland, D., *Key Eng. Mater.*, **99–100**, (1995), 203.

Holland, D., Hong, F., Logan, E. & Sutherland, S., *Proc. 2nd Int. Conf. New Materials and their Applications*, University of Warwick, UK, 10–12 April, 1990, Institute of Physics Conf. Series No. 111, edited by D. Holland, IOP (Bristol), 1990, pp 459–468.

Hong, F. & Holland, D., *J. Non-Cryst. Solids*, **112**, (1989), 357.

Holleran, L. M. & Martin, W., US Patent No. 4714687, 1987.

Hong, F & Holland, D., *Surf. Coat. Tech.*, **39/40**, (1989), 19.

Hong, F. & Holland, D., *J. Non-Cryst. Solids*, **112**, (1989), 357.

Hong, F., *Interactions between glass-ceramic coatings and metals*, PhD Dissertation, University of Warwick, 1991.

Hopkins, R. H., Kramer, W. E., Brandt, G. B., Hoffman, R. A., Steinbrugge, K. B. & Peterson, T. L., *J. Appl. Phys.*, **49**, (1978), 3133.

Horita, T., Sakai, N., Kawada, T., Yokokawa, H. & Dokiya, M., *Denki Kagaka*, **61**, (1993), 760.

Housekeeper, W. G., *J. Am. Inst. Elect. Eng.*, **42**, (1923), 870.

Huh, J.-W. & Kobayashi, H., *Nippon Kikai Gakki Ronbunshu*, **63A**, (1997), 57 (in Japanese).

Hutchinson, J. W. & Suo, Z., *Adv. Appl. Mech.*, **29**, (1992), 63.

Hyde, A. R. & Partridge, G., *Advanced Engineering with Ceramics*, British Ceramic Proceedings No. 46, edited by R. Morrell, Institute of Ceramics (Shelton, Stoke-on-Trent), 1990, pp 345–350.

Ikeda, Y., *Shin Nihon Denki Kiho*, **3**, (1968) (in Japanese).

Inglis, C. E., *Trans. Inst. Naval Arch.*, **55**, (1913), 219.

Ishiyama, M., Matsuda, T., Nagahara, S. & Suzuki, Y., *Asahi Garasu Kenkyu Hokoku*, **16**, (1966), 77.

Ison, P. J., Holland, D. & Bushby, R., *Phys. Chem. Glasses*, **41**, (2000), 267.

Jacobson, D. M., Ogilvy, A. J. W., & Leatham, A. G, Controlled expansion alloys offer manufacturing advantages for high-tech applications, Osprey Metals Ltd., 2005, www.smt.sandvik.com/

Jarrige, J., Joyeux, T., Labbe, J. C. & Lecompte, J. P., *Euro Ceramics VIII, Key Eng. Mater.*, **264–268**, edited by H. Mandal & L. Ovecoglu, Trans. Tech. Publications (Switzerland), 2004, pp. 675–678.

Jiansirisomboon, S., MacKenzie, K. J. D., Roberts, S. G. & Grant, P. S., *J. Non-Cryst.Solids*, **316**, (2003), 35–41.

Johnson, L. F. & Moran, M. B., in Window and dome technologies and materials VII, *Proc. SPIE*, **4375**, 2001, pp. 289–299.

Juve, D., Courbiere, M. & Treheux, D., in *Metal-Ceramic Interfaces*, edited by M. Ruhle, A. G. Evans, M. F. Ashby & J. P. Hirth, Acta-Scripta Metallurgica Proceedings Series 4, Pergamon Press (Oxford), 1990, pp. 152–158.

Kajiwara, M., *Glass Technol.*, **29**, (1988), 188.

Kara-Slimane, A. & Treheux, D., *Proc. 15th Int. Thermal Spray Conference*, vol. 2, ASM Thermal Spray Society/German Welding Society/High Temperature Society of Japan, 1998.

Kasuga, T., Mizuno, T., Watanabe, M., Nogami, M. & Niinomi, M., *Biomaterials*, **22**, (2001), 577.

Kasuga, T., Nogami, M. & Niinomi, M., *Key Eng. Mater.*, **192–195**, (2001), 223.

Katoh, K., *Surface and Interfacial Tension*, edited by S. Hartland, Marcel Dekker Inc. (New York), 2004, pp. 375–423.

Kataoka, N., Kawamoto, T. & Manabe, Y., Jpn. Patent No. 37-4029, 1962.

Kataoka, N. & Manabe, Y., *Osaka Kogyo Gijutsu Shikensho Kiho*, **23**, (1972), 204.

Kaufmann, E. N., *Characterisation of Materials*, John Wiley and Son, 2003.

Keding, R., Russel, C, Pascual, M. J., Pascual, L. & Duran, A., *Proc. Int. Congr. Glass*, Edinburgh, Volume 2. Extended Abstracts, 2001, pp. 922–923.

Kelly, A., *Strong Solids*, 2nd. edition, Clarendon Press (Oxford), 1973.

Kelly, J. R., Nishimura, I. & Campbell, S. D., *J. Prosthetic Dent.*, **75**, (1996), 18–32.

Kelsey, P. V., Siegal, W. T. & Miley, D. V., *Surfaces and interfaces in ceramic and ceramic-metal systems*, edited by J. A. Pask & A. Evans, Plenum Press (New York), 1981, pp 591-601.

Kilgo, R. D., Kovacic, L. & Brow, R. K., US Patent 5,965,469, Oct. 1999.

Kilgo, R. D., Kovacic, L. & Brow, R. K., US Patent 6,037,539, March 2000.

Kim, K. R., Suh, C. M., Murakami, R. I. & Chung, C. W., *Surf. Coat. Technol.*, **171**, (2003), 15–23.

Kim, Y.-S., Yu, S.-C., Lee, H., Tuan, A., Kim, C.-O. & Rhee, J.-R., *Phys. Status Solidi (A) Appl. Res.*, **201**, (2004), 1823.

King, R. M., *J. Am. Ceram. Soc.*, **16**, (1933), 232.

King, B. W., Tripp, H. P. & Duckworth, W. H., *J. Am. Ceram. Soc.*, **42**, (1959), 504.

Kingery, W. D., *Introduction to Ceramics*, John Wiley & Son (New York), 1960.

Klein, L. C., *Glass Ind.*, **62**, (1981), 14.

Klimonda, P., Lingstuyl, O., Lavelle, B. & Dabosi, F., *Surfaces and interfaces in ceramic and ceramic-metal systems*, edited by J. A. Pask & A. Evans, Plenum Press (New York), 1981, pp 477–486.

Klomp, J. T., *Ceramic Microstructures 86*, (1986), pp. 307–317.

Knorovsky, G. A., Brow, R. K., Watkins, R. D. and. Loehman, R. E., Interfacial debonding in stainless steel/glass-ceramic seals, in *Metal-ceramic joining*, edited by P. Kumar & V. A. Greenhut, The Mineral, Metals and Materials Society, 1991, pp. 237–246.

Kohl, W. H., Ceramic-to-Metal Sealing, in *Materials and Techniques for Electron Tubes*, Reinhold Pub. Corp., 1960.

Kohl, W. H., *Vacuum*, 14, (1964), 333.

Kobayashi, K., *Bull. Am. Ceram. Soc.*, **66**, (1987), 685.

Kohnle, C., Mintchev, O. & Schmauder, S., *Comput. Mater. Sci.*, **25**, (2002), 272–277.

Kolosoff, G., *Z. Math. Phys.*, **62**, (1914), 26.

Komori, D., Takashima, T. & Yamamoto, T., *Mem. Hokkaido Inst. Technol.*, **30**, (2002), 1–6. (in Japanese).

Konaicheva, N. V., Bilere, P. I. & Myakin, V. K., *J. Appl. Chem. USSR*, **63**, (1990), 643.

Kotomin, E. A., Maier, J., Stoneham, A. M., Zhukovski, Y. F., Fuks, D. & Dorfman, S., *Proc. 10th. Int. Ceramics Congr.*, Part C, 2003, pp. 669–682.

Krajewski, A., Ravaglioli, A., De Portu, G. & Visani, R., *Bull. Am. Ceram. Soc.*, **64**, (1985), 679.

Kramer, D. P., Harville, G. L., Buckner, D. A., McCarthy, J. P., Nease, A. B. & Sullenger, D. B, Physical property changes of a lithia-alumina-silica based glass as a function of composition, Monsanto Research Corporation Report MLM-3272, July 1985.

Kramer, D. P., & Massey, R. T., *Advances in Ceramics, vol. 9, Forming of Ceramics*, Am. Ceram. Soc. (Columbus), 1984, pp 265–273.

Kramer, D. P. & Massey, R. T., *Ceram. Eng. Sci.*, **5**, (1984), 739.

Kramer, D. P., & Massey, R. T. In-situ vacuum-assisted molding of glass-metal electrical components, Monsanto Research Corporation Report MLM-3267, July 1985.

Kramer, D. P., Massey, R. T., & Halcomb, D. L., Injection moulding sealing on glass to low melting metals, Monsanto Research Corporation Report MLM-3268, July 1985.

Kramer, D. P., Massey, R. T. & Halcomb, D. L., *Technology of Glass, Ceramic, or Glass-Ceramic to Metal Sealing*, edited by W. E. Moddeman, C. W. Merten & D. P. Kramer, Am. Soc. Mech. Eng. (New York), 1987, pp 31–37.

Kramer, D. P. & Moddeman, E., Chemistry of glass-ceramic bonding for header applications, Mound Applied Technologies Report MLM-3556, Nov. 1988.

Kramer, D. P. & Osborne, N. R., *Proc. 7th. Annual Conference on Composites and Advanced Ceramic Materials*, Am. Ceram. Soc., 1983, pp. 740–750.

Kramer, D. P., Salerno, R. F., & Egleston, E. E. Effect of several surface treatments on the strength of a glass ceramic-to-metal seal, Monsanto Research Corporation Report MLM-2893, February 1982.

Kramer, D. P., Spangler, E. M. & Beckman, T. M., *Am. Ceram. Soc. Bull.*, **72**, (1993), 78.

Kreidl, N. J., *Glass-forming systems*, edited by D. R. Uhlmann & N. J. Kreidl, Academic Press (New York), 1983, pp 105–299.

Krohn, D. A. & Cooper, A. R., *J. Am. Ceram. Soc.*, **52**, (1969), 661.

Krohn, M. H, Hellman, J. R., Shelleman, D. L., Green, D. J., Sakoske, G. E. & Salem, J. A., *J. Test. Eval.*, **30**, (2002), 470.

Kruzic, J. J., McNaney, J. M., Cannon, R. M. & Ritchie, R. O., *Mater. Res. Soc. Symp. Proc.*, vol. 654, 2001, pp. AA4.10.1–AA4.10.6.

Kuckert, H., Wagner, G. & Eifler, D., *Adv. Eng. Mater.*, **3**, (2001), 903.

Kumta, P. N. & Sriram, M. A., *J. Mater. Sci.*, **28**, (1993), 1097–1106.

Kunz, R., *Increase productivity with proper sensor selection*, www.flowbiz.com/VA/sensorZ.html

Kunz, S. C. & Loehman, R. E., *Adv. Ceram. Mater.*, **2**, (1987), 69.

Kurumada, A., Imamura, Y., Tomota, Y., Oku, T., Kubota, Y. & Noda, N., *J. Nucl. Mater.*, **313–316**, (2003), 245–249.

Langlet, M., Saltzberg, M. & Shannon, R. D., *J. Mater. Sci.*, **27**, (1992), 972.

Larin, V., Torcunov, A., Baranov, S., Vazquez, M. & Zhukov, A., Spanish Patent No. P9601993, 1996.

Lau, C. W., Rahman, A. & Delale, F., *Technology of Glass, Ceramic, or Glass-Ceramic to Metal Sealing*, edited by W. E. Moddeman, C. W. Merten & D. P. Kramer, ASME (New York), 1987, pp 89–98.

Laucher, J. H., Cook, R. L. & Andrews, A. I., *J. Am. Ceram. Soc.*, **39**, (1956), 288.

Lee, S.-B. & Han S.-M., *J. Korean Ceram. Soc.*, **24**, (1987), 227 (in Korean).

Lee, J.-S., Peng, J.-C. & Huang, C-W., *Glass Technol.*, **31**, (1990), 77.

Leedecke, C. J., Baird, P. C., & Orphanides, K. D., *Glass-to-Metal Seals*, HCC Aegis, Inc., 2004.

Leichtfried, G., Thurner, G. & Weirather, R., *Int. J. Refract. Met. Hard Mater.*, **16**, (1998), 13.

Lewandowski, H. J., Harber, D. M., Whitaker, D. L. & Cornell, E. A., www.colorado.edu/bec/Publications/JLTP_Lewandowski2003.pdf

Lewis, M. H., ed., *Glasses and Glass-Ceramics*, Chapman and Hall (London), 1989.

Ley, K. L., Krumpelt, M., Kumar, R., Mciser, J. H. & Bloom, I., *J. Mater. Res.*, **11**, (1997), 1489.

Li, X., Abe, T., Liu, Y. & Esashi, M., *IEEE Robotics & Autom. Sec.*, IEEE, (2001), 98.

Li, H., Khor, K. A. & Cheang, P., *Surf. Coat. Technol.*, **155**, (2002), 21–32.

Lidong, T., Xiangsen, M., Fuming, W. & Wenchao, L., *Proc. Int. Congr. Glass*, Edinburgh, Volume 2. Extended Abstracts, 2001, pp. 214.

Liu, Y. & Brunner, D., *Z. Metallkd.*, **93**, (2002), 444.

Liu, Y., Persson, C. & Melin, S., *J. Thermal Spray Technol.*, **13**, (2004), 377.

Liu, H. S., Shih, P. Y. & Chin, T. S., *Phys. Chem. Glasses*, **37**, (1996), 227.

Lo, T-N., Lui, T-S. & Chang, E., *Mater. Trans.*, **45**, (2004), 3065.

Loehman, R. E., *J. Met.*, **38**, (1986), 42.

Loehman, R. E., *Technology of Glass, Ceramic, or Glass-Ceramic to Metal Sealing*, edited by W. E. Moddeman, C. W. Merten & D. P. Kramer, Am. Soc. Mech. Eng. (New York), 1987, pp 39–45.

Loehman, R. E., *Bull. Am. Ceram. Soc.*, **68**, (1989), 891.

Loehman, R. E., Joining Engineering Ceramics, Technical Report SAND-90-0966C, 1990.

Loehman, R. E., Dumm, H. P. & Hofer, H., *Ceram. Eng. Sci. Proc.*, **23**, 2002, pp. 699–710.

Loehman, R. E. & Headley, T. J., *Mater. Sci. Res.*, vol. 21. Ceramic Microstructures '86, edited by J. A. Pask & A. G. Evans, Plenum Press (New York), 1987, pp 33–43.

Loehman, R. E.,. Kunz, S. C & Watkins, R. D., *Ceram. Eng. Sci. Proc.*, **7**, (1986), 721.

Loehman, R. E. & Pask, A. P., *Bull. Am. Ceram. Soc.*, **67**, (1988), 375.

Loewenstein, in *The Manufacturing Technology of Continuous Glass Fibres, Glass Science and Technology*, vol. 6, 2nd. edition (Elsevier, Amsterdam).

Lowrie, F. L. & Rawlings, R. D., *J. Eur. Ceram. Soc.*, 20, (2000), 751.

Lu, C. J., & Hesse, D., *Phil. Mag. Lett.*, **82**, (2002), 167.

Mackenzie, R. C. (editor), *Differential Thermal Analysis*, Academic Press (New York), 1972.

Mackerle, J., *Eng. Comput.*, **16**, (1999), 510.

Maeda, M. & Ikeda, T., *J. Phys. Chem. Solids*, **49**, (1988), 35.

Maeda, M., Igarashi, O., Shibayanagi, T. & Naka, M., *Mater. Trans.*, **44**, (2003), 2701.

Majumdar, A. & Jana, S., *Am. Ceram. Soc. Bull.*, **24**, (2001), 69.

Malmendier, J. W., US Patent No. 3883358, 1975.

Malmendier, J. W., & Sojka, J. E., US Patent No. 3885975, 1975.

Manfre, G. & Vianello, D., Italian Patent 930409, (1972).

Mantel, M., *J. Non-Cryst. Solids*, **273**, (2000), 294.

Malzbender, J., Stenbrecht, R. W. & Singheiser, L., *J. Mater. Res.*, **18**, (2003), 929.

Mardare, C. C., Mardare, A. I., Fernandes, J. R. F., Joanni, E., Pina, S. C. A., Fernandes, M. H. V. & Correia, R. N., *J. Eur. Ceram. Soc.*, **23**, (1003), 1027.

Marshall, D. B. & Evans, A. G., *J. Appl. Phys.*, **56**, (1984), 2632.

Martin, F. W., Japanese Patent No. 45-19982, 1970.

Martinez, E., Romero, J., Lousa, A. & Esteve, J., *Surf. Coat. Technol.*, **163–164**, (2003), 571–577.

Maskall, K. A. & White, D., *Vitreous Enamelling: a guide to modern enamelling practice*, Institute of Ceramics, Pergamon Press (London), 1986.

Matthewson, M. J. & Field, J. E., *J. Phys. E: Sci. Instrum.*, **13**, (1980), 355.

Matusita, K., Koide, M., Nishiyuki, T. & Hiroi, A., *Proc. Int. Congr. Glass*, Edinburgh, Volume 2. Extended Abstracts, 2001, pp. 663.

Mayer, P., Topping, J. A.., & Murthy, M. K., *J. Can. Ceram. Soc.*, **43**, (1974), 43.

McCollister, H. L. & Reed, S. T., US Patent No. 4414282, 1983.

McColm, I. J. & Dimbylow, C., *J. Mater. Sci.*, **9**, (1974), 1320.

McLean, J. W. & Sced, I. R., *Brit. Ceram. Soc. Trans.*, **72**, (1973), 235.

McKenzie, H. W. & Hand, R. J. *Basic optical stress measurement in glass*, Society of Glass Technology (Sheffield), 1999.

McKeown, J. T., Sugar, J. D., Gronsky, R. & Glaeser, A. M., *Welding J.*, **84**, (2005), 41.

McLellan, W. W. & Shand, E. B., editors, *Glass Engineering Handbook*, McGraw Hill (New York), 3rd edition, 1984.

McMillan, P. W., *Glass Ceramics*, Academic Press (London), 2nd. edition, 1979.

McMillan, P. W. & Hodgson, B. P., *Engineering*, **196**, (1963), 366.

McMillan, P. W. & Hodgson, B. P., British Patent No. 944571, 1963.

McMillan, P. W. & Hodgson, B. P., British Patent No. 1023480, 1966.

McMillan, P. W. & Hodgson, B. P., British Patent No. 1063291, 1967.

McMillan, P. W., Hodgson, B. P., Crozier, D. S. & Wells, S. T., British Patent No. 1174474, 1969.

McMillan, P. W., Hodgson, B. P. & Partridge, G., US Patent No. 3220815, 1965.

McMillan, P. W., Hodgson, B. P. & Partridge, G., *Glass Technol.*, **7**, (1966b), 121.

McMillan, P. W. & Partridge, G., US Patent No. 3170805, 1965.

McMillan, P. W. & Partridge, G., British Patent No. 1028871, 1966.

McMillan, P. W. & Partridge, G., British Patent No. 1151860, 1969.

McMillan, P. W. & Partridge, G., British Patent No. 1306727, 1973.

McMillan, P. W., Partridge, G. & Ward, F. R., British Patent No. 1205652, 1970.

McMillan, P. W. & Partridge, G., British Patent No. 924996, 1963.

McMillan, P. W., Partridge, G., Hodgson, B. P. & Heap, H. R., *Glass Technol.*, **7**, (1966c), 128.

McMillan, P. W. & Partridge, G., *Proc. Br. Ceram. Soc.*, **3**, (1965), 241.

McMillan, P. W. & Partridge, G., *J. Mater. Sci.*, **7**, (1972), 847.

McMillan, P. W., Phillips, S. V., & Partridge, G., *J. Mater. Sci.*, **1**, (1966), 269.

McNaughton, J. L. & Mortimer, C. T., *Differential Scanning Calorimetry*, IRS Chemistry Series No. 2, Butterworths (London), 1975.

McPhilmy, S. A., Baluk, M. J., Frankel, H. E. & Evans, J. W., *Proc. ISTFA 89*, ASM International, 1989, pp. 361–367.

Meek, T. T. & Blake, R. D., US Patent No. 4529856, 1985.

Meek, T. T. & Blake, R. D., *J. Mater. Sci. Lett.*, **5**, (1986), 270.

Mennig, M., Endres, K., Anschutz, D., Gier, A. & Schmidt, H., *Proc. Int. Congr. Glass*, Edinburgh, Volume 1. Extended Invited Papers, 2001, pp. 112–115.

Metcalfe, B. L. & Donald, I. W., in *New Materials and their Applications*, edited by D. Holland, Inst. Phys. Conf. Series No. 111, IOP (Bristol), 1990, pp. 469–478.

Metcalfe, B. L. & Donald, I. W., *Silic. Ind.*, (5–6), (1991), 99.

Metcalfe, B. L., Donald, I. W. & Bradley, D. J., *Institute of Ceramics Proceedings No. 48*, edited by R. Morrell & G. Partridge, ICS (Stoke), 1991, pp 177–188.

Mevrel, R., *High temperature surface interactions*, AGARD Conf. Proc., No. 461, NATO, 1989, pp 12-1–12-10.

Miao, H., Peng, Z., Pan, W., Qi, L., Yang, S. & Liu, C., *Euro Ceramics VIII*, Key Engineering Materials, vols. 264–268, edited by H. Mandal & L. Ovecoglu, Trans. Tech. Publications (Switzerland), 2004, pp. 605-608.

Michalske, T. A. & Freiman, J. W., *J. Am. Ceram. Soc.*, **66**, (1983), 284.

Michalske, T. A. & Bunker, B. C., *J. Am. Ceram. Soc.*, **70**, (1987), 780.

Minagawa, K. & Suzuki, K., Japanese Patent No. 49-36807, 1974.

Miriam, W. J., British Patent No. 1026178, 1966.

Miroshnichenko, I. S., Bashev, V. F., Pokrovskiy, Yu. K. & Spektor, E. Z., *Russ. Metall.*, **1**, (1980), 105.

Mitachi, S., Nagase, R., Takeuchi, Y. & Honda, R., *Glass Technol.*, **39**, (1998), 98.

Miyata, N. & Jinno, H., *J. Mater. Sci. Lett.*, **1**, (1982), 156.

Mizuhara, H., in *Advanced Brazing and Soldering Technologies*, IBSC 2000, edited by P. T. Vianco & M. Singh, American Welding Society (Miami, Florida), 2000, pp. 278–283.

Mizuhara, H. & Oyama, T., in *Engineered Materials Handbook, Vol. 4, Ceramics and Glasses*, Technical Chairman S. J Schneider, ASM International, 1991, pp. 502–510.

Mizuno, Y., Ikeda, M. & Yoshida, A., *J. Mater. Sci. Lett.*, **11**, (1992), 1653.

Moddeman, W. E., Birkbeck, J. C., Bowling, W. C., Burke, A. R. & R. Cassidy, T., *Ceram. Eng. Sci. Proc.*, **10**, (1989), 1403.

Moddeman, W. E., Craven, S. M. & Kramer, D. P., *J. Am. Ceram. Soc.*, **68**, (1985), C-298.

Moddeman, W. E., Pence, R. E., Massey, R. T., Cassidy, R. T. & Kramer, D. P., *Ceram. Eng. Sci. Proc.*, **10**, (1989), 1394.

Mollart, T. P., Lewis, K. L., Wort, C. J. H. & Pickles, C. S. J., *Window and dome technologies VIII*, SPIE, vol. 4375, 2001, pp. 199.

Moore, D. G., Pitts, J. W., Richmond, J. C. & Harrison, W. N., *J. Am. Ceram. Soc.*, 37, (1954), 1.

Moorehead, A. J., *Adv. Ceram. Mater.*, **2**, (1987), 159.

Moorehead, A. J. & Kim. H-E., in *Engineered Materials Handbook, Vol. 4, Ceramics and Glasses*, Technical Chairman S. J Schneider, ASM International, 1991, pp. 511–522.

Morena, R. & Francis, G. L., *Proc. XVIII Int. Congr. Glass*, edited by M. K. Choudhary, N. T. Huff and C. H. Drummond, American Ceramic Society (Waterville), 1998, CD-ROM.

Moriguchi, T., Miwa, K. & Shibuya, T., US Patent No. 3900330, 1975.

Mutoh, Y., Xu, J-Q., Miyashita, Y., Bernardo, G. G. & Takahashi, M., *JSME Int. Jnl, Series A (Solid Mech. Mater. Eng.)*, **46**, (2003), 403.

Nash, T. R., Pashby, E. M. & Collett, R. L., *Glass Technol.*, **24**, (1983), 298.

Nazri, G-A., & Pistoia, G., editors, *Lithium Batteries*, Kluwer Academic Publishers (Boston), 2004.

Nebot-Diaz, J., Bakali, J., Irun, M., Cordoncillo, E. & Escribano, P., *Proc. Int. Congr. Glass*, Edinburgh, Volume 2. Extended Abstracts, 2001, pp. 939–940.

Nelson, W., *Accelerated Testing*, John Wiley & Sons (New York), 1990.

Neyer, B. T., *Proc. 38th. Joint Propulsion Conference*, Indianapolis, July 2002.

Nicholls, J. R., *MRS Bull.*, **28**, (2003), 659-670.

Nicholas, M. G., *J. Mater. Sci.*, **21**, (1986), 3292.

Nicholas, M. G., *Ceramic Microstructures 86*, (1986), pp. 349–357.

Nicholas, M. G. & Mortimer, D. A., *Mater. Sci. Technol.*, **1**, (1985), 657.

Nikolic, L. & Radonjic, L., *Ceram. Int.*, **24**, (1998), 547–552.

Nixdorf, J. *Drahtwelt*, **53**, (1967), 696 (in German).

NMAB, High temperature electronic packaging, in *Materials for High-Temperature Semiconductor Devices*, Chapter 5, National Materials Advisory Board, (1995a).

NMAB, Device testing for high temperature electronic materials, in *Materials for High-Temperature Semiconductor Devices*, Chapter 6, National Materials Advisory Board, (1995b).

Noda, N., Ishihara, M., Yamamoto, N. & Fujimoto, T., *Mater. Sci. Forum*, **423–425**, (2003), 607–612.

Noda, N., Fujimoto, T. & Ishihhara, M., in *Proc. Int. Symp. of Young Scholars on Mechanics and Material Engineering for Science and Experiments*, 2003, pp. 811–816.

Nolte, H. J. & Spurek, R. F., US Patent 2,667,427, Jan. 1954.

Nolte, H. J. & Spurek, R. F., US Patent 2,667432, Jan. 1954.

Notis, M. R., *J. Amer. Ceram. Soc.*, 45, (1962), 412.

O'Hara, T. J., *Proc. 33rd Int. Power Sources Symposium*, the Electrochemical Society, 1988, pp. 262–273.

Okabe, N., Zhu, X., Takahashi, M. & Nakahashi, M., *Mater. Sci. Res. Int.*, **8**, (2002), 101–108.

Oliver, W. C. & Pharr, G. m., *J. Mater. Res.*, **7**, (1992), 1564.

Omar, A. A., El-Shennawi, A. W. A. & El-Ghannam, A. R., *J. Mater. Sci.*, **26**, (1991), 6049.

Onoki, T., Hosoi, K. & Hashida, T., *Scripta Mater.*, **52**, (2005), 767.

Oruganti, R. K. & Ghosh, A. K., *JOM*, **55**, (2003), 21–27.

Palmour, H., *J. Electrochem. Soc.*, **102**, (1955), 160C.

Pardoe, G. W. F., Butler, E. & Gelder, D., *J. Mater. Sci.*, **13**, (1978), 786.

Park, J.-H., Lee, D.-Y., Oh, K.-T, Lee, Y.-K. & Kim, K.-N., *J. Am. Ceram. Soc.*, **87**, (2004), 1792.

Parola, S., Verdenelli, M., Sigala, C., Scharff, J-P., Velez, K., Veytizou, C. & Quinson, J-F., *J. Sol-Gel Sci. Technol.*, **26**, (2003), 803–806.

Partridge, J. H., *Glass-to-Metal Seals*, Soc. Glass Technol. (Sheffield), 1949.

Partridge, G., *Inst. Phys. Conf. Ser. No. 89*, IOP Publishing (Bristol), 1987, pp 161–170.

Partridge, G., *Joining of Ceramics*, edited by M. G. Nicholas, The Institute of Ceramics, Chapman & Hall (London), 1990, pp 31–55.

Partridge, G. & Elyard, C. A., *Brit. Ceram. Proc.*, **34**, (1984), 219.

Partridge, G., Elyard, C. A. & Budd, M. I., *Glasses and Glass-Ceramics*, edited by M. H. Lewis, Chapman & Hall (London), 1989, pp 226–271.

Partridge, G., Elyard, C. A. & Keatman, H. D., *Glass Technol.*, **30**, (1989), 215.

Pascual, M. J., Duran, A. & Pascual, L., *Proc. Int. Congr. Glass*, Edinburgh, Volume 2, Extended Abstracts, 2001, pp. 659-660.

Pascual, M. J., Duran, A. & Pascual, L., *J. Non-Cryst. Solids*, **306**, (2002), 58.

Pascual, M. J., Pascual, L., Valle, F. J. & Duran, A., *J. Am. Ceram. Soc.*, **86**, (2003), 1918.

Pask, J. A. & Fulrath, R. M., *J. Am. Ceram. Soc.*, **45**, (1962), 592.

Pask, J. A., *Proc. Porcelain Enamel Inst. Tech. Forum*, **33**, (1971), 1.

Pask, J. A., & Tomsia, A. P., *Surfaces and interfaces in ceramic and ceramic-metal systems*, edited by J.

A. Pask & A. Evans, Plenum Press (New York), 1981, pp 411–419.

Pask, J. A., *Bull. Am. Ceram. Soc.*, **66**, (1987), 1587.

Pask, J. A., *Technology of Glass, Ceramic, or Glass-Ceramic to Metal Sealing*, edited by W. E. Moddeman, C. W. Merten & D. P. Kramer, Amer. Soc. Mech. Eng. (New York), 1987, pp 1–7.

Passerone, A., Valbusa, G. & Biagini, E., *J. Mater. Sci.*, **12**, (1977), 2465.

Pattee, H. E., *Joining Ceramics to Metals and Other Materials*, WRC Bulletin 178, Welding Research Council, 1978.

Paul, A., *Chemistry of Glasses*, Chapman and Hall (London & NY), 1982, pp. 41–50.

Pearsall, C. S., *Mater. Methods*, **70**, (1949), 61.

Peng, Y. B. & Day, D. E. , *Glass Technol.*, **32**, (1991), 166.

Peng, Y. B. & Day, D. E., *Glass Technol.*, **32**, (1991), 200.

Penkov, I. & Gutzow, I., *Silikattechnik*, **42**, (1991), 60.

Perez, N., *Fracture Mechanics*, Kluwer Academic Publishers (Boston), 2004.

Pevzner, B. Z. & Klyuev, V. P., *Phys. Chem. Glasses*, **41**, (2000), 384.

Pevzner, B. Z. & Klyuev, V. P., *Glass Technol.*, **44**, (2003), 94.

Pevzner, B. Z. & Niunin, G. I., *Phys. Chem.. Glasses*, **41**, (2000), 387.

Phan, M-H., Kim, Y-S., Chien, N-X., Yu, S. C., Lee, H. & Chou, N., *Jpn. J.. Appl. Phys. Part 1*, **42**, (2003), 5571.

Pirooz, P. P., US Patent No. 3088833, 1963.

Pirooz, P. P., US Patent No. 3088834, 1963.

Pirouz, P. & Ernst, F., in *Metal-Ceramic Interfaces*, edited by M. Ruhle, A. G. Evans, M. F. Ashby & J. P. Hirth, Acta-Scripta Metallurgica Proceedings Series 4, Pergamon Press (Oxford), 1990, pp. 199–221.

Polmear, I. J., *Metallurgy of Light the Metals*, Edward Arnold (London), 1989.

Pope, M. I. & Judd, M. D., *Differential Thermal Analysis*, Heyden (London), 1973.

Potter, A. & Henning, J. E., Continuous glass coating of fine metal wires, in *Energy Mater: Proc. Mater. Process Eng. Nat. Sym. Exhibition*, Chicago, 1968, pp 417.

Powers, M., in *Advanced brazing and soldering technologies*, IBSC 2000, edited by P. T. Vianco & M. Singh, American Welding Society (Miami, Florida), 2000, pp. 296–302.

Price, J. W., *Brit. Soc. Sci. Glass Blowers J.*, **22**, (1984), 34.

Quinn, G. D. & Morrell, R., *J. Am. Ceram. Soc.*, **74**, (1991), 2037.

Rabinovich, E. M., *J. Mater. Sci.*, **20**, (1985), 4259.

Rajaram, M. & Day, D. E., *J. Am. Ceram. Soc.*, **69**, (1986), 400.

Rangaraj, S. & Kokini, K., *Acta Mater.*, **51**, (2003), 251–267.

Rao, K. J., *Structural Chemistry of Glasses*, Elsevier (Oxford), 2002.

Rao, B. B., Reddy , N. K. & Jaleel, M. A., *J. Mater. Sci. Lett.*, **9**, (1990), 1159.

Rats, D., VonStebut, J. & Augereau, F., *Thin Solid Films*, **355**, (1999), 347.

Rawson, H., *Properties and Applications of Glass*, Elsevier (Amsterdam), 1980.

Ray, N. H., Lewis, C. J., Laycock, J. N. C. & W. D. Robinson, *Glass Technol.*, **14**, (1973), 50.

Ray, N. H., Laycock, J. N. C. & Robinson, W. D., *Glass Technol.*, **14**, (1973), 55.

Ray, N. H., Plaisted, R. J. & Robinson, W. D., *Glass Technol.*, **17**, (1976), 66.

Rey, M. C., Kramer, D. P., Henderson, W. R. & Abney, L. D., *Welding Research Supplement*, (May 1984), 162.

Richmond, J. C., Moore, D. G., Kirkpatrick, H. B.and Harrison, W. N., *J. Am. Ceram.Soc.*, **36**, (1953), 410.

Richter, H. R., *Glastech. Ber.*, **58K**, (1983), 402.

Rickerby, D. S. & Matthews, A., editors, *Advanced surface coatings*, Blackie & Son (London & Glasgow), 1991.

Rincón, J. M., *Key Eng. Mater.*, **206–213**, (2002), 2039.

Rincón, J. M., González-Peña, J. M. & Bosch, V. A., *Bull. Am. Ceram. Soc.*, **66**, (1987), 1124.

Risbud, S. H., Allen, G. D. & Poetzinger, J. E. *Ceramic Microstructures '86*, edited by J. A. Pask & A. G. Evans, Plenum Press (New York), 1986, pp 359–367.

Ritter, J. E., Jakus, K., Batakis, A. & Bandyopadhyay, N., *J. Non-Cryst. Solids*, **38&39**, (1980), 419.

Rulon, R. M., *Introduction to glass science*, edited by L. D. Pye, H. J. Stevens & W. C. LaCourse, Plenum Press (New York), 1972, pp 661–704.

Sack, W., Scheidler, H., & Petzoldt, J., *Glastech. Ber.*, **41**, (1968), 138.

Saiz, E., Tomsia, A. P., Fujino, S., & Gomez-Vega, J. M., *Graded coatings for metallic implant alloys*, internet article.

Sakka, S., *Treatise on Materials Science and Technology, vol. 22, Glass III*, edited by M. Tomozawa & R. H. Doremus, Academic Press (New York), 1982, pp 129–167.

Salamah, M. A. & White, D., *Surfaces and interfaces in ceramic and ceramic-metal systems*, edited by J. A. Pask & A. Evans, Plenum Press (New York), 1981, pp 467-476.

Salerno, R. F., Dichiaro, J. V., Egleston, E. E. & Koons, J. W., *An investigation of nonchlorinated substitute cleaning agents for methylene chloride*, Mound Applied Technologies Report MLM-3621, Jan. 1990.

Salman, S. M., Salama, S. N. & Darwish, H., *Proc. Int. Congr. Glass*, Edinburgh, Volume 2. Extended Abstracts, 2001, pp. 788–789.

Samsonov, G. V., *Handbook of the Physicochemical Properties of the Elements*, Oldbourne Book Co. (London), 1968.

Sanderson, K. D. & Knowles, J. A., European Patent EP1453771, Sept. 2004.

Santella, M. L., *Ceram. Bull.*, **71**, (1992), 947.

Sarkisov, P. D., *Proc. Glass '89, XV Int. Congr. Glass*, 1989, pp. 411–441.

Scherer, G. W., *Relaxation in Glass and Composites*, John Wiley & Sons (New York), 1986.

Scherer, G. W. & Rekhson, S. M., in *Treatise on Materials Science and Technology, vol. 26, Glass IV*, edited by M. Tomozawa & R. H. Doremus, Academic Press Inc. (London), 1985, pp 245–318.

Schiepers, R. C. J., Van Beek, J. A., De Giacomoni, E., Valla, B., Van Loo, F. J. J. & De With, G., in *Metal-Ceramic Interfaces*, edited by M. Ruhle, A. G. Evans, M. F. Ashby & J. P. Hirth, Acta-Scripta Metallurgica Proceedings Series 4, Pergamon Press (Oxford), 1990, pp. 138–143.

Schneider, D., Schwartz, T., Buchkremer, H. P. & Stoever, D., *Thin Solid Films*, **244**, (1993), 177.

Schneider, J. A., Guthrie, S. E., Kriese, M. D., Clift, W. M. & Moody, N. R., *Mater. Sci. Eng.*, **A259**, (1999), 253–260.

Schrooten, J. & Helsen, J. A., *Biomaterials*, **21**, (2000), 1461.

Schulz-Harder, J., *Microelectron. Reliability*, **43**, (2003), 359–365.

Schwickert, T., Geasee, P., Janke, A., Dickmann, U. & Conradt, R., in *Advanced brazing and soldering technologies, IBSC 2000*, edited by P. T. Vianco & M. Singh, American Welding Society (Miami, Florida), 2000, pp. 310–314.

Seager, L. W., Kokini, K., Trumble, K. & Krane, M. J. M., *Scripta Mater.*, **46**, (2002), 395.

Sedmale, G., Kubyakov, V., Sedmalis, U. & Sithinova, M., *Br. Ceram. Proc.*, **60** (vol. 1), 1999, pp. 221–222.

Sedmale, G., Vulfson, Y. & Sedmalis, U., *Proc. Int. Congr. Glass*, Edinburgh, Volume 2. Extended Abstracts, 2001, pp. 794–795.

Selcuk, A., Merere, G. & Atkinson, A., *J. Mater. Sci.*, **36**, (2001), 1173.

Selsing, J., *J. Am. Ceram. Soc.*, **44**, (1961), 419.

Selverian, J. H. & Kang, S., *Am. Ceram. Soc. Bull.*, **71**, (1992), 1511.

Selverian, J. H., O'Neill, D. & Kang, S., *Am. Ceram. Soc. Bull.*, **71**, (1992), 1403.

Ser, C. C., Zhang, X., Mohanraj, S., Premachandran, C. S. & Ranganathan, N., *Proc. 5th Electronics Packaging Technology Conf.*, 2003, pp. 307–310.

Serier, B., Bouiadjra, B. B. & Treheux, D., *Euro Ceramics VIII, Key Engineering Materials*, vols. 264-268, edited by H. Mandal & L. Ovecoglu, Trans. Tech. Publications (Switzerland), 2004, pp. 667–670.

Sesták, J., *Comprehensivs Analytical Chemistry, volume 12, Thermal Analysis, part D, Thermophysical Properties of Solids*, edited by G. Svehla, Elsevier (Amsterdam), 1984, pp 303–343.

Seufert, M. & Lenhart, A., *Proc. Int. Congr. Glass*, Edinburgh, Volume 2. Extended Abstracts, 2001, pp. 796–797.

Shapiro, A. A., Kubota, N., Yu, K. & Mecartney, M. C., *J. Electron. Mater.*, **30**, (2001), 386.

Shaw, M. C., *Eng. Fract. Mech.*, **61**, (1998), 49.

Shepeleva, L., Medres, B., Kaplan, W. D., Bamberger, M., McCay, M. H., McCay, T. D. & Sharp, M., *Surf. Coat. Technol.*, 125, (2000), 40–44.

Shetty, D. K., Rosenfield, A. R. & Duckworth, W. H., *J. Am. Ceram. Soc.*, **68**, (1985), C-282.

Shi, W. & James, P. F., in *The Physics of Non-Crystalline Solids*, edited by L. D. Pye, W. C. LaCourse & H. J. Stevens, Society of Glass Technology, Taylor & Francis (London, Washington DC), 1992, pp. 401–405.

Schilz-Harder, J., in *Proc. European Microelectronics Packaging and Interconnection Symposium, IMAPS Europe*, 2002, pp. 43–48.

Shim, D-J., Sun, H-W., Vengallotore, S. T. & Srikar, S. M., *MRS Symposium: Proc. Micro- and Nanosystems*, vol. 782, 2003, pp. 1201–1203.

Sieber, H., Hesse, D. & Werner, P., *Phil. Mag. A.*, **75**, (1997), 889.

Simon, S. & Nicula, A. L., *Phys. Status Solidi*, **81**, (1984), K1.

SinghDeo, N. N. & Schukla, R. K., *Glass Science and Technology, volume 2, Processing*, edited by D. R. Uhlmann & N. J. Kreidl, Academic Press (Orlando), 1984, pp 169–207.

Sitnikova, A. Ya., Appen, A. A., Anitov, I. S., Fedorov, V. N., Kalina, A. M. & Piryutko, M. M., *J. Appl. Chem. USSR*, **47**, (1974), 1981.

Smith, G. P., *Mater. Des.*, **10**, (1989), 54.

Smith, W. E., US Patent No. 3380818, 1968.

Smith, R. W., in *Advanced brazing and soldering technologies, IBSC 2000*, edited by P. T. Vianco & M. Singh, American Welding Society (Miami, Florida), 2000, pp. 310–314.

Smith, G. L., Neilson, G. F. & Weinberg, M. C., *Phys. Chem. Glasses*, **28**, (1987), 257.

Smith, S. M. & Scattergood, R. O., *J. Am. Ceram. Soc.*, **75**, (1992), 305.

Smithells, C. J., Glass-to-Tungsten Seals, in *Tungsten*, Chapman & Hall (London), 1952, pp. 278.

Soerensen, B. F. & Primdahl, S., *J. Mater. Sci.*, **33**, (1998), 5291.

Sohn, S-B., Choi, S-Y., *J. Am. Ceram. Soc.*, **87**, (2004), 60.

Sohn, S-B., Choi, S-Y., Kim, G-H., Song, H-S. & Kim, G-D., *J. Non-Cryst. Solids*, **297**, (2002), 103–112.

Son, S. J., Park, K. H., Katoh, Y. & Kohyama, A., *J. Nucl. Mater.*, **329–333**, (2004), 1549.

Stalcy, I I. F., *J. Am. Ceram. Soc.*, **17**, (1934), 163.

Stanton, K. T., Healy, N., Vanhumbeeck, J. F., Newcomb, S. & Hill, R. G., *7th. World Biomaterials Congress Transactions*, Sydney, Australia, 2004.

Stark, D., in *Proc. of the SPIE - The International Society for Optical Engineering*, vol. 4980, 2003, pp.289–300.

Steinberg, J., in *Engineered Materials Handbook, Vol. 4, Ceramics and Glasses*, Technical Chairman S. J Schneider, ASM International, 1991, pp. 542–545.

Stephens, J. J. & Hoskinh, F. M., *Proc. Symp. Materials Design Approaches and Experiences*, edited by J-C. Zhao, M. Fahrmann & T. M. Pollock, Metals and Materials Soc., 2001, pp. 227–239.

Stookey, S. D., British Patent No. 752243, 1956.

Stookey, S. D., British Patent No. 829447, 1960.

Stookey, S. D., US Patent No. 2933857, 1960.

Stott, F. H., *The role of active elements in the oxidation behaviour of high temperature metals and alloys*, edited by E. Lang, Elsevier Applied Science (London), 1989, pp 3–21.

Strnad, Z., *Glass-Ceramic Materials*, Elsevier (Amsterdam), 1986.

Strnad, Z., Spirkova, B. & Dusil, J., *Silikaty*, **20**, (1976), 225.

Stroda, M. & Stoch, L., *Proc. Int. Congr. Glass*, Edinburgh, Volume 2. Extended Abstracts, 2001, pp. 978–979.

Sturgeon, A. J., Holland, D., Partridge, G. & Elyard, C. A., *Glass Technol.*, **27**, (1986), 102.

Suga, T. & Miyazawa, K., in *Metal-Ceramic Interfaces*, edited by M. Ruhle, A. G. Evans, M. F. Ashby & J. P. Hirth, Acta-Scripta Metallurgica Proceedings Series 4, Pergamon Press (Oxford), 1990, pp. 189–195.

Suganuma, K., in *Engineered Materials Handbook, Vol. 4, Ceramics and Glasses,* Technical Chairman S. J Schneider, ASM International, 1991, pp. 523–531.

Suh, C. M., Lee, M. H., Kim, S. H., Kim, D. G., Chol, Y. G. & Lee, T. Y., *Int. J. Mod. Phys. B*, **17**, (2003), 1335.

Suzuki, Y., Nagahara, S. & Ichimura, N., US Patent No. 3674520, 1972.

Suzuki, Y., & Ichimura, N., *Asahi Garasu Kenkyu Hokoku*, **18**, (1968), 49.

Takahashi, M., Okabe, N., Zhu, X. & Kagawa, K., *Mater. Sci. Res. Int.*, **8**, (2002), 109–115.

Takahashi, M., Okabe, N., Zhu, X. & Kagawa, K., *Fatigue Fract. Eng. Mater. Struct.*, **26**, (2003), 391–398.

Takahashi, S., Yoshiba, M. & Harada, Y., *Materials at High Temperatures*, vol. 18, 2001, pp. 125–130.

Takamori, T., *Treatise on materials science and technology*, vol. 17, edited by M. Tomozawa & R. H. Doremus, Academic Press (New York), 1979, pp 173–255.

Tang, Z, Wang, F. & Wu, W., *Surf. Coat. Technol.*, **110**, (1998), 57–61.

Tao, W. & Dorner-Reisel, A., *J. Mater. Sci.*, **39**, (2004), 4309.

Taswell, L. F. & Jones, M. W., *Glass Technol.*, **31**, (1990), 44.

Tavgen, V. V., *Glass Ceram.*, **57**, (2000), 183.

Taylor, G. F., *Phys. Rev.*, **23**, (1924), 655.

Taylor, G. F., US Patent 1793529, 1931.

Terry, M. T., Edgemon, G. L., Mickalonis, J. I. & Mizia, R. E., *Proc. WM'02*, 2002, 1.

The Institute of Porcelain Enamellers, 2004.

Thomas, I. M., *Sol-gel technology for thin films, fibers, preforms, electronics and speciality shapes*, edited by L. C. Klein, Noyers Publications (Pakk Ridge, NJ), 1982, pp 2–15.

Thomas, W. B., *Solid State Technol.*, **29**, (1986), 73.

Thorp, J. S., Akhtaruzzaman, M., Logan, E. A. & Holland, D., *J. Mater. Sci.*, **26**, (1991), 5373.

Timoshenko, S., *J. Opt. Soc. Am.*, **11**, (1925), 233.

Tindyala, M. A. & Ott, W. R., *Bull. Am. Ceram. Soc.*, **57**, (1978), 432.

Tomsia, A. & Pask, J. A., *J. Am. Ceram. Soc.*, **69**, (1986), C-239.

Tomsia, A. P., Feipeng, Z. & Pask, J. A., *J. Am. Ceram. Soc.*, **68**, (1985), 20.

Tomsia, A. P. & Pask, J. A., *Joining of Ceramics*, edited by M. G. Nicholas, The Institute of Ceramics, Chapman & Hall (London), 1990, pp 7–30.

Tomsia, A. P., Pask, J. A. & Loehman, R. E., in *Engineered Materials Handbook, Vol. 4, Ceramics and Glasses,* Technical Chairman S. J Schneider, ASM International, 1991, pp. 493–501.

Trumble, K. P. & Ruhle, M., in *Metal-Ceramic Interfaces*, edited by M. Ruhle, A. G. Evans, M. F. Ashby & J. P. Hirth, Acta-Scripta Metallurgica Proceedings Series 4, Pergamon Press (Oxford), 1990, pp. 79–92.

Tufescu, F. M., Chiriac, H., Ovari, T-A. & Stancu, A., *J. Magn. Magn. Mater.*, **242–245**, (2002), 254–256.

Tufescu, F. M., Ovari, T. A., Chiriac, H. & Stancu, A., *J. Optoelectron. Adv. Mater.*, **5**, (2003), 273.

Turner, C. W., *Bull. Am. Ceram. Soc.*, **70**, (1991), 1487.

Turwith, M., Elssner, G. & Petzov, G., *Ceramic Microstructures 86*, 1986, pp. 969–979.

Ulitovsky, A. V., *Pribory Tech. Eksperimenta*, **3**, (1957), 115.

Vaben, R., Traeger, F. & Stover, D., *J. Thermal Spray Technol.*, **13**, (2004), 396.

Vakhula, Ya. I., *Glass Ceram.*, **57**, (2000), 101.

Valentini, R., Solina, A., Paganini, L. & DeGregorio, P., *J. Mater. Sci.*, **27**, (1992), 6579-6582.

Valentini, R., Solina, A., Paganini, L. & DeGregorio, P., *J. Mater. Sci.,* **27**, (1992), 6583-6589.

Van Helvoort, A. T. J., Knowles, K. M. & Fernie, J. A., *Euro Ceramics VIII, Key Engineering Materials,* vols. 264-268, edited by H. Mandal & L. Ovecoglu, Trans. Tech. Publications (Switzerland), 2004, pp. 649–654.

Van Houten, G. R., *Ceram. Bull.,* **38**, (1959), 301.

Vargin, V. V. & Miklyukov, E. M., *Zh. Prikl. Khim.,* **11**, (1968), 194

Varshneya, A. K., *Treatise on materials science and technology,* vol. 22, edited by M. Tomozawa & R. H. Doremus, Academic Press (New York), 1982, pp 241–306.

Verne, E., Valles, C. F., Brovarone, C. V., Spriano, S. & Moisescu, C., *J. Eur. Ceram. Soc.,* **24**, (2004), 2699.

Vitman, F. F. & Pukh, V. P., *Ind. Lab.,* **29**, (1963), 925.

Verganelakis, V., Nicolaou, P. D. & Kordas, G., *Glass Technol.,* **41**, (2000), 22.

Vogel, W. & Höland, W., *Angew. Chem. Int. Ed. Engl.,* **26**, (1987), 527.

Volf, M. B., *Chemical approach to glass,* Elsevier (Amsterdam), 1984.

Wachtman, J. B., Capps, W. & Mandrel, J., *J. Mater.,* **7**, (1972), 188.

Wachtman, J. B. and Haber, R. A. (editors), *Ceramic Films and Coatings,* William Andrew Publishing/ Noyes, 1993.

Wada, M., & Kawamura, S., *Bull. Inst. Chem. Res., Kyoto Univ.,* **59**, (1981), 256.

Wagner, G., Walther, F., Nebel, T., Eifler, D. & Roeder, E., *Proc. Int. Congr. Glass,* Edinburgh, Volume 2. Extended Abstracts, 2001, pp. 666–667.

Wagner, G., Walther, F., Nebel, T. & Eifler, D., *Glass Technol.,* **44**, (2003), 152.

Walton, J. D. & Sweo, B. J., *J. Am. Ceramic. Soc.,* **36**, (1953), 335.

Wang, T. H. & James, P. F., *Proc. 2nd Int. Conf. New Materials and their Applications,* University of Warwick, UK, 10–12 April, 1990, Institute of Physics Conf. Series No. 111, edited by D. Holland, IOP (Bristol), 1990, pp 401–410.

Wang, R. & Kido, M., *Mater. Res. Bull.,* **38**, (2003), 1401.

Wang, R. & Mitsuo, K., *Mater. Res. Bull.,* **38**, (2003), 1401–1411.

Wang, T. & Dorner-Reisel, A., *J. Mater. Sci.,* **39**, (2004), 4309.

Wange, P., Vogel, J., Horn, L., Holand, W. & Vogel, W., *Silic. Ind.,* (7–8), (1990), 231.

Wasynczuk, J. A. & Ruhle, M., *Ceramic Microstructures 86,* 1986, pp. 341–348.

Watanabe, A., Mitsudou, M., Kihara, S. & Abe, Y., *J. Am. Ceram. Soc.,* **72**, (1989), 1499.

Watkins, R. D., in *Engineered Materials Handbook, Vol. 4, Ceramics and Glasses,* Technical Chairman S. J Schneider, ASM International, 1991, pp. 478–481.

Watkins, R. D. & Loehman, R. E., *Adv. Ceram. Mater.,* **1**, (1986), 77.

Watanabe, M., Manabu, E. & Teruo, K., *Mater. Sci. Eng. A,* **359**, (2003), 368–374.

Wei, J., Xie, H., Nai, M. L., Wong, C. K. & Lee, L. C., *J. Micromech. Microeng.,* **13**, (2003), 217–222.

Weast, R. C., *Handbook of Chemistry and Physics,* CRC Press (Cleveland), 55th edition, 1974.

Weiderhorn, S. M. & Bolz, L. H., *J. Am. Ceram. Soc.,* **53**, (1970), 543.

Weiderhorn, S. M., *J. Am. Ceram. Soc.,* **50**, (1967), 407.

Weil, K. S. & Paxton, D. M., *Ceram. Eng. Sci. Proc.,* **23**, (2002), 785–792.

Weisner, H. & Schneider, J., *Phys. Status Solidi,* **26**, (1974), 71.

Wen, L., Guorong, C., Jijian, C. & Meifang, Z., *Glass Technol.,* **40**, (1999), 184.

Wessel, E. & Steinbrecht, R. W., *Key Engineering Materials* vol. 223, 2002, pp. 55–60.

West, A. R. & Glasser, F. P., *J. Mater. Sci.,* **5**, (1970), 557.

West, A. R. & Glasser, F. P., *J. Mater. Sci.,* **5**, (1970), 676.

Weil, K. S. & Paxton, D. M., *Ceram. Eng. Sci. Proc.,* vol. 23, 2002, pp. 785–792.

White, G. S., in *Mechanical testing methodology for ceramic design and reliability,* edited by D. C. Cranmer & D. W. Richerson, Marcel Dekker Inc. (New York), 1998, pp. 17–42.

Wiggin, H. & Co. Ltd. Hereford, UK, alloy data sheets.

Wiggins, J., Srikanth, H., Wang, K-Y., Spinu, L. & Tang, J., *J. Appl. Phys.,* **83**, (2000), 4810–4812.

Wilder, J. A., *J. Non-Cryst. Solids*, **38&39**, (1980), 879.

Wilder, J. A., Healey, J. T. & Bunker, B. C., *Advances in Ceramics 4*, edited by J. H. Simmons, Amer. Ceram. Soc. (Columbus), 1982, pp 313–326.

Witte, R., Herfurth, H. & Heinemann, S., In *Photon processing in microelectronics and photonics*, edited by K. Sugioka, M. C. Gower, R. F. Haglund, A. Pique, F. Trager, J. J. Dubowski & W. Hoving, Proc. SPIE, vol. 4637, 2002, pp. 487–495.

Wolterdorf, J., Pippel, E., Roeder, E., Wagner, G. & Wagner, J., *Phys. Status Solidi A: Appl. Res.*, **150**, (1995), 307.

Woodward, R., *Mater. Des.*, **10**, (1989), 248.

Wratil, J., *Vitreous Enamels*, Borax Holdings Ltd. (London), 1984.

Xiong, H-P., Kawasaki, A., Kang, Y-S., and Watanabe, R., *Surf. Coatings Technol.*, **194**, (2005), 203.

Yamamoto, T., Itoh. H., Mori, M., Mori, N. & Watanabe, T., *Denki Kagaku*, **64**, (1996), 575.

Yamanaka, T. & Takagi, Y., Japanese Patent No. 50-55612, 1975.

Yang, X., Jha, A., Brydson, R. & Cochrane, R. C., *Thin Solid Films*, **443**, (2003), 33.

Yang, Z., Stevenson, J. W., & Meinhardt, K. D., *Solid State Ionics*, **160**, (2003), 213.

Yiwang, B. & Yanchun, Z., *Mater. Res. Soc. Proc.*, vol. 750, 2002, pp. 47–52.

Young, T., *Trans. Roy. Soc. Lond.*, **95**, (1805), 65.

Zachariasen, W. H., *J. Am. Chem. Soc.*, **54**, (1932), 3841.

Zhenguo, Y., Meinhardt, K. D. & Stevenson, J. W., *J. Electrochem. Soc.*, **150**, (2003), 1095–1101.

Zhenguo, Y., Stevenson, J. W. & Meinhardt, K. D., *Solid State Ionics, Diffus. React.*, **160**, (2003), 213.

Zhiting, G., Ma, J., Ning, H. & Huang, F., *Rare Met. Mater. Eng.*, **32**, (2003), 650.

Zhou, H., Hu, X., Bush, M. & Lawn, B. R., *J. Mater. Res.*, **15**, (2000), 676.

Zhou, B. & Kokini K., *Mater. Sci. Eng. A*, A348, (2003), 271–279.

Zhu, D., Choi, S. R., and Miller, R. A., *Surf. Coatings Technol.*, **188–189**, (2004), 146.

Zhukov, A., *J. Magn. Magn. Mater.*, **242–245**, (2002), 216–223.

Zhukov, A., Garcia-Beneytez, J. M. & Vazquez, M., *J. Phys. IV*, **8**, (1988), Pr2-763.

Zhukov, A., Lazareva, E. A. & Kapelyvzhnaya, N. P., *Glass Ceram.*, **58**, (2001), 360.

Zhuravlev, G. I. & Borisenko, A. A., *Strength Mater.*, **22**, (1990), 1426.

Zubekhin, A. P., Lazareva, E. A. & Kapelyuzhnaya, N. P., *Glass Ceram.*, **58**, (2001), 360.